AF676528

Environmental Policy Design and Implementation: Toward Sustainable Society

Environmental Policy Design and Implementation: Toward Sustainable Society

Editors

Adam P. Hejnowicz
Jessica P. R. Thorn

MDPI • Basel • Beijing • Wuhan • Barcelona • Belgrade • Manchester • Tokyo • Cluj • Tianjin

Editors

Adam P. Hejnowicz
Newcastle University
UK
University of York
UK

Jessica P. R. Thorn
University of St. Andrews
UK
University of York
UK
University of Cape Town
South Africa

Editorial Office
MDPI
St. Alban-Anlage 66
4052 Basel, Switzerland

This is a reprint of articles from the Special Issue published online in the open access journal *Sustainability* (ISSN 2071-1050) (available at: https://www.mdpi.com/journal/sustainability/special_issues/environmental_policy_design_implementation).

For citation purposes, cite each article independently as indicated on the article page online and as indicated below:

LastName, A.A.; LastName, B.B.; LastName, C.C. Article Title. *Journal Name* **Year**, *Volume Number*, Page Range.

ISBN 978-3-0365-5185-2 (Hbk)
ISBN 978-3-0365-5186-9 (PDF)

© 2022 by the authors. Articles in this book are Open Access and distributed under the Creative Commons Attribution (CC BY) license, which allows users to download, copy and build upon published articles, as long as the author and publisher are properly credited, which ensures maximum dissemination and a wider impact of our publications.
The book as a whole is distributed by MDPI under the terms and conditions of the Creative Commons license CC BY-NC-ND.

Contents

About the Editors

Adam Hejnowicz

Adam Hejnowicz PhD is a Research Associate in the School of Engineering, Newcastle University, and part of the GCRF Living Deltas Hub, a Visiting Researcher in the Department of Biology at the University of York, a member of ESRC Centre for the Evaluation of Complexity Across the Nexus based at the University of Surrey, and a contributing author to the IPBES Values Assessment. An interdisciplinary environmental scientist, his applied, policy-facing, and transdisciplinary research concerns the sustainability and governance of complex social–ecological systems. His work draws on mixed methods and plural epistemologies. Currently, his research focuses on examining the intersection between climate change adaptation, policy, and governance in Asian mega-deltas in Vietnam, India, and Bangladesh. Other work has focused on peri-urban resilience, climate migration and COVID-19 impacts on informal settlements; new economics; emerging manufacturing environmental impacts; the water–energy–food nexus; the role of evaluation in transformational change; value systems and organizational cultures; post-Brexit UK environmental and agricultural policy; and ecosystem services and market-based mechanisms for natural resource management. All throughout, his work has involved working with local, regional, and national government agencies, departments, and ministries; community-based organizations; local, national, and international NGOs as well as multilateral donors and intergovernmental organizations. He believes in the importance of freely available and inclusive knowledge for a better society.

Jessica Thorn

Jessica Thorn PhD is a social–ecological systems scientist with a background in ecology and human geography. She is a Lecturer in Sustainable Development at the School of Geography and Sustainable Development at the University of St Andrews, African Research Initiative for Scientific Excellence (ARISE) fellow at the Department of Environmental Sciences at the University of Namibia, and contributing author to the International Panel on Climate Change AR6 WGII. Her work broadly concentrates on climate change adaptation, social–ecological systems, biodiversity conservation, the sustainability of mega infrastructure and green transitions in peri urban, smallholder and mountain systems. Hailing from Namibia, she has 15 years of experience working on various ESRC, EU, IDRC, GCRF, CR4D, NSF, NERC, NRF, DFID, CGAIR, IDRC, ESRC, UNEP, UNDP, and USAID funded projects affiliated with the UN, World Bank, governments, and NGOs, working closely with local stakeholders across sectors and travelling to 57 countries. The scale of her research varies from individual households to national comparative surveys as well as synthetic analyses, participatory scenario planning, ecosystem service assessments, and systematic reviews. She has taught on international courses and supervises PhD and MSc students and contributed to United Nations Global Environmental Outlook and The Economics of Ecosystems and Biodiversity reports. Prior to coming to St Andrews, she completed her BSocSci (Hons) at the University of Cape Town, MSc and DPhil at Oxford, postdoctoral studies at Colorado State University, University of York, and ETH Zurich, and held a fellowship at the African Climate and Development Initiative.

sustainability

Editorial

Environmental Policy Design and Implementation: Toward a Sustainable Society

Adam P. Hejnowicz [1,2,*] and Jessica P. R. Thorn [3,4,5]

1 School of Engineering, Newcastle University, Newcastle-Upon-Tyne NE1 7RU, UK
2 Department of Biology, University of York, York YO10 5DD, UK
3 School of Geography and Sustainable Development, University of St. Andrews, St. Andrews KY16 9AL, UK; jprt1@st-andrews.ac.uk
4 Department of Environment and Geography, University of York, York YO10 5NG, UK
5 African Climate and Development Initiative, University of Cape Town, Cape Town 7001, South Africa
* Correspondence: adam.hejnowicz@newcastle.ac.uk

Citation: Hejnowicz, A.P.; Thorn, J.P.R. Environmental Policy Design and Implementation: Toward a Sustainable Society. *Sustainability* **2022**, *14*, 3199. https://doi.org/10.3390/su14063199

Received: 9 February 2022
Accepted: 10 February 2022
Published: 9 March 2022

Publisher's Note: MDPI stays neutral with regard to jurisdictional claims in published maps and institutional affiliations.

Copyright: © 2022 by the authors. Licensee MDPI, Basel, Switzerland. This article is an open access article distributed under the terms and conditions of the Creative Commons Attribution (CC BY) license (https://creativecommons.org/licenses/by/4.0/).

1. Introduction

> "No matter how complex global problems may seem, it is we ourselves who have given rise to them. They cannot be beyond our power to resolve." —Daisaku Ikeda, Japanese Buddhist philosopher

The science could not be more emphatic, as a spate of hard-hitting assessments have indicated, achieving prosperous societies, climate stability and a flourishing biosphere requires urgent global action across scales and sectors [1–9]. Meeting the ambitions of the post-2015 Sustainable Development agenda, Paris Climate Agreement, and the Convention of Biodiversity's post-2020 Biodiversity Framework will require radical change in the design and implementation of environmental policies (SDG 17.14), especially those that intersect key goals of economic development (SDG 8, 9, 11) and production and consumption (SDG 2, 7, 12). Such policies must enable transitions towards knowledge-based economies grounded in evidence-based policy making. Here, the cooperation of city governments, the private sector, development practitioners, conservationists, communities, urban planners, and others will be key.

The question then becomes how, under these conditions, can policies be effectively designed and implemented in a way that will steer societies towards more sustainable, inclusive outcomes in the short- and long-term future? This is a complex question and an enormous challenge, and in this Special Issue we only begin to scratch the surface. We do so in two substantive ways by advancing our understanding of: (a) the present state and effectiveness of local, national, and regional policies engaging with, and transforming, the climatic, environmental, social, and economic impacts of development activities; and (b) how environmental policies might be designed and embedded into future development planning to encourage coordination and coherence across policy domains.

To accomplish this, we present a collection of ten papers (Table 1) focusing predominantly on sub-Saharan Africa, including two papers from Europe and Asia for wider relevance. Collectively, these studies work across spatial and temporal scales from local communities, to municipal, national, regional, and international levels and from recent decades up to the mid-21st century. These studies are situated in a range of systems, from urban to rural, dryland to tropical climates, and employ both qualitative and quantitative methodologies. While covering diverse themes, all studies relate to policy implementation, participation and equitable representation in decision making. Crucially, in an era where decolonization is an increasingly important matter to address in academia and beyond, many article authors primarily live and work in the contexts they are researching.

Table 1. Brief description of the papers included in the special issue.

Paper	Issue	Focus	Approach	Methodology	Thematic Areas	Scale	Country
Mbidzo et al., 2021 [10]	Efficacy of community-based conservation for sustainable resource management and local benefits	National natural resource management policy implementation and local-level impacts	Appraisal of common pool resource management under different governance regimes (i.e., conservancies and communtiy forests)	Policy analysis using Ostrom's CPR framework and key informant interviews	Land use, biodiversity protection, common pool resources, institutional interactions and decision making	Sub-national (regional)	Namibia
Sigalla et al., 2021 [11]	Participatory effectiveness of multi-stakeholder platforms for cross-sector water governance	Participatory representation and engagement of stakeholder groups in decision making platforms to inform water resource management	Examination of gender equality, stakeholder composition and managerial mainstreaming in MSPs	Literature policy review, key informant interviews and focus groups	Water governance, participatory decision making, representative pluralism	National and sub-national (basin)	Tanzania
Angula et al., 2021 [12]	The role and extent of gender responsiveness in international climate change financing instruments	Green Climate Fund and gender responsiveness	Programme level analysis of gender inequalities and differentiated impacts in Green Climate Fund ecosystem-based adaptation initiatives	In-depth interviews and groups discussions	Communtiy-based natural resource management, ecosystem-based adaptation, climate financing, gender representation	National	Namibia
Beauchamp et al., 2021 [13]	Appropriate incorporation of resilience and human wellbeing concepts and indicators into evaluation frameworks	Climate adaptation interventions to build resilience and enhance human wellbeing of local communities	Assessing the interplay of local predictors of resilience and human wellbeing and indicators of resilience in relation to the Devolved Climate Finance mechanism	household survey, participatory discussions and statistical analysis	Local-led adaptation, resilience and human-wellbeing, policy evaluation, local compatibility	Sub-national (regional)	Tanzania

Table 1. *Cont.*

Paper	Issue	Focus	Approach	Methodology	Thematic Areas	Scale	Country
Johnson 2021 [14]	Unravelling the policy discourses embedded in the translation and implementation of international green market mechanisms	National implementation of REDD+ practices and impacts	Narrative assessment of national policy discourse and relationship with local policy implementation	Discourse analysis (critical political ecology), qualitative interviews, and focus groups	Climate mitigation financing, policy implementation, State power, neoliberalism	National	Ghana
Wijesinghe and Thorn 2021 [15]	Governance of urban climate resilience	Integration of urban green infrastructure into local urban governance and spatial planning in informal settlements to enhance resilience and human wellbeing	Assessing the benefits and trade-offs of urban green infrastructure, analysing the institutional arrangements for urban green infrastructure governance, and highlighting desirable pathways for urban governance	Case studies, key informant interviews, focus groups, and participant observation	Urban governance, climate resilience, informality, green infrastructure, local participatory stewardship	Municipal	Namibia
Wakdok and Bleischwitz et al., 2021 [16]	Climate change consequences for human security in resource insecture regions	Examining the climate–security–resource nexus and implications for the implementation of the SDGs	Explore land use and grazing policies in relation to security and conflict pathways concerning shared resources between farmers and migrating herdsmen, and the role of climate change as a stress multiplier	Literature review, comparative case study, and scenario approach	Resource conflict, climate change, migration, sustainable development, land use and livelihoods	International	Nigeria and Chad
Scheren et al., 2021 [17]	Balancing development and ecological integrity futures	Highlighting sustainable pathways for Africa's ecological resource base over a 50 year time horizon	Develop collectively owned scenarios that can be used to explore key drivers of change and system level responses to steer sustainable social-ecological futures	Participatory scenario planning, DPSIR Framework	Sustainable development, participatory planning, decision-support, social-ecological futures	International	Pan-Africa
Franco et al., 2021 [18]	Utility and effectiveness of spatial modelling of national sustainability metric	Mapping municipal level sustainability indices based on the supply and demand of natural resources (i.e., their footprint)	Assessment of the "ecological balance" per capita for each municipality based on calculating the "ecological footprint" and "biocapacity" per capita to determine municipal scale sustainability	Ecological Footprint framework	Resource use and demand, decision-support tool, spatially explicit sustainability modelling	National and sub-national (municipal)	Italy
Xing et al., 2020 [19]	Efficacy of policy to green industry via exploring the relationship between environmental policies (regulation), sustainable innovation and financial performance	Assessing the mediating impact of "green dynamic capability" on sustainability exploration/exploitation innovation of industrial businesses	Measuring the capacity of specific environmental policies to stimulate sustainable innovation, conceptualised as "sustainability exploration innovation" and "sustainability exploitation innovation", in polluting manufacturing industries by introducing "green dynamic capability" as a mediating mechanism—extending the Porter Hypothesis framework.	Survey, Multiple mediation model	Environmental policy regulation, sustainability innovation, business responsiveness	National	China

2. Background

Over the past few decades, many low- and middle-income countries, especially in sub-Saharan Africa, have witnessed rapid socioeconomic developments that have resulted in significant transformations of local and national social-ecological environments [20–24]. What these references demonstrate is that on the one hand, these transformations have realized considerable economic growth, inward investment, improvements in infrastructure and basic amenities, poverty reduction and livelihood diversification. In turn, nurturing the education and skills of many young populations and catalyzing the growth of dynamic urban centers. However, on the other hand, these benefits remain unevenly distributed between and within countries, often leading to natural resource exploitation, habitat loss, and even species extinction, while poorer or marginalized communities across the rural-urban continuum are regularly alienated from decision making processes. Moreover, these developments frequently occur against the backdrop of weak governance, institutional bureaucratic backlogs, operational silos, and lack of transparency and accountability.

The continuing COVID-19 pandemic, which resulted in an estimated 3.2% contraction of the global economy in 2020 and according to the World Health Organization has so far claimed over 5.78 million lives, has further spotlighted the stark geopolitical disparities and unequal power relations that frame interactions within and between countries and their populations [25]. These asymmetries underpin the profound structural, social, and economic inequalities that exacerbate adverse impacts of global environmental change, where the most vulnerable communities continue to carry the greatest burden of these unfolding events.

Examples are all around to see. Altered La Niña rainfall patterns in 2020/2021 severely impacted global agricultural production and livelihoods, while drought impacted large areas of Africa, Asia, and Latin America, and tropical storms and cyclones caused widespread damage. Many nations witnessed devastating increases in the frequency and severity of wildfires, such as South Africa, Greece, Russia, Turkey, Greece, India, Israel, and across North America in 2021, releasing 6450 megatons of CO_2 (or 148% more than the total European Union fossil fuel emissions in 2020) [26]. The global costs of natural disaster damage in 2020 totaled US$210 billion according to a report by reinsurance company Munich Re [27]. Meanwhile, global land transformation continues apace. The sheer magnitude and extent of land appropriation for development and extraction meant over 12 million hectares of tropical tree cover was lost in 2020 [28], which continues to imperil the land rights, livelihoods, and cultures of rural and Indigenous communities [29]. At the same time, rates of unplanned urbanization are growing at an extraordinary pace (particularly in Asia and Africa) leading to some 1 billion people living in informal settlements [6]. Furthermore, the levels of consumption of food, water, and energy resources are vastly outstripping what is considered sustainable [1,30–32].

Within the environmental policy domain, questions of how to improve the efficacy, legitimacy, and efficiency of local to global governance has been a long-standing debate, as have the internal and external integration and streamlining of institutions, organizations, and their bureaucracies. Without doubt, the existential threats posed by the confluence of the climate, biodiversity and pollution crises require collective, multi-scalar policy action, as recognized by recent high-level pledges made by world leaders at the 2021 United Nations Climate Change Conference of Parties (COP26) in Glasgow. Time will tell if these pledges remain grandiose political rhetoric or become actualized into action. Nonetheless, the COP26 underscored the many and varied voices resolutely calling for the re-orientation of the world towards a pathway of equitable, sustainable development.

Even with such a global drive, there is no one 'best' path towards achieving sustainable societies nor necessarily a consensus on what those societies would be like. Instead, there are multiple avenues that may be appropriate to pursue. Geographies of development matter; populations are heterogeneous differing along wealth, ethnic and religious lines, and face different types and magnitudes of social-ecological challenges. Moreover, communities, countries and regions are not all starting from the same position, nor do they have the same technical, physical, financial, human, and natural capital endowments or capacities. There are also historical legacies and path dependencies associated with conflict or colonialism that are important to acknowledge. This means reaching a so-called 'sustainable society' is harder and longer for some compared to others. Furthermore, sustainable development as a concept and its mainstreaming in the form of the SDGs is itself contested [33,34]. Questions remain regarding how local perspectives can be embedded in national policy frameworks and planning processes, the ability of sustainability metrics to support decision making, the role of dominant discourses in shaping policy narratives and implementation, and how environmental policies can stimulate industrial and public innovation now and in the future. There also remains limited empirical evidence of the efficacy of community-based natural resource management and multistakeholder platforms, gendered considerations of international climate financing initiatives, governing green infrastructure in peri-urban systems, and appropriate policy responses to human security implications of resource constraints under climate change.

These issues are explored by authors within this collection.

3. Key Insights

In this section, rather than provide a sequential synopsis of each paper, we highlight eight cross-cutting insights from our collection of articles that indicate how policy (broadly conceived) can engage with and inform the transition towards sustainable societies.

3.1. Develop Appropriate Coordinated, Integrated Institutional Arrangements

By comparing conservancies and community forests in the Zambezi region of Namibia, ref. [10] argue that matching actors, resources, and legal and administrative arrangements across scales is critical for the effective management of common pool resources. In analyzing the effective roll-out of multi-stakeholder platforms for water basin governance in Tanzania, ref. [11] similarly note that institutional matching (i.e., the correspondence between institutions and the level of authority, scale, or issues which they are addressing) is a common challenge necessary to overcome. Likewise, ref. [15] argue that to progress towards urban climate resilience in Namibia that forming coordinated governance systems that clarify mandates, roles, and modalities is mainstreaming urban green infrastructure and ecosystem services into municipal urban spatial planning and policy is essential. Equally, ref. [14] advances that enhancing the efficacy of reducing emissions from deforestation and forest degradation (REDD+) in Ghana requires individual programs to be coordinated or integrated into the 'national forestry governance landscape'. Fully aligning with these sentiments, arising from a multi-year scenario planning process exploring pan-African ecological futures, ref. [17] emphasize that holistic governance (i.e., the combining and streamlining of legislation, regulation, and informal rules across scales) is essential to move towards effective natural resource governance.

The consequences of inappropriate or inadequate governance are also clear. Prescriptive, top-down measures can diminish customary governance arrangements and undermine local social-ecological resilience [10]; structural institutional barriers can alienate or exclude individuals or groups from actively contributing to governance processes [11]; while insufficient governance resources and capacities are prohibitive to advancing informal settlement upgrading [15] or conservancies [10]. This strongly suggests that governance and institutional structures and processes need to be carefully crafted and considered, and critically evaluated. In this regard, ref. [17] underscore the importance of strengthening institutions, building capacity, ensuring consistent and long-term financing, and harnessing new technologies.

3.2. Ensure Inclusive, Pluralistic Stakeholder Engagement and Meaningful Participation

Several papers deal with the issue of stakeholder participation in decision making processes. Ref. [11] point to the need for the 'right mix' of actors (in terms of how stakeholders behave in response to rules and assigned roles), suggesting the need for more extensive private and public, as well as social, political, environmental, and economic and other sector participation to enable the proper functioning of multistakeholder platforms. In advocating for collaborative governance, ref. [15] call for broader community-based consultation, involvement, and stewardship in urban green infrastructure installation, restoration, and maintenance. The authors put forth the case that revitalizing multistakeholder platforms could provide a mechanism to enhance "inclusivity and accessibility in the planning, design and management, while improving local stewardship and valuing of green spaces" (p. 17). Further, the work of [12] in Namibia indicates that participation in decision making and leadership may be enhanced through aligning climate adaptation governance with community based natural resource management institutional arrangements. Extending participation and enhancing stakeholder engagement is also about building and co-constructing new partnerships, which enable the establishment of new initiatives. As ref. [17] state in relation to the challenge of appropriately managing important ecological assets, to be effective this requires the reconfiguration and creation of new roles and relationships between public, private and civil society sectors and actors. Ref. [17] further highlight the need to establish new partnerships, especially those that heavily engage with the private sector through corporate social responsibility.

At the same time, participation is not an unalloyed good, and indeed can be counterproductive if it is approached in a cursory manner, lacks meaningful engagement, and fails to account for meaningful inclusivity. As ref. [11] reveal, participation in water sector multistakeholder platforms in Tanzania primarily occurs at a technical level or below, resulting in discussions that focus on technicalities at the expense of wider, longer-term, and strategic deliberations. A lack of diverse stakeholder engagement erodes the breadth and quality of discussions, reducing the likelihood of effective multi-scalar integrated water resource management. In addition, ref. [14] points out that participation does not necessarily endow decision making capability or power, even if it provides a space for discussion. In the case of REDD+ implementation in Ghana, whilst government, private, civil society, research, and development sectors are part of the conversation, local communities are frequently absent, and especially so in high-level decision making forums. Instead, ref. [14] notes, local communities are often represented by proxies who purport to speak on their behalf, construed as homogenous collectives with singular perspectives and aligned common interests.

3.3. *Improve Gender Representation, Responsiveness and Reduce Inequalities*

Across all sectors and levels of decision making responsibility, historical and current, women, Indigenous, impoverished, and other marginalized groups are frequently marginalized or excluded from core policy, governance, or management arenas. This exclusion is not only to the detriment of these groups, but also to the detriment of structures and processes of decision making. Ref. [17] remark, for instance, that there is a widespread gendered division of labor within water institutions, with women commonly relegated to administrative and non-decision making roles. Authors also find that multistakeholder platforms do not adhere to gender equality despite this principle being part of their founding guidelines. Examining the gender responsiveness of Green Climate Fund projects in Namibia, ref. [12] demonstrate that social and cultural factors work together to prohibit the participation of women in the implementation of Green Climate Fund programs, particularly in the form of patriarchal dominance, which constrains the ability of women to take leadership roles, contribute meaningfully to decisions, and undermines livelihood diversification, for instance to working as wildlife game guards. On the other hand, the authors argue that considering gender at the outset of community-driven adaptation projects can reduce gender inequities and build capacity, while improving the chances of achieving climate resilient outcomes. They go on to suggest that climate-financed interventions should focus on engaging both men and women of all ages, promote women to leadership roles, collapse income disparities, and fully acknowledge the value women's work and their reproductive rights.

3.4. Develop an Integrated and Coherent Multi-Scale Policy Landscape

Fragmented, incoherent, or contradictory policies are not only less effective, but can actively undermine sustainable solutions or perpetuate challenges. Assessing climate change impacts on human security in Northern Nigeria and the Lake Chad region, ref. [16] assert the centrality of land grazing policy failure and implicate it as the primary driver of human displacement. They propose that a nexus approach to policy formulation, design and implementation can help provide a holistic mechanism to address agricultural and pastoral land scarcity, ecosystem service degradation and navigate conflicts. A nexus policy approach further leverages a systems-based, multi-scale, multistakeholder appraisal capable of negotiating the tensions between environmental impacts, land use policy, and wider social and cultural factors influencing human migration.

Relatedly, ref. [17] advocate integrated planning capabilities to deliver more strategic evidence-based decision making, suggesting that such an approach can be deployed across scales, combine different forms of data and evidence, be used to assess the social-ecological costs and benefits of development projects, improve stakeholder collaboration, and thus better manage and steer the large-scale social-ecological transformations of Africa's landscapes, ecology, and natural capital. To build effective urban climate resilience for informal settlement communities, ref. [15] additionally make the case that this rests on improving policy coherence. As an example, the authors cite the need to include urban green infrastructure in integrated development plans such as the city of Windhoek's human settlements upgrading policy.

3.5. Understand the Politics and Power Dynamics of Policy and Be Sensitive to Local Needs and Conditions

A frequent barrier to achieving legitimate, consensual sustainable development is state misappropriation of power and the lack of awareness of local perceptions of policy interventions. In the case of Ghana's REDD+ strategy, ref. [14] asserts how its focus is intentionally in the 'wrong' direction. Rather than addressing the macro-level market and policy forces that enable the development of illegal markets for forest products and maintain their demand, instead the policy focuses on eliminating the micro-level illegal activities in rural areas that contribute to deforestation. The consequence of this is that this renders interventions both apolitical and technical. As ref. [14] goes on to argue, this enables the state to accrue decision making powers and financial resources under the guise of social-ecological responsibility, while simultaneously expanding its power and control into rural areas and over forest resources at the expense of local communities. This latter point chimes strongly with the recommendation by [10] that, in some (although not all) contexts, if communities have secure land tenure rights, whether communal or individual, overall natural resources on their land, this can improve land management.

Approaching the issue of localization of monitoring and evaluating climate change adaptation program in Tanzania, ref. [13] show how higher-level policy proxies for wellbeing and resilience can be misaligned and affected by different factors, while being rooted in local dynamics. They argue that from the outset programs must be attuned to, and fully acknowledge, local social and cultural norms and power dynamics (even though this is not the majority practice). This is critical to avoid unintended outcomes that lead to maladaptation. The authors also argue, as part of so-called 'locally-led adaptation', that researchers and practitioners need to be sensitive to the lived experiences and subjective perceptions of communities. On a similar note, ref. [12] emphasize that ecosystem-based adaptation policies and planning should encourage household level adaptation responses, and in cases where this is not fully sufficient provide public support for planned adaptation.

3.6. Encourage Environments That Stimulate Innovation and Support Leadership

Providing enabling environments that support green innovation is crucial to helping the private sector move in a sustainable direction. In their analysis of environmental regulation policy on Chinese manufacturing companies, ref. [19] show that regulatory policy can positively affect firm financial performance via 'green dynamic capability' (i.e., encouraging businesses to reconfigure resources to develop greener capabilities). The authors demonstrate that innovation is largely via the 'sustainability exploitation innovation' (i.e., as a result of incremental changes made by firms through improving performance such as by purchasing patents and outsourcing production) not through product development. In the context of peri-urban green infrastructure, ref. [15] raise the prospect of innovation via strategies such as 'safe-to-fail' pilot schemes, or learning-by-doing, collaborative, experimental spaces in the form of urban living labs. Innovation in adaptation is also reliant on the galvanizing and convening power of local leaders, which can support for instance training and extension programs [12].

3.7. Employ Novel Methods to Provide Effective Decision-Support Tools for Sustainable Development

Methods can offer invaluable decision support tools to inform evidence-based policy making for sustainable transitions. Increasingly, these include strategic environmental and social impact assessments, conservation planning tools, and natural capital accounting [17].

One method which has gained traction in policy circles over recent years is scenarios. Scenarios are articulations of plausible future system states. Scenarios help policy makers and other stakeholders move beyond their normal restrictive political, business, and economic short- and medium-term time horizons, and to think strategically over the long-term about potential development pathways. Scenarios have the potential to move policy away from being reactive towards being adaptive and iterative. In their paper, ref. [17] report on a participatory scenario planning exercise underpinning the African Ecological Futures initiative. Participatory tools of this kind support knowledge exchange, social learning, transdisciplinary practice, and co-production. Bringing together diverse stakeholders, the process generated four scenarios that were entitled: 'Going Global', 'Helping hands', 'All in Together' and 'Good Neighbors'. The authors stress the power of these narratives by exploring how different policy actions could influence national and continental development pathways and environmental outcomes. Pragmatically mobilizing narratives in this way can provide tangible ways to feed into policy decision making processes. Ref [16] use scenarios in a more theoretical exploratory manner, based on literature review and expert judgement, to explore the risks and opportunities of how capital investment, technology and partnership building could transform the Lake Chad region into a sustainable finance and development hub, benefiting as many as 50 million people living in the region.

Another method is highlighted by [18] who use an ecological footprint approach to develop a municipal scale sustainability index. Their analysis shows that 60% of municipalities are unsustainable, covering virtually 95% of the Italian population. Despite the acknowledged shortcomings of the ecological footprint framing, the authors argue that their sustainability index can feed into different stages of decision making, particularly the early warning and monitoring phases, to improve the targeting of policy interventions and their adaptation over time.

3.8. Ensure Consistent Financing That Supports Local Communities, Social-Ecological Systems, and Institutions

The availability, distribution, and use of international and national finance channels to support environmentally sustainable policy interventions is central to deliver and sustain local change. Equally important are the checks and balances of these funds and their responsiveness to local contexts. Taking a macro perspective, ref. [17] urge lending institutions and market investors to create appropriate investment safeguards, legal and regulatory frameworks, and long-term social-ecological impact strategies that mainstream ecological and social benefits into financial risk assessments and minimize the ecological damage. Directing their attention to climate financing, refs. [12,16] argue for the need to repurpose the Green Climate Fund so it is more effective in accounting for livelihood practices, land use policies, conflict, and interactions between ecosystem-based adaptation, gender, and other socially differentiated divisions of labor. Finally, ref. [10] argue that ensuring financial benefits flow directly to communities (e.g., in return for their labor efforts) is key to supporting local natural resource management institutions.

4. Looking Forward

Crafting effective environmental policies that enable societies to move towards greater sustainability is an ongoing challenge, but one which is ever more urgent to address if we are to successfully confront the multiple crises of the Anthropocene. In reflecting on the contributions to this special issue, we hope to offer a small snapshot of the diversity and depth of research that is addressing eight cross-cutting themes fundamental to environmental policy issues. Individually, none of these themes is new. However, considered together, they not only align with current political, civil society and scientific discussions at meetings such as COP26, but (we hope) offer a more holistic pathway to realize transformative change.

In closing, as we look to the future and the progressive alignment of environmental policy and sustainable development for the benefit of all peoples, societies, and the natural world, it is worth contemplating the words of Wangari Maathai and the responsibility we all have in contributing to that vision:

> "Today we are faced with a challenge that calls for a shift in our thinking, so that humanity stops threatening its life-support system. We are called to assist the Earth to heal her wounds and, in the process, heal our own—indeed to embrace the whole of creation in all its diversity, beauty, and wonder."

Author Contributions: Conceptualization, writing, drafting and editing—A.P.H. and J.P.R.T. All authors have read and agreed to the published version of the manuscript.

Funding: We acknowledge funding by The Centre for the Evaluation of Complexity Across the Nexus, a UK Economic and Social Research Council large centre (project number: ES/N012550/1) and the UK Research Innovation's Global Challenges Research Fund (UKRI GCRF) Living Deltas Hub. Plus, the UK Research and Innovation's Global Challenges Research Fund (UKRI GCRF) through the Development Corridors Partnership project (project number: ES/P011500/1).

Conflicts of Interest: The authors declare no conflict of interest.

References

1. IPBES. *Summary for Policymakers of the Global Assessment Report on Biodiversity and Ecosystem Services Of the Intergovernmental Science-Policy Platform on Biodiversity and Ecosystem Services*; Díaz, S., Settele, J., Ngo, H.T., Guèze, M., Agard, J., Arneth, A., Balvanera, P., Brauman, K.A., Butchart, S.H.M., Chan, K.M.A., et al., Eds.; IPBES Secretariat: Bonn, Germany, 2019.
2. IPCC. Summary for Policymakers. In *Global Warming of 1.5 °C. An IPCC Special Report on the Impacts of Global Warming of 1.5 °C above Pre-Industrial Levels and Related Global Greenhouse Gas Emission Pathways, in the Context of Strengthening the Global Response to the Threat of Climate Change, Sustainable Development, and Efforts to Eradicate Poverty*; Masson-Delmotte, V., Zhai, P., Pörtner, H.-O., Roberts, D., Skea, J., Shukla, P.R., Pirani, A., Moufouma-Okia, W., Péan, C., Pidcock, R., et al., Eds.; World Meteorological Organization: Geneva, Switzerland, 2018.
3. IPCC. Summary for Policymakers. In *Climate Change 2021: The Physical Science Basis. Contribution of Working Group I to the Sixth Assessment Report of the Intergovernmental Panel on Climate Change*; Masson-Delmotte, V., Zhai, P., Pirani, A., Connors, S.L., Péan, C., Berger, S., Caud, N., Chen, Y., Goldfarb, L., Gomis, M.I., et al., Eds.; Cambridge University Press: Cambridge, UK, 2021; *in press*.
4. FAO. *The State of the World's Biodiversity for Food and Agriculture*; Bélanger, J., Pilling, D., Eds.; FAO Commission on Genetic Resources for Food and Agriculture Assessments: Rome, Italy, 2019.
5. FSIN. *2020 Global Report on Food Crises: A Joint Analysis for Better Decisions. Food Security Information Network*; FAO; WFP; IFPRI: Rome, Italy; Washington DC, USA, 2020.
6. UN. *World Urbanization Prospects: The 2018 Revision*; United Nations Division of Economic and Social Affairs, United Nations: New York, NY, USA, 2018.
7. UN. *Report of the Secretary-General on SDG Progress 2019 Special Edition*; United Nations: New York, NY, USA, 2019.
8. UN. *World Social Report 2020: Inequality in a Rapidly Changing World*; United Nations Division of Economic and Social Affairs, United Nations: New York, NY, USA, 2020.
9. WWF. *Living Planet Report 2020-Bending the Curve of Biodiversity Loss*; Almond, R.E.A., Grooten, M., Petersen, T., Eds.; WWF: Gland, Switzerland, 2020.

10. Mbidzo, M.; Newing, H.; Thorn, J.P.R. Can Nationally Prescribed Institutional Arrangements Enable Community-Based Conservation? An Analysis of Conservancies and Community Forests in the Zambezi Region of Namibia. *Sustainability* **2021**, *13*, 10663. [CrossRef]
11. Sigalla, O.Z.; Tumbo, M.; Joseph, J. Multi-Stakeholder Platform in Water Resources Management: A Critical Analysis of Stakeholders' Participation for Sustainable Water Resources. *Sustainability* **2021**, *13*, 9260. [CrossRef]
12. Angula, M.N.; Mogotsi, I.; Lendelvo, S.; Aribeb, K.M.; Iteta, A.-M.; Thorn, J.P.R. Strengthening Gender Responsiveness of the Green Climate Fund Ecosystem-Based Adaptation Programme in Namibia. *Sustainability* **2021**, *13*, 10162. [CrossRef]
13. Beauchamp, E.; Sainsbury, N.C.; Greene, S.; Chaigneau, T. Aligning Resilience and Wellbeing Outcomes for Locally-Led Adaptation in Tanzania. *Sustainability* **2021**, *13*, 8976. [CrossRef]
14. Johnson, S. Discourse and Practice of REDD+ in Ghana and the Expansion of State Power. *Sustainability* **2021**, *13*, 11358. [CrossRef]
15. Wijesinghe, A.; Thorn, J.P.R. Governance of Urban Green Infrastructure in Informal Settlements of Windhoek, Namibia. *Sustainability* **2021**, *13*, 8937. [CrossRef]
16. Wakdok, S.S.; Bleischwitz, R. Climate Change, Security, and the Resource Nexus: Case Study of Northern Nigeria and Lake Chad. *Sustainability* **2021**, *13*, 10734. [CrossRef]
17. Scheren, P.; Tyrrell, P.; Brehony, P.; Allan, J.R.; Thorn, J.P.R.; Chinho, T.; Katerere, Y.; Ushie, V.; Worden, J.S. Defining Pathways towards African Ecological Futures. *Sustainability* **2021**, *13*, 8894. [CrossRef]
18. Franco, S.; Pancino, B.; Martella, A. Mapping National Environmental Sustainability Distribution by Ecological Footprint: The Case of Italy. *Sustainability* **2021**, *13*, 8671. [CrossRef]
19. Xing, X.; Liu, T.; Shen, L.; Wang, J. Linking Environmental Regulation and Financial Performance: The Mediating Role of Green Dynamic Capability and Sustainable Innovation. *Sustainability* **2020**, *12*, 1007. [CrossRef]
20. Jayne, T.S.; Chamberlin, J.; Benfica, R. Africa's Unfolding Economic Transformation. *J. Dev. Stud.* **2018**, *54*, 777–787. [CrossRef]
21. Laurance, W.F. Why environmental impact assessments often fail. *THERYA* **2022**, *13*, 67–72. [CrossRef]
22. AUC/OECD. *Africa's Development Dynamics 2018: Growth, Jobs and Inequalities*; OECD Publishing: Paris, France; AUC: Addis Ababa, Ethiopia, 2018. [CrossRef]
23. AUC/OECD. *Africa's Development Dynamics 2019: Achieving Productive Transformation*; OECD Publishing: Paris, France; AUC: Addis Ababa, Ethiopia, 2019. [CrossRef]
24. AUC/OECD. *Africa's Development Dynamics 2021: Digital Transformation for Quality Jobs*; OECD Publishing: Paris, France; AUC: Addis Ababa, Ethiopia, 2021. [CrossRef]
25. McCann, G.; Malenga, C. Covid-19 and Global Inequality. In *COVID-19 in the Global South: Impacts and Responses*; Carmody, P., McCann, G., Colleran, C., O'Halloran, C., Eds.; Bristol University Press: Bristol, UK, 2020.
26. Copernicus Atmosphere Monitoring Service. Wildfires Wreaked Havoc in 2021, CAMS Tracked Their Impact. Available online: https://atmosphere.copernicus.eu/wildfires-wreaked-havoc-2021-cams-tracked-their-impact (accessed on 11 November 2021).
27. Munich, R.E. Record Hurricane Season and Major Wildfires—The Natural Disaster Figures for 2020. Available online: https://www.munichre.com/en/company/media-relations/media-information-and-corporate-news/media-information/2021/2020-natural-disasters-balance.html (accessed on 11 February 2022).
28. WRI. *Global Forest Review, 2021, Update 2*; World Resources Institute: Washington, DC, USA, 2021; Available online: https://research.wri.org/gfr/global-forest-review (accessed on 11 November 2021).
29. UN. *State of the World's Indigenous Peoples 5th Volume: Rights to Lands, Territories, and Resources*; United Nations Division of Economic and Social Affairs, United Nations: New York, NY, USA, 2021.
30. Biggs, E.M.; Bruce, E.; Boruff, B.; Duncan, J.M.; Horsley, J.; Pauli, N.; McNeill, K.; Neef, A.; Van Ogtrop, F.; Curnow, J.; et al. Sustainable development and the water–energy–food nexus: A perspective on livelihoods. *Environ. Sci. Policy* **2015**, *54*, 389–397. [CrossRef]
31. Schlör, H.; Venghaus, S.; Fischer, W.; Märker, C.; Hake, J.-F. Deliberations about a perfect storm—The meaning of justice for food energy water-nexus (FEW-Nexus). *J. Environ. Manag.* **2018**, *220*, 16–29. [CrossRef]
32. CBD. *First Draft of the Post-2020 Global Biodiversity Framework*; CBD/WG2020/3/3; United Nations Environment Programme: Nairobi, Kenya, 2021.
33. Connelly, S. Mapping Sustainable Development as a Contested Concept. *Local Environ. Int. J. Justice Sustain.* **2007**, *12*, 259–278. [CrossRef]
34. Hickel, J. The contradiction of the sustainable development goals: Growth versus ecology on a finite planet. *Sustain. Dev.* **2019**, *27*, 873–884. [CrossRef]

sustainability

MDPI

Article

Can Nationally Prescribed Institutional Arrangements Enable Community-Based Conservation? An Analysis of Conservancies and Community Forests in the Zambezi Region of Namibia

Meed Mbidzo [1,*], Helen Newing [2] and Jessica P. R. Thorn [3,4]

1 Department of Agriculture and Natural Resource Sciences, Namibia University of Science and Technology, 13 Jackson Kaujeua Street, Windhoek 13388, Namibia
2 Interdisciplinary Centre for Conservation Science (ICSS), Department of Zoology, University of Oxford, 11a Mansfield Road, Oxford OX1 3SZ, UK; h.s.newing@gmail.com
3 York Institute of Tropical Ecosystems, Department of Environment and Geography, University of York, Wentworth Way, York YO10 5NG, UK; jessica.thorn@york.ac.uk
4 African Climate and Development Initiative (ACDI), University of Cape Town, Geological Sciences Building Level 6, 13 Library Road, Cape Town 7700, South Africa
* Correspondence: mmbidzo@yahoo.com

Abstract: Community-based conservation is advocated as an idea that long-term conservation success requires engaging with, providing benefits for, and establishing institutions representing local communities. However, community-based conservation's efficacy and impact in sustainable resource management varies depending on national natural resource policies and implications for local institutional arrangements. This paper analyses the significance of natural resource management policies and institutional design on the management of common pool resources (CPRs), by comparing Namibian conservancies and community forests. To meet this aim, we reviewed key national policies pertinent to natural resource governance and conducted 28 semi-structured interviews between 2012 and 2013. Key informants included conservancy and community forest staff and committee members, village headmen, NGO coordinators, regional foresters, wildlife officials (wardens), and senior government officials in the Ministry of Environment and Tourism and the Ministry of Agriculture, Water and Forestry. We explored the following questions: how do national natural resource management policies affect the operations of local common pool resource institutions? and how do external factors affect local institutions and community participation in CPRs decision-making? Our results show that a diversity of national policies significantly influenced local institutional arrangements. Formation of conservancies and community forests by communities is not only directly linked with state policies designed to increase wildlife numbers and promote forest growth or improve condition, but also formulated primarily for benefits from and control over natural resources. The often-assumed direct relationship between national policies and local institutional arrangements does not always hold in practice, resulting in institutional mismatch. We aim to advance theoretical and applied discourse on common pool resource governance in social-ecological systems, with implications for sustainable land management policies in Namibia and other landscapes across sub-Saharan Africa.

Keywords: common pool resource governance; design principles; forestry policy; institutions; Namibia; wildlife policy

Citation: Mbidzo, M.; Newing, H.; Thorn, J.P.R. Can Nationally Prescribed Institutional Arrangements Enable Community-Based Conservation? An Analysis of Conservancies and Community Forests in the Zambezi Region of Namibia. *Sustainability* **2021**, *13*, 10663. https://doi.org/10.3390/su131910663

Academic Editor: Brent Jacobs

Received: 31 May 2021
Accepted: 19 September 2021
Published: 25 September 2021

Publisher's Note: MDPI stays neutral with regard to jurisdictional claims in published maps and institutional affiliations.

Copyright: © 2021 by the authors. Licensee MDPI, Basel, Switzerland. This article is an open access article distributed under the terms and conditions of the Creative Commons Attribution (CC BY) license (https://creativecommons.org/licenses/by/4.0/).

1. Introduction

The relationship between resource degradation and common property systems has been the subject of intensive research for many years. The underlying debate is based on the assumption that when resources are limited and accessible to multiple users, everyone over-uses resources [1]. To avoid this 'tragedy of the commons', it was initially postulated that the commons should be privatised or, if kept as public property, exclusive rights to entry and use should be allocated by public authorities [1]. However, this position has been widely critiqued (e.g., [2–4]) because it overlooks the existence of local norms, rules, and governance systems. There is now substantial evidence demonstrating that resource users effectively conserve and sustain their natural resources over the long term through local institutions [2,5–8]. In some countries, this understanding has led to the devolution of natural resource management, but with varying degrees of success [9]. Clearly, understanding the conditions affecting success is important for informing developments in international and national policy.

Common pool resource (CPR) theory has emerged as a strong analytical framework for understanding the local characteristics of successful community-based conservation [1,5,6,10,11]. Here, we refer to community-based conservation as an approach where long-term conservation success requires engaging with and providing benefits for local communities [12]. There is substantial literature that applies the CPR framework to common resources in different ecosystems, including forests, rangelands, and marine or freshwater systems [13]. Central to CPR theories are Ostrom's eight design principles, which outline common characteristics of successful long-enduring, self-governing CPR institutions [2,14,15] (Table 1). Whilst there are many variations on this set of principles which overlap considerably with conditions identified by other scholars [e.g., 10,11], they remain a useful starting point to assess various local institutional arrangements that are present in CPR systems.

Table 1. Design principles for long-enduring and self-governing CPR institutions [2,15].

Design Principle	Explanation
1. Clearly defined boundaries	1A: User boundaries—individuals or households with rights to withdraw resources from the CPR, and 1B: Resource boundaries—the boundaries of the CPR itself.
2. Congruence	2A: Congruence with local conditions—appropriation and provision rules are congruent with local social and environmental conditions. 2B: Appropriation and provision—the distribution of benefits from appropriation rules is roughly proportionate to the costs imposed by provision rules.
3. Collective choice arrangements	Most individuals affected by operational rules can participate in modifying those rules.
4. Monitoring	4A: Monitoring users—monitors who are accountable to the users monitor the appropriation and provision levels of the users. 4B: Monitoring the resource-monitors who are accountable to the users monitor the condition of the resource.
5. Graduated sanctions	Violators of rules are sanctioned depending on the seriousness and context of the offence by other users, by officials accountable to these users or from both.
6. Conflict resolution mechanisms	Users and their officials have rapid access to low-cost, local means to resolve conflict among users or between users and officials.
7. Minimal recognition of rights to organise	The rights of users to devise their own institutions are not challenged by external authorities.
8. Nested enterprises (for CPRs that are part of larger systems)	Appropriation, provision, monitoring, enforcement, conflict resolution and governance activities are organized in multiple layers of nested enterprises.

In this paper, we use Ostrom's principles to explore the relationship between nationally prescribed institutional arrangements and devolved natural resource governance in Namibia. The Namibian devolution process was implemented through the community-based natural resource management (CBNRM) program, drawing on international experience, especially Zimbabwe's Communal Areas Programme for Indigenous Resource Management (CAMPFIRE) [16]. CAMPFIRE demonstrated that in order to influence people's behavior, management authority and benefit rights need to be devolved to the lowest possible unit [16,17]. However, in this case, the actual level of devolution-associated success was limited [18,19]. Thus, devolving authority to a higher local level is not sufficient to ensure efficient CBNRM by itself. Institutional arrangements are also critical, including how and by whom rules are made and sanctions are applied [3]. The presence of both community conservancies (focusing on wildlife management and tourism) and community forests (focusing on managing natural vegetation, forests, and woody vegetation, excluding wildlife) in Namibia, with contrasting nationally prescribed institutional arrangements, offers a natural experiment to explore the effects of these arrangements on local institutions and participation in decision making.

Many approaches have been suggested to deal with the challenges of community management of natural resources. These include CPR frameworks, CBNRM, integrated natural resource management, co-management, and institutional design principles, among others. CPRs are resource systems that are "... sufficiently large as to make it costly (but not impossible) to exclude potential beneficiaries from obtaining benefits" from their use [2] (p. 30). Therefore, at the heart of CPR theory is an understanding that there is a tradeoff between the costs and the benefits of excluding potential users, and the outcomes of this tradeoff, both for the community as a whole and for individuals. Although not conceived with this objective, Ostrom's framework also overlaps with the social-ecological systems concept which emphasizes the complex interactions and outcomes of different social, biophysical, policy, and economic systems [18–21], and allows for analyses of systems which are multi-layered at different internal and external levels, geographic scales, or nested systems [21]. It also involves the active integration of local inhabitants' voices, knowledge, and expectations [22]. Although we do not apply the first and second tier concepts of Ostrom's social-ecological framework [23,24], we acknowledge these nested, multiscale interactions in our analysis.

There is mounting evidence to show collaborative, community-based conservation as important in helping national governments meet international policy targets, while resulting in effective biodiversity outcomes [25]. At least 50% of the global land area is under customary ownership and management; the livelihoods of 2–3 billion people are directly dependent on the landscapes; and most of this land is rich in biodiversity [26]. The participation and rights of Indigenous peoples is listed as the first of 13 enabling conditions in the CBD's draft post-2020 Global Biodiversity Framework [27]. Despite considerable research on enabling conditions for effective governance of CPR, there is limited empirical understanding in particular contexts, including Namibia, of under what conditions institutional arrangements perform best [28].

This paper analyses the influence of national institutional prescriptions on local management of CPR institutions of communal conservancies and community forests. To meet this aim, we explore the following questions: how do natural resource management policies affect the operations of local CPR institutions and participation in decision making? and how do external factors affect local institutions and community participation in CPRs decision making? Results can contribute to an emerging research agenda on CPR theory [4] and concepts of enabling conditions.

2. Namibian Context

Namibia is well known for its recent work in community conservation—including community forests and communal conservancies [29]. Prior to 1960, all natural resources belonged to the colonial or South African government, and it was not until 1968 that freehold farmers' rights over wildlife were recognized [30]. This property regime was reinforced in 1975, when rights for white freehold farmers over wildlife were recognized through the Nature Conservation Ordinance (No. 4 of 1975). This legislative reform contributed to major increases in wildlife numbers on commercial farms for controlled hunting, managed through a permitting system [31]. However, local people on communal land had no legal rights to use natural resources or take action against any illegal hunting.

During the early 1980s, in response to declining wildlife numbers [32], conservationists started building trust with traditional leaders on communal lands, through the appointment of renumerated game guards for patrolling and reporting. By the late 1980s, wildlife populations substantially recovered [33]. This era marked the origin of CBNRM in Namibia. However, it had no legal backing. After independence in 1990, to redress past inequalities in land distribution and rights over wildlife, Namibia decentralized natural resource management on communal land, with regional government offices still involved. The creation of a mechanism for communal conservancies in 1996 marked a paradigm shift. Communal conservancies enabled communities to manage the resources where they live, and acknowledged the challenges of centralized government enforcement due to the large distances from central offices and limited resources [34]. Following this, in 2001, a similar mechanism for community forests was created.

By 2019, communal conservancies and community forests had been created on about 58.7% of all communal land, with an estimated 227,802 residents [35]. They covered 21.9% of the national territory, compared to 17.6% in National Parks and state-owned concessions and 6.1% in private conservancies. Thus, conservancies and community forests remain a significant component of the overall national conservation estate. The numbers continue to grow, with financial, technical, and political support from the national government, civil society, and multilateral donor agencies including USAID, UNDP, GEF, and the World Bank. However, there is an 84% overlap between conservancies and forests, which means that the same geographical areas are subject to two different sets of policy prescriptions and accountable to different government Ministries.

3. Materials and Methods

The presence of two contrasting nationally prescribed institutional mechanisms for CBNRM offers a unique opportunity for a comparative analysis of their role in enabling CBNRM on the ground.

The research consisted of two parts: a policy analysis and a set of key informant interviews. The policy analysis involved the identification of key laws and policy instruments through desk searches followed by a comparative content analysis focusing on institutional aspects and using CPR theory as a framework.

Subsequently, 28 in-depth semi-structured interviews were conducted between July 2012 and May 2013 using an interview guide to elicit information about the operations of institutions involved in natural resource management. Interviews assessed the functioning of local CPR institutions and determined the degree to which these institutions met conditions regarded as important for successful CPR institutions: boundaries, decision making, rules, monitoring, sanctions, and conflict resolution [2]. Further information about local and external institutions was collected related to the following: origins and development of the organization in terms of historical context and interests; institutional capacity in terms of skills, personnel, and financial resources; and institutional linkages in terms of levels of collective actions and information exchange.

The respondents were identified through snowball sampling. Individuals were targeted by virtue of their institutional position or experience and were expected to have in-depth knowledge of conservancies and community forests. At the national level, they included senior staff in the Ministry of Environment and Tourism, the Ministry of Agriculture, Water and Forestry, and one relevant NGO Integrated Rural Development and Nature Conservation (IRDNC). Regional and local respondents were selected from the Zambezi region, where there was a cluster of overlapping conservancies and community forests. They included government officials, NGO staff, regional foresters, chief control wardens, and others including conservancy/community forest staff, and committee members and community leaders. Interviews were conducted with three members of the same institution wherever possible but, other than this, the sample size was not set in advance but was reviewed during data collection using the principle of triangulation to determine whether a point of saturation had been reached. Saturation is defined as the point at which additional data collection "produces little important new information or understanding that is relevant" [36] (p. 75). A full list of informants is presented in Table 2. A more in-depth field study was undertaken in this region and is reported elsewhere [37].

Table 2. List of informants interviewed.

Informant No.	Operational Level	Sector	Institution	Position
1	National	Government	Ministry of Environment and Tourism (MET)	Deputy director, scientific services
2	National	Government	Ministry of Environment and Tourism	Director, environmental affairs
3	National	Government	Ministry of Environment and Tourism	National director, Regional services and parks management
4	National	Government	Ministry of Agriculture, Water and Forestry/Deutscher Entwicklungsdienst (DED)	Community forestry in Namibia programme officer
5	National	Government	Ministry of Agriculture, Water and Forestry/Deutsche Gesellschaft für Internationale Zusammenarbeit (GIZ)	Senior management advisor
6	National	Government	Ministry of Environment and Tourism	Coordinator, Namibia Protected Landscape Conservation Areas Initiative (NAMPLACE) project
7	National	NGO	Integrated Rural Development and Nature Conservation (IRDNC)	Co-Director
8	Regional	NGO	Integrated Rural Development and Nature Conservation (IRDNC)	Regional Assistant director
9	Regional	Government	Kavango Zambezi Transfrontier Conservation Area (KAZA TFCA)	Regional Liaison officer
10	Regional	Government	Ministry of Land and Resettlement	Deputy director, regional programme implementation
11	Regional	Government	Ministry of Environment and Tourism	CBNRM warden, regional services
12	Regional	Government	Ministry of Agriculture, Water and Forestry	Senior forestry technician
13	Regional	Government	Ministry of Agriculture, Water and Forestry	Community forestry technician
14	Regional	Government	Ministry of Environment and Tourism	Landscape specialist, Namibia Protected Landscape Conservation Areas Initiative (NAMPLACE) project
15	Regional	Government	Zambezi regional council	Chief regional officer
16	Local	Communities	Sobbe communal conservancy	Acting conservancy manager
17	Local	Communities	Sobbe communal conservancy	Senior community resource monitor
18	Local	Communities	Kwandu communal conservancy	Conservancy chairperson

Table 2. *Cont.*

Informant No.	Operational Level	Sector	Institution	Position
19	Local	Communities	Kwandu communal conservancy	Manager
20	Local	Communities	Kwandu communal conservancy	Field officer
21	Local	Communities	Kwandu traditional authority	Headmen
22	Local	Communities	Mashi communal conservancy	Chairperson
23	Local	Communities	Mashi communal conservancy	Manager
24	Local	Communities	Sachona community forest	Chairperson
25	Local	Communities	Masida community forest	Chairperson
26	Local	Communities	Masida community forest	Vice chairperson
27	Local	Communities	Kwandu community forest	Honorary forester
28	Local	Communities	Lubuta community forest	Chairperson

All interviews were conducted in person by the first author in either English or in Silozi and lasted 1.5–2 h. Interviews were recorded in detailed handwritten notes, audio recordings or both. All audio data were transcribed. The principles of thematic analysis were used to organize the qualitative data by creating and applying codes to the data [38,39]. The development of the coding protocol (categories) was informed by the conceptual framework of CPR design principles [40] and codes (free nodes) based on the research questions. Six broad 'operational' themes or codes were created, several of which included two or more sub-codes. The six broad themes included the following: (1) community characteristics; (2) rules in use; (3) rule enforcement; (4) support; (5) conflicts; and (6) interactions. Coding was completed using NVIVO v.10, first according to the broad predetermined themes and subsequently using 'free' nodes. Following coding, material was extracted on each theme and synthesized into a summary.

4. Results

4.1. Desk Analysis: Key National Policies Influencing CPR Institutional Arrangements in Namibia

The regulations that affect CBNRM in Namibia are wide-ranging, dispersed across various legal and policy frameworks and ministries, and have changed over time. This makes implementation and coordination of CBNRM challenging. Here, we discuss five key policies.

First, the policy on Wildlife Management, Utilisation and Tourism in Communal Areas was enacted by the Ministry of Environment and Tourism in 1995. In this policy, past discriminatory provisions of the Nature Conservation Ordinance (1975) (the second key policy) were removed for communal farmers to gain the same recognition of rights over wildlife as freehold farmers. This paved the way for the formation of conservancies on communal land.

Following this, legislative reform allowed conservancies to be registered through the Nature Conservation Amendment Act 5 of 1996. The act recognized the right to the consumptive and non-consumptive use (typically for tourism) and sustainable management of wildlife in conservancies [41]. Many saw this as an important step for communities to have greater control and benefit from resources; to provide for wildlife damages offsets; to reduce uncontrolled harvesting; and to prevent harassment from illegal hunters [42]. Under this act, clearly defined boundaries, membership, a committee, a constitution, and a plan for the equitable distribution of benefits to members are all required to be a registered conservancy. Membership is voluntary rather than prescribed; and is based on how members of the conservancy are defined by the adult (≥18 years) communal area representatives. This enables communities to use existing institutions, including traditional institutions, as the basis for their conservancy committee (see next section). An individual or community can sell or lease the rights of management and exclusion or both, as outlined by the right of alienation. Property can be transferred from the Ministry of Environment and Tourism or the Ministry of Agriculture, Water and Forestry to the communities in two ways. The first is that property rights are awarded de jure, whereby the government explicitly grants such rights to communities living in these areas by formal law. The second is recognition of de facto rights, where land is communally owned and falls under customary law [43].

A third important policy is the Forest Act 12 of 2001. Like the conservancy legislation, the act recognizes the rights of communities over forest resources, with the twin goals of CBNRM in mind: conserving biodiversity and improving rural livelihoods. Communities enter into a written forest management agreement with the government based on defining boundaries, developing management and benefit-sharing plans, cost sharing arrangements, and appointing a management authority. To date, however, the establishment of community forests has been somewhat slow compared to conservancies, partly due to lack of funding.

The fourth important policy is the Communal Land Reform Act 5 of 2002, which is implemented under the Ministry of Lands and Resettlement (MLR). This act recognizes customary law and makes provision for traditional authorities to administer, allocate, and be involved in the registration of communal land rights. The act also determines the conditions of grazing rights on communal land, including allocation to non-residents [44]. Customary land rights to areas under 20 hectares can be allocated to individuals for up to 99 years and transferred to descendants of the rights-holder.

Fifth, the Traditional Authority Act 25 of 2000 makes provision for traditional authorities to apply customary law in the allocation of communal land, harvesting forest resources and other matters related to CPR. The Traditional Authority Act is implemented by the Ministry of Regional, Local Government and Housing and Rural Development (MRLGHRD). Section 16 of the Traditional Authority Act requires traditional authorities *'to support policies of government, regional councils and local authority councils and refrain from any act which undermines the authority of those institutions'* (p. 13). Thus, most conservancies and community forests are directly linked to a traditional authority [43,45].

Under this act, the traditional authority court (khuta), is the governing body in each district (Figure 1). Each village has a headman (induna) and a senior headman (induna silalo) who represents several villages. In most cases, the principal advisor (ngambela) does not directly communicate with the chief (litunga), but instead information is conveyed through the deputy advisor (natamoyo). Disputes are first considered at the village level by the village indunas. If a solution is not found, the matter is escalated to the district khuta where village and senior headman discuss the matter, overseen by the induna silalo that presides over the district. If a solution is still not found, the matter is escalated by the induna silalo to the higher traditional authority khuta to the ngambela. If the ngambela is unable to settle the matter, it is referred to the litunga who will hear witness statements privately and publicly with the concerned communities before a verdict. The last resort would be to refer the matter to the magistrate court.

However, three areas prove challenging to the implementation of the Traditional Authority Act: (i) it is open to interpretations when there are power struggles or legal cases between traditional authorities and government [45]; (ii) customary laws (i.e., norms, rules of procedure, traditions, and usage) that are not written down can be difficult for outsiders to ascertain and subject to diverse interpretations; and (iii) enforcement is dependent on the traditional authority's legitimacy.

Table 3 summarizes the key features of the national prescribed institutional frameworks for community conservancies and community forests that are relevant to CPR theory. These relate to access and withdrawal, management decision-making, powers of exclusion, and transfer of rights, including alienability.

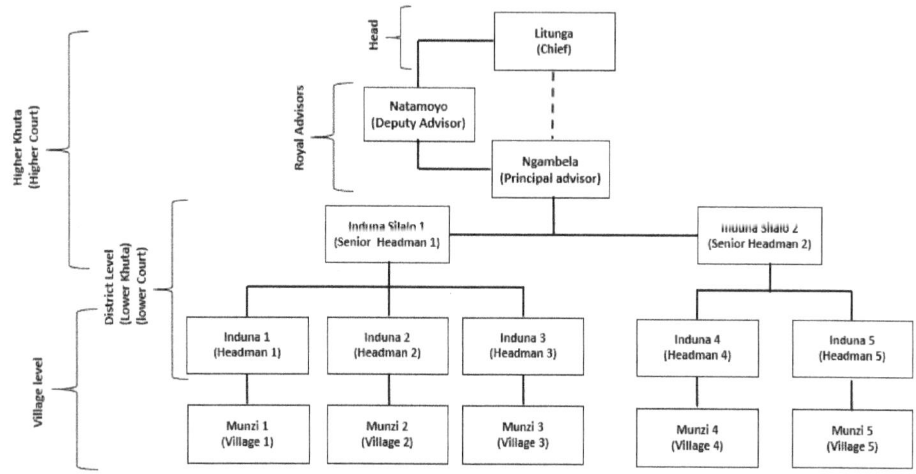

Figure 1. Simplified traditional rural governance structure in the Zambezi Region. Typically, khutas comprise of the chief (litunga), principal advisor (ngambela), deputy advisor (natamoyo), senior headman (induna silalo), headmen (indunas, or councilor) and secretaries. The ngambela is the administrator, while the natamoyo is the personal advisor to the litunga.

Table 3. Comparison of key features of nationally prescribed institutional arrangements for conservancies and community forests.

Access and Withdrawal	Management Powers	Powers of Exclusion	Transfer of Rights, Including Alienability
	Communal conservancy		
Geographical boundaries are legally defined. Conservancy members are legally registered and their rights to use and benefit from certain wildlife resources are legally recognised. Wildlife quotas are set by the Ministry of Environment and Tourism. The traditional authority, not the conservancy, may grant grazing rights to non-residents.	Communal conservancy management and executive committees make management decisions about wildlife. Community game guards monitor wildlife and report violations to the Ministry of Environment and Tourism. Conservancies may apply for permits for live capture and sale of wildlife. They can also ask for permission to reduce numbers of certain wildlife species.	The management committee can cancel an individual's membership. The community decides on which villages may be part of the communal conservancy. However, they have no powers to exclude outsiders.	Rights to sell or lease the resources are very limited. A conservancy can enter into a contract with an investor granting them permission to develop a tourism facility such as a lodge. Customary land rights may be transferred to descendants of the rights-holder. Only the Ministry of Environment and Tourism can dissolve the communal conservancy.
	Community forest		
Geographical boundaries are legally defined. Any person with traditional rights to the area has rights to harvest and benefit from forest resources. Annual allowable cut for tree species is determined by the Directorate of Forestry. Grazing rights can be granted by the government in consultation with the traditional authority.	Management powers over a specified area are devolved to the community level. The community shares responsibility with the Directorate of Forestry regarding the control of forest use.	Powers to exclude outsiders from encroaching the forest are limited. The traditional authority may grant grazing rights to non-residents.	Rights to sell or lease the resources are very limited. Customary land rights may be transferred to descendants of the right holder. Only the Ministry of Agriculture, Water and Forestry can dissolve the community forest.

In the rest of this section, we explore and discuss the implications of the above features for enabling implementation on the ground, informed by the results from the semi-structured interviews and using Ostrom's eight design principles as a framework (Table 1).

4.2. Clearly Defined Boundaries

Design principle 1 is concerned with defining the boundaries both of users; the set of individuals or households who have rights to withdraw the natural resources (principle 1A) and of the natural resources included or excluded under the CPR regime (principle 1B). As outlined above, user boundaries are precisely defined under the legal frameworks for conservancies, which should keep a register of all members. All conservancy constitutions defined members by their residence status within the conservancy boundaries and in most cases the register included information on residence status. However, the interviews revealed contradictions between formal registration and customary systems of resource governance. One elderly community member expressed this as follows:

> *I don't have to register to become a member of this conservancy, Everyone knows I was born here, even my parents were born here, this is my area, why should I register?*

In contrast, members of community forests are defined by law as anyone with customary land rights to an area, even if they do not currently reside in the area. This means that user boundaries are not as clearly defined as for conservancies and there is no list of members. On the other hand, there is more flexibility for decisions about which individuals or households are included—made according to local customary governance systems. According to the chairperson of one conservancy committee that also administers an overlapping community forest, the reasoning behind this approach is to ensure that no-one with customary rights is excluded from benefiting directly from the use of forest resources.

In relation to boundaries of the natural resources, the geographical boundaries of both conservancies and community forests are legally defined and well understood by local community members. In community forests, which are concerned only with the management of stationary resources (plants), this is sufficient to fully define the boundaries of the natural resources concerned. However, conservancies are concerned with wildlife, some species of which move between conservancies, across landscapes and even between countries, which greatly complicates efforts to set quotas and to monitor resources and their use. According to a landscape specialist, this has been partially addressed through joint game counts, patrols, and post-translocation monitoring of wildlife over a wider area.

4.3. Congruence between Appropriation and Provision Rules and Local Conditions

Design principle 2A concerns congruence between rules for the appropriation, as well as use and provision of resources with local social and environmental conditions. A crucial aspect of environmental conditions is the state of the natural resources concerned. In both conservancies and community forests, this was considered through the use of ecological monitoring to set quota allowances. Monitoring in conservancies took the form of annual game counts of different species, which was regarded as largely effective despite the limitations related to wildlife mobility outlined in the previous section. Monitoring in community forests took the form of forest inventories which, according to an official from the Directorate of Forestry, were used in setting timber harvesting quotas. However, the accuracy of forest inventories was raised as a potential limiting factor in the effectiveness of the timber quota system.

There was little evidence of flexibility to adapt rules about appropriation and use to local social and cultural conditions. Conservancies can request hunting quotas for their own use for religious and cultural festivals from the Ministry of Environment and Tourism. Subsistence hunting is not permitted unless individuals have paid for a permit, which most members cannot afford runs counter to customary norms and practices

Design principle 2B is concerned with the distribution of costs and benefits from resource management and use and the balance between them. In both conservancies and community forests, members who were employed or who were in management committees (incurring a cost in terms of time and effort) received monetary benefits in return. Committee members stated that others received non-monetary benefits, such as shared meat in conservancies and the rights to use forest resources including non-timber forest products (NTFPs) in community forests. One option for conservancies was to offer short-term positions so that more people would receive benefits over time. However, according to one NGO official, this created challenges in terms of continuity and abuses of power:

> *"Weak institutional memory is a challenge if you have a new committee that is enacted every two to three years … Those* [conservancy] *positions are very vulnerable because people see that* [having a position in the conservancy] *as the* [main] *benefit".*

4.4. Collective Choice Arrangements

Principle 3 states that most members who are affected by the rules within a CPR regime should have a say in formulating the rules. This depends both on the extent to which the rules are nationally prescribed or can be formulated locally, and second on who has a say in decisions at the local level. In relation to the first of these, there are many nationally prescribed rules for both conservancies and community forests over which the resource users have no say. Members of conservancies raised this as problematic, saying that the prescribed rules were too restrictive and that they were powerless to oppose them. One example was the nationally prescribed ban on traditional subsistence hunting without payment. Committee members expressed the view that at the least they should be able to hunt birds (such as guinea fowl) and small animals (such as South African springhare) for subsistence use, and also that they should be allowed to walk with dogs in the forest for protection against wild animal attacks. During 2013 and 2014, the Directorate of Forestry placed a national moratorium on the harvest and trade of timber in Namibia, which affected income generation in community forests. The moratorium was triggered by concerns about unsustainable use of forest resources, particularly in the north-eastern regions of the country. The chairperson of one community forest expressed frustration over the moratorium:

> *"They* [the Directorate of Forestry] *came to stop us from cutting timber because of some other people outside the community forest that were cutting timber without permits, why is that?"*

NGOs have also played a crucial role in defining the rules for conservancies, and while the intention may be to empower communities to formulate the rules, this may not translate in practice. For example, one senior NGO official described the process as:

> *"Guiding conservancies to make sure they have good governance"* but was quick to admit that *" … we cannot pretend that we are not influencing them* [conservancies]*".*

In terms of local participation in rule-making, some conservancies promoted the direct participation of all members (e.g., in attending annual general meetings, voting on conservancy matters, speaking at any meeting) whereas others allowed only a small number of representatives to participate. In community forests, the scope for member participation was restricted both by the lack of regular meetings of the whole community and by a lack of information about the occasional meetings that did take place. According to the chairperson of one community forest, in many instances the management committee took decisions in consultation with just the traditional authority.

4.5. Monitoring

According to principle 4, monitoring should be carried out regularly both of user behavior (4A) and of the condition of the natural resources (4B), and monitors should be accountable to the user group. In conservancies, monitoring took the form of regular patrols by community game guards who were employed by and answerable to the management committees, whereas in community forests the committee members had to carry out monitoring themselves. Conservancy monitoring tended to be carried out on a regular basis, while community forest monitoring activities tended to be less regular due to the lack of paid personnel and incentives. As was said by the chairperson of a community forest:

> *'People are not willing to work for free, that is why some committee members prefer not to be active'.*

Both local and external respondents pointed out the lack of funds in community forests as a major setback to the success of forest management. Interviews with forestry officials indicated that funding to community forests had come mainly from international donors. Most conservancies, on the other hand, secured substantial operational funds from trophy hunting and other wildlife related activities such as joint-venture tourism. An analysis of financial reports from a sample of conservancies in the study area show that conservancy income varied between years and conservancies. Conservancies that had diverse sources of income (e.g., trophy hunting and tourism joint ventures) showed a steady increase in income over three years while those that relied only on trophy hunting showed no clear patterns. However, the results further indicate that most of the funds generated by conservancies went towards operational costs, leaving very little to benefit the wider community.

Monitoring of natural resources is included in the policy prescriptions for both conservancies and community forests. In both cases, responsibility for monitoring rests with the user group, with assistance from other stakeholders (e.g., NGOs, ministries). According to MET officials, conservancies are required to conduct annual game counts in order to be allocated hunting quotas. Monitoring activities are described in Section 4.5 above. Monitoring in conservancies tended to be much more regular and robust than in community forests, because of the existence of financial resources and monetary payments for community monitors. One government official from the Directorate of Forestry also indicated that local forest monitors were prone to bribes from timber dealers and that therefore the Department of Forestry needed to be involved in monitoring forest resources whenever possible.

4.6. Graduated Sanctions

Principle 5 states that violators of the rules are sanctioned according to a graduated system, depending on the seriousness and context of the offence. This principle also states that sanctions are applied by users, officials accountable to these users, or both. Interviewees at all levels indicated that the principle of graduated sanctions was strongly adhered to in both conservancies and community forests. The procedures for handling cases of rule-breaking and types of conflict resolution or sanction depended on the severity of the case. For example, sanctions for illegal hunting in conservancies varied according to species. However, sanctions in community forests were generally mild unless the offence involves harvesting high-value timber. Sanctions were commonly perceived as harsher for first offences in conservancies than in community forests. However, severity of offence was defined differently by different actors. For instance, one ministry official considered illegal hunting involving protected and high-value species as a severe offence, while some conservancy committee members regarded all types of hunting including possession of game meat without permission as severe offences.

The extent of adherence to the second part of this principle, which relates to who has the authority to apply sanctions, is less clear-cut. One conservancy field officer stated that cases of illegal hunting of any wildlife species are reported directly to the Ministry of Environment and Tourism, who then decide whether to fine the rule violator through the court system or directly. In community forests, the users had greater powers of discretion in relation to sanctioning. Community forestry chairpersons indicated that local violators are usually just warned in the first instance and the illegally harvested product is confiscated. However, persistent and serious violations are reported to the Directorate of Forestry.

Collaboration between local CPR institutions and state institutions is particularly important in relation to enforcement and sanctioning where rules are broken by outsiders. In relation to conservancies, it was reported that communities cannot deal with serious cases of illegal hunting by outsiders themselves due to their limited mandate, decision making power, and equipment. The role of the traditional authority has diminished over time, as reported by one conservancy manager:

> "*They* [the traditional authority] *don't deal with natural resource crimes anymore, although in the past the conservancy would report to them*".

Legally, conservancies can apprehend but not arrest offenders and they must report them to the relevant ministry (e.g., Ministry of Environment and Tourism, Directorate of Forestry) or police. In contrast, illegal harvesting of forest resources was still usually dealt by the community forest management committee and the traditional authority.

4.7. Conflict Resolution Mechanisms

Principle 6 states that users and their officials should have rapid access to low-cost, local means to resolve conflict, whether it is among users or between users and officials. This principle is adhered to in community forests than conservancies because, in community forests, local conflict resolution mechanisms are in place and are often considered more responsive and effective than those involving external authorities. Wildlife-related conflicts, which fall under the responsibility of conservancies, are mainly handled externally by government institutions.

Although several key informants agreed that the traditional authority plays an important role in the governance of CPR in both institutions, there were different views about the role of the traditional authorities and their relationship to the nationally prescribed institutional structures. On the one hand, some informants argued that although the traditional authorities played a critical role in the initial formation of conservancies, they interfered with the later operations of the conservancy. On the other hand, respondents perceive that the power of traditional authorities has been weakened by the new conservancy structures, as voiced by one traditional leader:

> "*The conservancy is dominating us. Even now we have papers from the khuta* [traditional authority court] *saying we must deal with issues of natural resource use. We are supposed to charge people and get money out of it, but the conservancy has now taken over and dominates the khuta*".

4.8. Minimal Recognition of Rights to Organise

Principle 7 states that the rights of users to devise their own institutions should not be challenged by external authorities. In Namibia, all communal land belongs to the state but communities that apply and register their areas as conservancies and community forests have conditional recognition of rights to manage and benefit from CPRs in their areas. However, they have limited flexibility to devise their own institutions because of the detailed prescriptions in national policy. Therefore, the principle is only partially met. One particular issue arising from this study is that some aspects of the prescriptions are incompatible with customary institutions and norms. Moreover, communities still need to seek permission from external government ministries to use natural resources.

4.9. Nested CPR Systems

Design principle 8 states that the different aspects of CPR systems should be organized in multiple, nested layers. This is the crux of the current paper, which focuses on the relationship between the national and local levels. This is an area where CPR theory intersects closely with social-ecological systems theory, based on the principle that higher-level institutional structures and prescriptions should support local communities in order to increase resilience.

In the case of community forests and communal conservancies, the principle of nested systems is adhered to in the superficial sense that there is more than one layer of organization. There are substantial prescriptions for institutional structures and rules at the national level, whereas some details can be defined locally. There are also some aspects that are defined internationally and are reflected in national policy (e.g., in relation to trade in endangered species through the Convention of International Trade in Endangered Species of Fauna and Flora). However, as the preceding sections have demonstrated, the extent to which national components support local governance or give flexibility to adjust to local contexts is very limited. In some instances, national policy is in direct conflict with customary governance systems and has disenfranchised customary authorities. The overall long-term effect may thus be to weaken, rather than strengthen, community coherence and governance. Although legislation allows communities to manage resources, they must do so within the specific prescriptions laid down at the national level and are still substantially dependent on government decision-makers to develop, implement, and enforce national policies.

The existence of the two contrasting institutional frameworks at the national level creates further complexity and confusion, especially where they are applied over the same area of land. This was commented upon not only by community members but also by government officials. One ministry official described the problem as follows:

> *'I don't think it is proper to give people an area to manage animals within it, but they don't have the right to use and manage the trees and plants around them. Would one take the conservancy to court if the elephant kept on destroying the forest? Should a forester say to the conservancy your elephants are destroying my forest? This is why all resources should be inclusive and belong to one target group in a specific area'.*

Interviews with government officials revealed that the ministries responsible for implementing conservancy and community forest policies tended to make decisions independently. This separation was further apparent at the local level through the formation of separate committees for the two types of resources—often weakening pre-existing traditional systems of natural resource governance. Some regional government officials indicated that integrated decision-making system at the local level is difficult to achieve because of the lack of cross-sectoral cooperation at higher levels and an NGO official expressed the same sentiment:

> *"On the ground, yes, there is some sort of collaboration. But at national level, there is no collaboration between the stakeholders within CBNRM. We need to sit together and chat a future together, and do integrated planning, implementation and monitoring".*

The desk review of policies and other government documents revealed clearly that the barriers to integrated natural resource decision making stemmed from segregated national governmental structures and legal frameworks governing different types of natural resources, and particularly inadequate coordination between the separate agencies responsible for management of wildlife and forests.

A simplified summary of the extent to which Ostrom's design principles are met in national prescriptions for conservancies and community forests is presented in Table 4.

Table 4. Strength of alignment between CPR design principles and institutional arrangements for conservancies and community forests.

Design Principle	Conservancies	Community Forests
1a. Clearly defined user boundaries	Strong. A full list of registered users is required in order to register a conservancy.	Medium. Criteria for defining users are set nationally but their application is left to the local level.

Table 4. *Cont.*

Design Principle	Conservancies	Community Forests
1b. Clearly defined resource boundaries	Medium. Geographical boundaries are defined but wildlife are highly mobile.	Strong. Geographical boundaries are defined and the boundaries fully define the resources to be managed (plants).
2a. Congruence with local conditions	Social: Weak. Little evidence of flexibility to adapt to local social conditions. Environmental. Medium. quotas are based on annual game counts, but game move over large areas, limited effectiveness of this approach	Social: strong. Few restrictions on most subsistence forest resource use. Some involvement of traditional authorities. Environmental: medium. Timber harvesting quotas are based on forest inventories, but concerns were raised about their accuracy.
2b. Proportionality of costs and benefits	Strong. Those who take on specific activities such as monitoring are paid.	Weak. Very little financial benefits to those who are involved in specific activities.
3. Collective-choice arrangements	Medium. Many aspects are nationally prescribed, but community members do have a say in locally prescribed aspects.	Low. Many aspects are nationally prescribed and for those that are locally prescribed there is little opportunity for most members to have a say.
4a. Monitoring of users	Medium. Regular patrols by paid community game guards answerable to management committee.	Low. Monitoring irregular due to lack of incentives.
4b. Monitoring of the resource	Strong. Regular game counts.	Weak. Forest inventories of variable quality.
5. Graduated sanctions	Medium. Sanctions varied according to the type of offence and whether it was a first offence, but the powers of conservancy to apply sanctions were limited and the powers of traditional authorities had diminished.	Medium. Sanctions varied according to the type of offence and whether it was a first offence but tended to be mild. Community forest management committees and traditional authorities had greater powers to apply sanctions.
6. Low-cost local conflict resolution mechanisms	Weak. Conflicts related to wildlife were handled mainly by government institutions.	Strong. Local conflict resolution mechanisms in place.
7. Minimal recognition of rights to organise	Medium. Both policy mechanisms (conservancies and community forests) enable communities to gain recognition of rights to resource use in defined areas of land but they must organize themselves according to rigid nationally prescribed institutional requirements.	
8. Nested enterprises	Medium. Both policy mechanisms (conservancies and community forests) involve more than one layer of institutional structures but there are inconsistencies between the different layers and further contradictions where conservancies and community forests overlap.	

5. Discussion and Conclusions

The above analysis provides some preliminary insights into the effects of nationally prescribed institutional arrangements for conservancies and community forests and the broader implications in terms of enabling conditions for community-based conservation.

First, it demonstrates the critical importance of adequate coordination across sectors, scales, types of resources, and between different legal and administrative systems. This is particularly needed considering the large dispersal area of wildlife which move across conservancy boundaries (similar to fisheries, see [46]) and require a nested approach involving both local and external management. In Namibia, sectoral policies and legislation have created competing and overlapping national, regional, and local community institutions for management of different kinds of natural resources (wildlife, under communal conservancies, and timber and NTFPs under community forests). The need for more integrative whole-of-government approaches across sectors is recognized as an enabling condition for effective conservation in the draft post-2020 Global Biodiversity Framework [27].

Second, it highlights the risk that overly rigid national prescriptions may be disenfranchising traditional authorities and weakening the resilience of local social-ecological systems. This is an issue that has been widely documented in other parts of the world (for example see [47]). Ostrom's design principles refer to nested systems that operate at different levels and imply a balance between the creation of a framework at the higher levels to enable local actions and the need to leave flexibility for the details to be designed locally according to the local social and environmental context.

Third, financial independence is critical to any institution that hopes to produce results and perform administrative functions. Conservancies were able to generate income through tourism, whereas no community forests were reported to have secured long-term funds for their operations, and this was recognized as a major constraint. The identification of sustainable sources of conservation finance is a key issue in the post-2020 Global Biodiversity Framework, including finance for community contributions to conservation [48]. We recommend the introduction of measures to ensure that communities receive direct benefits in return for their efforts, which could motivate them to sustain active participation in forest management and hold positions for a longer period to avoid rent-seeking behavior and loss of institutional memory and effectiveness [49]. Where conservancies and community forests overlap, opportunities for joint monitoring exist. Forest inventories need to be regularly updated to ensure that allocated harvesting quotas are still within sustainable limits. The allocation of wildlife hunting quotas to a cluster of conservancies in the same vicinity could improve congruence with resource conditions.

Fourth, as early as 1998 [34], Jones called for a change in policy to embed secure land tenure in natural resource management systems to ensure that communities hold secure rights over all natural resources on their land. Although progress has been made (e.g., through the Flexible Land Tenure Act 2012 and CBNRM policy which promotes integrated land and natural resource planning and decision making), more than two decades later, the strengthened tenure rights have not materialized for most communities in spite of widespread recognition in academic and global conservation policy fora of the fundamental requirement for secure tenure rights as a precondition for the long-term effectiveness of community-based natural resource management [27,50]. There is a pressing need to revisit the current legislation to strengthened tenure rights and ensure conducive environment for the development of community conservation in Namibia.

The analysis in this paper of conservancies and community forests from Namibia demonstrates the value of Ostrom's design principles in highlighting how national policies may enable or disable community conservation in particular contexts. However, it also indicates the complexity and plurality of the relationships between national policies and local institutional arrangements. An overarching conclusion is that over-prescription at the national level can be counterproductive, weakening customary governance systems and local social-ecological resilience. The extensive body of research on CPRs has demonstrated that it is not just the nature of the rules and other institutional arrangements that is important, but also that the resource users are fully involved in their design and implementation, including defining resource boundaries, enhancing the monitoring, enforcement, conflict resolution and prosecution capacity, and meeting regularly with a wide range of actors to continually evaluate operation rules and norms, overcome ambiguities in handling violations, and enhance compliance [2,51].

Our research aligns with a recent focus on how environmental policies might be designed and transformed to improve the outcomes of community conservation. When the study was conducted, officials working in the different ministries responsible for implementing conservancy and community forest policies and legislation were making decisions independently. In 2020, the Department of Forestry was incorporated into the Ministry of Environment and Tourism; however, it is too early to evaluate the effect of such a change to the governance of wildlife and forest resources. This analysis provides a basis on which a detailed analysis of the actual performance of CPR institutions can be conducted; to determine under which conditions certain CPR systems would perform best. This study could also be used as a basis for follow up to see if conditions have changed and therefore provides a baseline which could be used to assess further changes in policy frameworks. We hope to have contributed to the global debate on theories about CPRs and environmental policy analysis.

Author Contributions: Conceptualization, formal analysis, investigation, visualisation, and project administration: M.M.; writing: M.M., H.N. and J.P.R.T.; supervision, H.N.; funding acquisition, M.M. (PhD funding) and J.P.R.T. (APC funding). All authors have read and agreed to the published version of the manuscript.

Funding: This research was funded by the Commonwealth Scholarship Commission in the United Kingdom, NACS-2011-288; and the National Council of Higher Education in Namibia (NCHE).

Institutional Review Board Statement: This study was approved by the Ethics Review Committee of the University of Kent, and a research permit was issued by the Namibian Ministry of Environment and Tourism (Permit number 1747/2012). At the local level, permission to conduct research was granted by the traditional authority, mainly from the senior headman.

Informed Consent Statement: Informed consent was sought according to the University of Kent Research Ethics Guidelines and measures were put in place to prevent harm to the participants of the study. A brief description of the study and its purpose was presented to each informant before the start of the interview.

Data Availability Statement: The data presented in this study are available on request from the corresponding author. The data are not publicly available due to privacy restrictions.

Acknowledgments: We wish to thank all people from the various organizations who took time to attend to the interviews, including people in all the study villages who took time from their daily activities to provide us with valuable information about their interactions with wildlife and forest resources. Permission to conduct this research was granted by the Ministry of Environment and Tourism, and the Ministry of Agriculture, Water and Forestry. We are indebted to their staff for facilitating the research in the Zambezi Region. Staff of the Integrated Rural Development and Nature both national and regional provided valuable insights on conservancies. The generous financial support provided by the Commonwealth Scholarship Commission in the United Kingdom and the National Council of Higher Education in Namibia made the research possible.

Conflicts of Interest: The authors declare no conflict of interest. The funders had no role in the design of the study; in the collection, analyses, or interpretation of data; in the writing of the manuscript, or in the decision to publish the results.

References

1. Hardin, G. The tragedy of the Commons. In *Managing the Commons*; Hardin, G., Baden, J., Eds.; WH Freeman: San Francisco, CA, USA, 1968.
2. Ostrom, E. *Governing the Commons: The Evolution of Institutions for Collective Action*; Cambridge University Press: Cambridge, UK, 1990.
3. Bromley, D.W.; Cernea, M.M. *The Management of Common Property Natural Resources: Some Conceptual and Operational Fallacies*; World Bank Publications: Washington, DC, USA, 1989; Volume 57.
4. van Laerhoven, F.; Schoon, M.; Villamayor-Tomas, S. Celebrating the 30th anniversary of Ostrom's governing the Commons: Traditions and trends in the study of the Commons, revisited. *Int. J. Commons* **2020**, *14*, 208–224. [CrossRef]
5. Agrawal, A. Common property institutions and sustainable governance of resources. *World Dev.* **2001**, *29*, 1649–1672. [CrossRef]
6. Dolsak, N.; Ostrom, E. The Challenges of the Commons. In *The Commons in the New Millennium: Challenges and Adaptations*; The MIT Press: Cambridge, MA, USA, 2003; pp. 3–34.
7. Gibson, C.C.; McKean, M.A.; Ostrom, E. Explaining deforestation: The role of local institutions. In *People and Forests: Communities, Institutions, and Governance*; The MIT Press: Cambridge, MA, USA, 2000; pp. 1–26.
8. Gautam, A.P.; Shivakoti, G.P. Conditions for successful local collective action in forestry: Some evidence from the hills of Nepal. *Soc. Nat. Resour.* **2005**, *18*, 153–171. [CrossRef]
9. Larson, A.M.; Dahal, G.R. Introduction: Forest tenure reform: New resource rights for forest-based communities? *Conserv. Soc.* **2012**, *10*, 77–90. [CrossRef]

10. Wade, R. The management of common property resources: Collective action as an alternative to privatisation or state regulation. *Camb. J. Econ.* **1987**, *11*, 95–106. [CrossRef]
11. Baland, J.-M.; Platteau, J.-P. *Halting Degradation of Natural Resources: Is There a Role for Rural Communities?* Oxford University Press: Oxford, UK, 1996.
12. Brooks, J.; Waylen, K.A.; Mulder, M.B. Assessing community-based conservation projects: A systematic review and multilevel analysis of attitudinal, behavioral, ecological, and economic outcomes. *Environ. Evid.* **2013**, *2*, 2. [CrossRef]
13. Van Laerhoven, F.; Ostrom, E. Traditions and trends in the study of the Commons. *Int. J. Commons* **2007**, *1*, 3–28. [CrossRef]
14. Ostrom, E. *Understanding Institutional Diversity*; Princeton University Press: Princeton, NJ, USA, 2005.
15. Cox, M.; Arnold, G.; Tomás, S.V. A review of design principles for community-based natural resource management. *Ecol. Soc.* **2010**, *15*, 38. [CrossRef]
16. Jones, B.T. Community Management of Natural Resources in Namibia 1999. IIED, London. Available online: https://pubs.iied.org/7415iied (accessed on 15 March 2021).
17. Murphree, M.W. *Congruent Objectives, Competing Interests and Strategic Compromise: Concepts and Process in the Evolution of Zimbabwe's CAMPFIRE Programme*; Institute for Development Policy and Management, University of Manchester: Manchester, UK, 1998.
18. Murombedzi, J.C. Devolution and Stewardship in Zimbabwe's CAMPFIRE Programme. *J. Int. Dev.* **1999**, *11*, 287–293. [CrossRef]
19. Whande, W.; Kepe, T.; Murphree, M. Local communities, equity and conservation in Southern Africa: A synthesis of lessons learnt and recommendations from a Southern African technical workshop. In Proceedings of the Communities and Conservation in Southern Africa: Key Issues and Challenges Towards a More Equitable and Sustainable Future Workshop, Pretoria, South Africa, 26–28 February 2003.
20. Cumming, G.S.; Olsson, P.; Chapin, F.S.; Holling, C.S. Resilience, experimentation, and scale mismatches in social-ecological landscapes. *Landsc. Ecol.* **2013**, *28*, 1139–1150. [CrossRef]
21. Janssen, M.A.; Anderies, J.M. A multi-method approach to study robustness of social–ecological systems: The case of small-scale irrigation systems. *J. Inst. Econ.* **2013**, *9*, 427–447. [CrossRef]
22. Carpenter, S.R.; Arrow, K.J.; Barrett, S.; Biggs, R.; Brock, W.A.; Crépin, A.-S.; Engström, G.; Folke, C.; Hughes, T.P.; Kautsky, N. General resilience to cope with extreme events. *Sustainability* **2012**, *4*, 3248–3259. [CrossRef]
23. Hinkel, J.; Cox, M.E.; Schlüter, M.; Binder, C.R.; Falk, T. A diagnostic procedure for applying the social-ecological systems framework in diverse cases. *Ecol. Soc.* **2015**, *20*, 32. [CrossRef]
24. Ostrom, E. A general framework for analyzing sustainability of social-ecological systems. *Science* **2009**, *325*, 419–422. [CrossRef]
25. Oldekop, J.A.; Holmes, G.; Harris, W.E.; Evans, K.L. A global assessment of the social and conservation outcomes of protected areas. *Conserv. Biol.* **2016**, *30*, 133–141. [CrossRef]
26. RRI (Rights and Resources Initiative). *Rights-Based Conservation: The Path to Preserving Earth's Biological and Cultural Diversity?* Rights and Resources Initiative: Washington, DC, USA, 2020; Available online: https://rightsandresources.org/wp-content/uploads/Final_Rights_Conservation_RRI_07-21-2021.pdf (accessed on 23 September 2021).
27. CBD (Convention on Biological Diversity) Update of the Zero Draft of the Post-2020 Global Biodiversity Framework. CBD/POSTt2020/PREP/2/1. 2020. Available online: https://www.cbd.int/doc/c/3064/749a/0f65ac7f9def86707f4eaefa/post2020-prep-02-01-en.pdf (accessed on 23 September 2021).
28. Gari, S.R.; Newton, A.; Icely, J.D.; Delgado-Serrano, M.M. An analysis of the global applicability of Ostrom's design principles to diagnose the functionality of common-pool resource institutions. *Sustainability* **2017**, *9*, 1287. [CrossRef]
29. Community Forestry in Namibia. Available online: http://www.fao.org/forestry/37205-0bae550691151380 23f681a41e025cf08.pdf (accessed on 4 September 2021).
30. Weaver, L.C.; Skyer, P. Conservancies: Integrating wildlife land-use options into the livelihood, development and conservation strategies of Namibian communities. In Proceedings of the Conservation and Development Interventions at the Wildlife/Livestock Interface. Implications for Wildlife, Livestock and Human Health, Durban, South Africa, 14–15 September 2003; Volume 30, pp. 89–104.
31. Barnes, J.I.; De Jager, J.L.V. Economic and financial incentives for wildlife use on private land in Namibia and the implications for policy. *South African J. Wildl. Res.* **1996**, *26*, 37–46.
32. Nott, C.; Jacobsohn, M. Key issues in Namibia's communal conservancy movement. In *Rights, Resources and Rural Development: Community-Based Natural Resource Management in Southern Africa*; Earthscan Publications: London, UK, 2004; pp. 194–199.
33. Hoole, A. *2007. Lessons from the Equator Initiative: Common Property Perspectives for Community-Based Conservation in Southern Africa and Namibia*; University of Manitoba: Winnipeg, Canada, 2007; p. 32. Available online: https://idl-bnc-idrc.dspacedirect.org/bitstream/handle/10625/39004/128323.pdf (accessed on 10 May 2021).
34. Jones, B.T.B. Namibia's approach to community-based natural resource management (CBNRM): Towards sustainable development in communal areas. In *Paper Prepared for the Scandinavian Seminar College Project: Policies and Practices Supporting Sustainable Development in Sub-Saharan Africa*; J Am Acad Orthop Surg: Windhoek, Namibia, 1998.
35. *The State of Community Conservation in Namibia—A Review of Communal Conservancies, Community Forests and Other CBNRM Initiatives*; NACSO: Windhoek, Namibia, 2019; p. 88.
36. Newing, H. *Conducting Research in Conservation: Social Science Methods and Practice*; Routledge: Milton Park, UK, 2010.

37. Mbidzo, M. The Dynamics of Governing Natural Resources in Namibia's Conservancies and Community Forests: Implications for Empowerment, Equity and Sustainability. Ph.D. Thesis, University of Kent, Canterbury, UK, 2016.
38. Boyatzis, R.E. *Transforming Qualitative Information: Thematic Analysis and Code Development*; Sage: Thousand Oaks, CA, USA, 1998.
39. Braun, V.; Clarke, V. Using thematic analysis in psychology. *Qual. Res. Psychol.* **2006**, *3*, 77–101. [CrossRef]
40. Ostrom, E.; Gardner, R.; Walker, J. *Rules, Games and Common-Pool Resources*; University of Michigan Press: Ann Arbor, MI, USA, 1994.
41. Corbett, A.; Jones, B.T. *The Legal Aspects of Governance in CBNRM in Namibia*; Directorate of Environmental Affairs, Ministry of Environment and Tourism: Windhoek, Namibia, 2000.
42. Meinzen-Dick, R.; Mwangi, E. Cutting the web of interests: Pitfalls of formalizing property rights. *Land Use Policy* **2009**, *26*, 36–43. [CrossRef]
43. Hinz, M.O. Traditional Governance and African customary law: Comparative observations from a Namibian perspective. *Hum. Rights Rule Law Namib.* **2008**, *20*, 59–87.
44. Jones, B.T. *Synthesis of the CBNRM Policy and Legislation in Botswana, Malawi, Mozambique, Namibia, Zambia, and Zimbabwe*; WWF-SARPO: Harare, Zimbabwe, 2007; p. 97.
45. Hinz, M.O. Traditional Courts in Namibia–Part of the Judiciary? Jurisprudential Challenges of Traditional Justice. In *The Independence of the Judiciary in Namibia*; Horn, N., Bösl, A., Eds.; Macmillan: Windhoek, Namibia, 2008; pp. 149–173, ISBN 978-99916-0-807-5.
46. Fleischman, F.; Ban, N.C.; Evans, L.S.; Epstein, G.; Garcia-Lopez, G.; Villamayor Tomas, S. Governing large-scale social-ecological systems: Lessons from a comparison of five cases. *Int. J. Commons* **2014**, *8*, 428–456. [CrossRef]
47. Newing, H.S.; Wahl, L. Benefiting local populations? Communal reserves in Peru. *Cult. Surviv.* **2004**, *28*, 38–41.
48. Biodiversity and Finance: Building on Common Ground with Customary Rights-Holders. Available online: https://www.forestpeoples.org/en/convention-biological-diversity-cbd/briefing-paper/2021/biodiversity-and-finance-building-common (accessed on 4 September 2021).
49. Krause, T.; Collen, W.; Nicholas, K.A. Evaluating safeguards in a conservation incentive program: Participation, consent, and benefit sharing in Indigenous Communities of the Ecuadorian Amazon. *Ecol. Soc.* **2013**, *18*, 1. [CrossRef]
50. Díaz, S.; Settele, J.; Brondízio, E.S.; Ngo, H.T.; Guèze, M.; Agard, J.; Arneth, A.; Balvanera, P.; Brauman, K.; Butchart, S.H. *Summary for Policymakers of the Global Assessment Report on Biodiversity and Ecosystem Services of the Intergovernmental Science-Policy Platform on Biodiversity and Ecosystem Services*; IPBES: Bonn, Germant, 2019; p. 56.
51. Roe, D. *Conservation, Crime and Communities: Case Studies of Efforts to Engage Local Communities in Tackling Illegal Wildlife Trade*; IIED: London, UK, 2015.

Article

Multi-Stakeholder Platform in Water Resources Management: A Critical Analysis of Stakeholders' Participation for Sustainable Water Resources

Onesmo Z. Sigalla [1,2,*], Madaka Tumbo [3] and Jane Joseph [1]

1 Water Right (T) Ltd., 109 Regent Estate, Mikocheni, Dar es Salaam P.O. Box 8703, Tanzania; janejoseph83@gmail.com
2 Nelson Mandela—African Institution of Science and Technology, Arusha P.O. Box 447, Tanzania
3 Ministry of Water, Water Resources Institute, Dar es Salaam P.O. Box 35059, Tanzania; madaka.tumbo@gmail.com
* Correspondence: onesigalla@gmail.com; Tel.: +255-(754/713)-535-997

Abstract: Multi-stakeholder platforms (MSPs) have gained momentum in addressing contentious and cross-sectoral aspects of natural resources management. They have helped to enhance cross-learning and the inclusion of marginalized groups. Tanzania's water resources management sub-sector has championed these platforms as a means of breaking silos around planning, coordination, and resource mobilization. However, it is not uncommon to experience the occasional dominance of some influential sectors or groups due to their resources contribution to the process, contemporary influence, or statutory authority. Between 2013 and 2020, Tanzania has pioneered the establishment of MSPs at a national level and across the river and lake basins. This paper examines the representation of stakeholder groups in these platforms. Additionally, it establishes the baseline information that contributes to unlocking the current project-based platform design characterized by inherent limitations to potential changes in stakeholders' attitudes and actions. The research analyzed stakeholder's views, their representation, and the local and international literature to formulate opinions. Findings indicated that gender equality had not been adhered to despite being in the guidelines for establishing MSPs. The balance of public, private, and civil society organizations (CSOs) is acutely dominated by the public sector organizations, especially water-related ones. Finally, participation on the decision-making level is minimal, causing unsustainable platforms unless development partners continue to support operational costs.

Keywords: MSP; representation; stakeholders; stakeholders' engagement; water governance; pluralistic approach

Citation: Sigalla, O.Z.; Tumbo, M.; Joseph, J. Multi-Stakeholder Platform in Water Resources Management: A Critical Analysis of Stakeholders' Participation for Sustainable Water Resources. *Sustainability* **2021**, *13*, 9260. https://doi.org/10.3390/su13169260

Academic Editors: Adam P. Hejnowicz and Jessica P. R. Thorn

Received: 31 May 2021
Accepted: 20 July 2021
Published: 18 August 2021

Publisher's Note: MDPI stays neutral with regard to jurisdictional claims in published maps and institutional affiliations.

Copyright: © 2021 by the authors. Licensee MDPI, Basel, Switzerland. This article is an open access article distributed under the terms and conditions of the Creative Commons Attribution (CC BY) license (https://creativecommons.org/licenses/by/4.0/).

1. Introduction

Integrating diverse stakeholders in water resources management has been an essential part of sustainable water resources management. However, complexities that arise in the dynamism inherent in the human–water interactions are shaped by growth in population and urbanization, which modify the demand for water resources [1–3]. In addition, the use and management of water resources depend on economic growth, urbanization, land-use change, hydrological–climatic changes, technological advances, historical perspectives, politics, and complex, traditional practices based on religious and cultural beliefs and attitudes [4]. Water-related problems are, thus, interlinked and solvable only by interactions among diverse scientific disciplines and stakeholders in the auspice of integrated water resources management (IWRM), as is aided partly by implementing multi-stakeholder platforms [4,5].

Since the early 2000s, the concept of multi-stakeholder platforms (MSP) has gained traction in several sectors [6,7]. The concept adapts different names across sectors including multi-stakeholder forums, multi-stakeholder processes or partnerships, and multi-stakeholder initiatives. Essentially, this concept entails collective (sometimes "collaborative") governance, an innovative and solutions-oriented model focusing on public value. This is where diverse stakeholders can collaborate to improve public resources and deliver services [8–12]. The critical tenet of these platforms lies in bringing together government, civil society, and the private sector to address complex development challenges that no one party alone has the capacity, resources, and know-how to do so more effectively [13,14]. In addition, the uniqueness of platform is in learning by doing: using feedback mechanisms from the environment (biophysical and social) to shape policy, followed by further systematic experimentation, in a never-ending cycle [15,16]. In so doing, MSPs come to complement and not usurp the role of governments in achieving these ends. In water resources management, it comes as a logical companion to implement IWRM [6,17], which was introduced as part of Agenda 21 of the United Nations Conference on Environment and Development (UNCED) held at Rio de Janeiro in 1992 [18]. IWRM has been broadly defined as a process that promotes the coordinated development and management of water, land, and related resources in order to maximize the resultant economic and social welfare in an equitable manner without compromising the sustainability of vital ecosystems [19]. It is based on the three principles: social equity, economic efficiency, and environmental sustainability [20].

In this regard, IWRM and MSP help to achieve the UN 2030 Agenda, which requires multiple sectors and actors to work together seamlessly. Sustainable Development Goal (SDG) 17 explicitly recognizes multi-stakeholder partnerships as important vehicles for mobilizing and sharing knowledge, expertise, technologies, and financial resources to support countries' SDG commitments [21]. Further, SDG 17 seeks to encourage and promote effective public–private–civil society partnerships, building and capitalizing on their respective capacities and experience in resource mobilization and management. This provides an enabling function for the implementation of SDG 6 on water and sanitation, especially SDG 6.5 on water resources management and in the context of achieving water security for all [21]. In addition, when well-designed, these platforms may also help to achieve SDG 5 on gender equity and empowering of women and girls [21], the IWRM principle on social equity, the Dublin Statement on the role of women [22], and adherence to national water policy and legislation on the one-third gender principle in representation bodies [23,24]. However, experience shows that female participation remains limited, while general representative members in statutory bodies, i.e., catchment water committees and basin/national water boards, are limited to five and ten seats, respectively (Figure 1). MSP then expands the mechanism for broader stakeholder engagement, which helps to achieve adaptive management that features stakeholder input and knowledge generation, objective setting, management planning, monitoring implementation, and incremental plan adjustment in the face of uncertainty [9–12].

This paper examines sectoral representation and the inherent opportunities and barriers of the existing state of affairs. In addition, it establishes the baseline of the level of representation and its issues in these nascent stages of MSP evolution in Tanzania. In this regard, we address the following objectives in this paper:

(a) to assess the level of adherence to the one-third gender rule for all water-related institutions of representation as proposed in the water policy and legislation in Tanzania;
(b) to examine the balance of participation between different groups of stakeholders as envisaged in SDG 17 and government MSP regulations; and
(c) to evaluate the role of participation of the managerial level in the mainstreaming of MSP undertakings to respective partners.

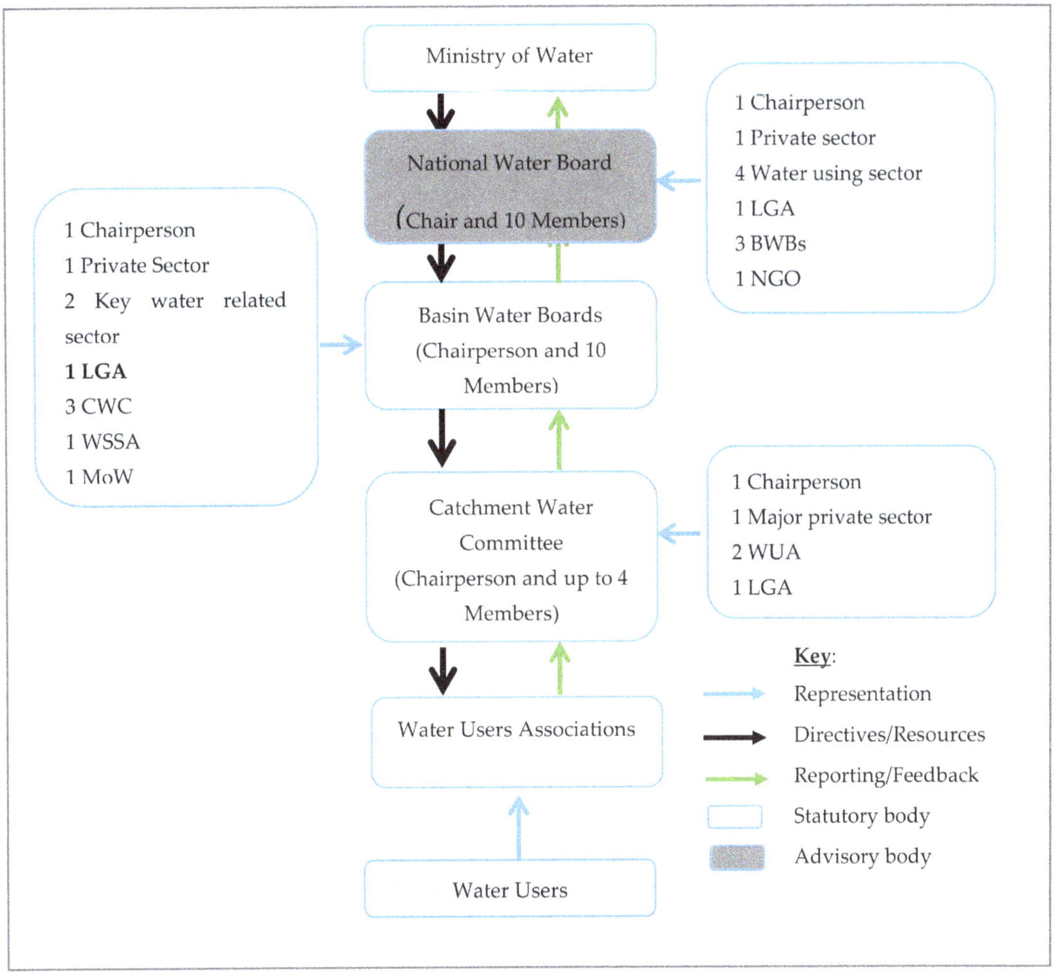

Figure 1. Institutional hierarchy for water resources management in Tanzania—modified after the national water sector development strategy. Adapted from Ref. [25]. (LGA—local government authorities, Mow—Ministry of Water, NGO—non-governmental organizations).

2. Materials and Methods

2.1. The Study Site

The MSPs in Tanzania are operationalized at the national, basin, and catchment levels. In the context of this paper, we selected two national- and basin-level MSPs for the analysis. The selected basins were Lake Rukwa and Lake Nyasa basins, as seen in Figure 2. These constitute most of Tanzania's southwestern highland block, which is famous as a critical food basket. The two basin MSPs are similar in that both are dominated by agrarian economies, contain mining hotspots and national parks, and are transboundary and influenced by the fast-growing city of Mbeya and the borders of Malawi and Zambia, with potential for unsustainable development if not well guided, as discussed in [26]. The case study MSPs were selected to compare and contrast participation issues and experiences at the national and sub-national or basin level. In addition, we studied the same between infant basin form (that have started as recent as 2019) and the relatively experienced national platform. The national platform has been formally in existence since

2017, so it is expected to have gathered substantial insights. The same is then expected to have been percolated to the basin or sub-national levels.

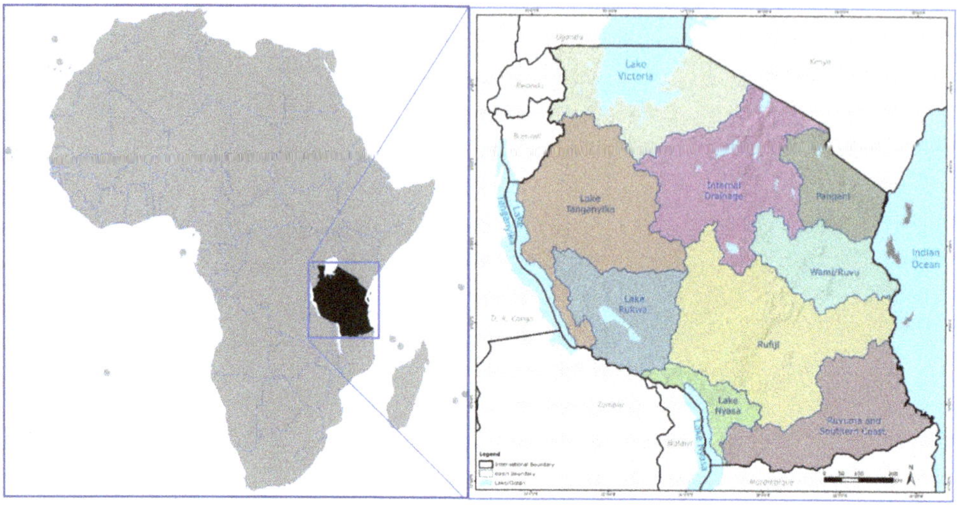

Figure 2. Jurisdiction of the selected multi-stakeholder platform—extracted from the national water atlas. Adapted from Ref. [27].

2.2. Data Collection

2.2.1. Literature Review

A review of national and international frameworks and literature supporting stakeholders' engagement in water resources management was performed. This benefited from open access electronic databases such as the Ministry of Water webpage, the World Bank Open Knowledge Repository (OKR), the 2030 Water Resources Group Repository, the Global Water Partnership Digital Library, and other internet searches. These were either by navigating through respective web pages or using search phrases on the subject matter in different search engines such as Google Scholar. A literature review was carried out to understand the subject matter and helped to augment and triangulate information gathered during focus group discussions and key informant interviews.

Table 1 below summarizes information used, such as national and international commitments and guidelines in specific aspects, e.g., affirmative action on gender balance, extracted from relevant strategic documents. Others included broadly agreed definitions and principles for different concepts, e.g., the IWRM and MSP, which were also crafted from the review of international frameworks. This also included a reference of what is considered a multi-stakeholder partnership, which in SDG 17 entailed a public–private–civil society partnership. Subsequently, the collected documents were filtered to get the following groups of required areas of this paper:

- widely accepted definition of concepts, i.e., IWRM and MSP;
- principles governing these concepts;
- national and international commitments, e.g., gender parity; and
- relevant experiences elsewhere that relate to these focus concepts.

2.2.2. Key Informant Interviews

Discussions with different stakeholders and forum secretariat have been conducted since early 2019 (Table 2). Questionnaires were drafted and used as a guide to collect data on representation, respective sectors, and the level of decision-making. The interviews engaged basin water officers, environmental experts from riparian administrative regions and districts, basin and national water board members, private sector members, civil society organizations, and the community through water users' associations, WUAs, and irrigation associations. These were purposefully sampled to capture individuals with knowledge and experience in water resources management, MSPs, and active engagement in the MSP deliberation process. Interviews were conducted in Swahili and English depending on interviewee preference. Responses were captured in questionnaire forms and additional explanations; key quotes and a general understanding of the responses were transcribed in notebooks.

Table 1. Description of some of the key national and international frameworks considered.

S/N	Description of Framework/Literature	Information Extracted	Source
1.0. National Frameworks			
1.1	National water policy of 2002	Role of stakeholders in water resources management, one-third gender principle	[23]
1.2	National water sector development strategy of 2006	Institutional framework for WRM	[25]
1.3	Water resources management act 2009	Provisions for implementation of water policy	[24]
1.4	Regulations on Multi Stakeholders Forum Gazette Notice No. 56 of 2021	Provisions for formal recognition of national and basin platforms	[28]
1.5	Basin platform proceedings report (national and basin level)	Respective working groups identified WRM issues and implementation strategies	[29,30]
2.0. International Frameworks			
2.1	UN 2030 Agenda	SDG 5 on women and girls, SDG 6.5 on Water resources management, and SDG 17 on partnerships	[21]
2.2	UNCED Agenda 21	IWRM framework	[18]
2.4	Dublin Statement	Role of women in IWRM	[22]
2.5	Multi-Stakeholder Platforms	MSP evolution, composition, and implementation in Tanzania and globally	[31]

Table 2. Key informant respondents and type of data collected.

S/N	KII Respondent	Department/ Section/Focus	Respondents	Key Information Gathered
1	Lake Rukwa Basin Water Board	Stakeholders engagement	2	platform proceedings reports, enabling statutory environment, sources of investment for MSP, uptake by stakeholders, emerging benefits of MSPs
2	Lake Nyasa Basin Water Board	Stakeholders engagement	1	
3	Ministry of Water	Division of water resources	1	
4	Water Users Association	Association leaders	4	Emerging benefits of MSPs
5	Civil Society	Advocacy, technical support	4	Uptake of MSP deliberations, sustainability issues
6	Private Sector	Beverage, agribusiness, and water bottling	4	Opportunities to influence policy and reputational risks
7	Development Partners	Natural resources management	2	Sustainable financing collaborations, e.g., with the private sector
8	Total		18	

2.2.3. Focused Group Discussions

The focus group discussions were conducted during the occasion of the respective platforms. The discussions involved twelve (12) working groups and helped to identify water resources management issues in the respective platforms, as summarized in Table 3. The groups involved between 7 to 15 members in each working group. Information gathered during discussions included water resources management issues that the working group seeks to address, respective drivers, strategies, and potential barriers at the implementation and strategic level. The groups also provided recommendations of measures to be taken to address water resources management challenges.

Table 3. Working group participants in the three platforms of interest.

Platform Description	Working Group	No. of Participants (Only Elected Group Members)
Lake Rukwa Basin Multi-Stakeholders Forum on Water Resources Management	Agriculture	15
	Environmental management	13
	Water supply	14
	Mining	8
Lake Nyasa Basin Multi-Stakeholders Forum on Water Resources Management	Agriculture	14
	Environmental management	15
	Water supply	12
	Mining	7
National Multi-Stakeholders Forum on Water Resources Management	Private sector (beverages, mining, and textiles)	9
	Knowledge management (research, policy, and practice)	13
	Resources mobilization (irrigation finance initiative and national water fund)	13

2.3. Data Analysis

Primary data collected from this study were descriptively analyzed using MS Excel software. Secondary data and literature reviews were synthesized and analyzed empirically. Both results were presented in tables, figures, or pie charts that offered a better way to compare and contrast results. Additional information was captured in the form of quotations from key informants. Finally, results were presented under three themes, namely (i) gender balance, which aimed at evaluating the level of adherence to or departure to national and international guidelines on gender; (ii) the balance of participation of sectors, which assessed the participation split from the public sector, private sector, and civil society; and (iii) the uptake of MSP deliberations, which aimed at assessing the uptake of MSP deliberations by the respective stakeholders.

3. Results

3.1. Identified Participants and Categories

Tables 4–6 below present different participant categories from the three platforms assessed in the current study. These are from the most recent MSPs at the Lake Rukwa platform attended by 83 participants, the Lake Nyasa platform attended by 63 participants, and the national platform with 150 attendees. For the purpose of this study, participants were further disaggregated by gender, hosting sector, and respective level of authority. In addition to these tables, responses from stakeholders are included in a narrative with insert quotations emphasizing results. Subsequent subsections present assessment results in three categories that capture the study objectives, i.e., the level of adherence to gender balance, the balance of participation between sectors, and mainstreaming MSP undertakings by participating sectors.

Table 4. Stakeholder participation data for Lake Rukwa basin platform indicating gender and authority levels (Source: Field data 2021).

Sector	Male	Female	Total	Mngmt.	Technical	Assistant Tech	Support
Community institutions	13	5	18	17	1	0	0
Civil society organizations	6	0	6	5	0	1	0
Ministry of Water	0	3	3	1	2	0	0
Ministry of Water implementing agency	18	6	24	6	9	7	2
Other public sector institutions	24	5	29	4	17	5	3
Private sector	3	0	3	0	3	0	0
Total 3	64	19	83	33	32	13	5

Table 5. Stakeholder participation data for Lake Nyasa Basin platform indicating gender and authority levels (Source: Field data 2021).

Sector	Male	Female	Total 1	Mngmt	Technical	Assistant Tech	Support
Community institutions	3	0	3	3	0	0	0
NGO/CSO	3	0	3	3	0	0	0
Ministry of Water	3	1	4	1	1	1	1
Ministry of Water implementing agency	23	11	34	5	19	2	7
Other public sector institutions	12	3	15	1	10	2	3
Private sector	2	2	4	1	3	0	0
Total 3	46	17	63	14	33	5	11

(NGO—non-governmental organizations; CSO—civil society organizations).

Table 6. Stakeholder participation data for the national platform, including virtual and physical participation (Source: Field data 2021).

Sector	Physical F	Physical M	Virtual F	Virtual M	Total
Ministry of Water	9	18	1	1	29
Ministry of Water IA	2	12	3	11	28
Public	5	11	2	12	30
Private	6	12	3	8	29
NGO/CSO	1	7	1	5	14
Development partners	5	4	5	6	20
Total	28	64	15	43	150

(NGO—non-governmental organizations; CSO—civil society organizations).

3.2. Identified Key Issues

The national- and basin-level platforms discussed a number of issues that they uniquely intend to address (Table 7). While it is not the intention of this study to analyze these issues, their interlinking and cross-sectoral nature helped to inform our opinions. This is in the context of (a) the unique role that women can play, (b) the different mandates and knowledge that sectors of interest have (the current study considers the public sector, private sector, and civil society, as per SDG 17), and (c) the potential that an appropriate mix of community members, technocrats, and decision/policy makers can bring in addressing these identified issues.

Table 7. Key issues identified by basin- and national-level platforms.

Platform Description	Working Group	Key Issues of Focus
Lake Rukwa Basin Multi Stakeholders Forum on Water Resources Management	Agriculture	Poor water use efficiency and poor productivity
	Environmental management	Inadequate law enforcement and coordination of actors
	Water supply	Limited access to clean and safe water
	Mining	Pollution of water sources
Lake Nyasa Basin Multi Stakeholders Forum on Water Resources Management	Agriculture	Poor water use efficiency, declining water flows, and illegal fishing
	Environmental management	Destruction of natural vegetation; siltation of water sources, and dwindling river flows
	Water supply	Insufficient non-Revenue water and declining water sources
	Mining	Deforestation and water pollution
National Multi Stakeholders Forum on Water Resources Management	Private sector (beverages, mining, and textiles)	Declining water availability
	Knowledge management (research, policy, and practice)	Limited dissemination of information
	Resources mobilization (irrigation finance initiative and national water fund)	Limited resources for WRM activities

The issues were captured during the development of the respective theory of change or work plans in [29,30], and they are also well-captured by other scholars, e.g., [23,32,33].

3.3. Adherence to Gender Balance

Based on the gender disaggregation of participants listed in the considered platforms, it was noted that none of the platforms adhered to the affirmative action embedded in national and international frameworks. While the Tanzania policy and legal framework for water resources management established a one-third gender rule in all representation institutions [23,24], the United Nations SDG 5 intends to improve gender equality and empowerment of women and girls [21]. Figure 3 illustrates this skewness in that Lake Rukwa had only 23% female participants, Lake Nyasa had 27%, and the national platform was attended by 29% female participants. This collaborated well with sentiments from female participants in Lake Nyasa Basin MSP who indicated that:

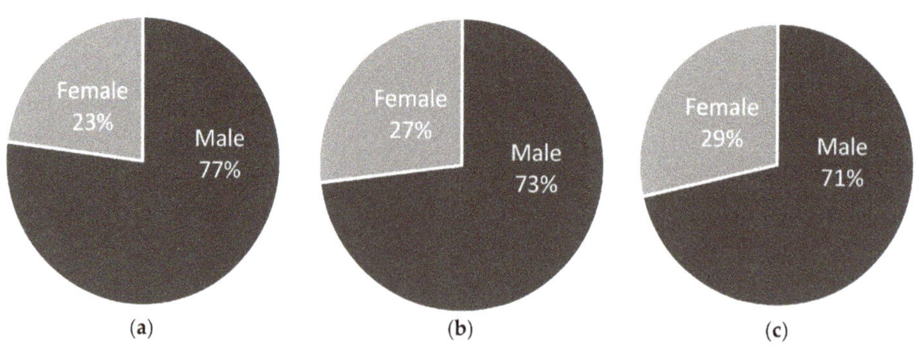

Figure 3. Percentage balance of gender in the participation of stakeholders in MSPs. (**a**) Lake Rukwa, (**b**) Lake Nyasa, and (**c**) the national platform.

"Whereas the Minister emphasizes the role of bureaucrats to off-shoulder water buckets from women, our voices are limited, starting with the way we take part in participation, discussions, and positions of leadership, which somehow owes to culture and numerous responsibilities in the homestead." (Sentiments captured from one of the platform members in Lake Nyasa).

Although the composition of all the assessed platforms did not adhere to the one-third gender rule, as pointed out above, all platforms had a female member as a vice-chairperson. While efforts are needed to encourage the representation of females, their presence (employment) in participating organizations adds another complexity that might be beyond the influence of MSP coordinating entities. The 2014 Integrated Labor Force Survey (ILFS) indicates that females in Tanzania form a larger share of the working-age population but a smaller share of the economically active population. Women account for 52% of the working-age population (15 years and over), but the labor force participation rate is higher among males (89.4%) than among females (forming 84.2%) [34].

Further to the general analysis of gender balance above, the study aimed to assess the same balance across the vital MSP sectors, i.e., public sector, private sector, and civil society organizations or non-governmental organizations. The interpretation of Figure 4 shows that, across the board, female members were fewer (or none) compared to their male counterparts. In addition, Lake Nyasa had the fewest female members coming from community institutions, civil society, and the Ministry of Water.

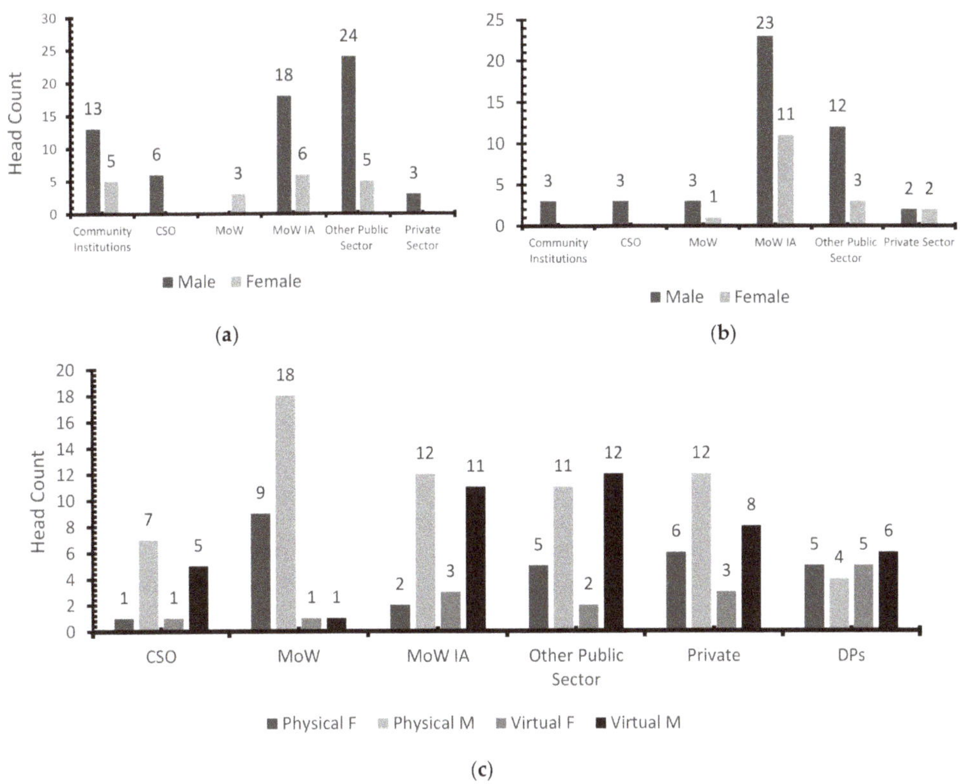

Figure 4. Gender disaggregation of participants at (**a**) Lake Rukwa, (**b**) Lake Nyasa, and (**c**) national platform. The national platform had a virtual participation facility as well.

3.4. Balance of Sector Representation

As observed in SDG 17 [21], the national water resources legislation [28] and initiatives such as the WWF water program [35] and 2030 water resources group [31,36,37], the public–private–CSO balance is paramount and must come equally for a robust platform. However, Figure 5 shows that sector participation is acutely skewed with the public sector being the dominant player at 67%, 84%, and 58% for Lake Rukwa, Lake Nyasa, and the national platform, respectively. Furthermore, the intended expansion of other sectors in WRM decision-making is undermined when the Ministry of Water and its implementing agencies form 32%, 60%, and 38% for the same stated forums. On the other hand, private sector and CSO participation are the narrowest, with the former standing at 4%, 6%, and 19% in the respective platforms. Similarly, the latter is 7%, 5%, and 10% for Lake Rukwa, Lake Nyasa, and National platforms, respectively. This state of affairs promotes a lack of hybrid sectors, which denies building more vital institutions and MSP sustainability. This was also observed by the chairperson of the national forum, who gave a narrative of the growing interest of the private sector in water resources management and these platforms in particular,

> "*I have worked in the sector for many years to a level of Permanent Secretary in the Ministry; we never used to have a push towards engaging the private sector. However, although slow, it is encouraging to see this shift where matters get to be discussed and picked up from here, which transforms platforms from merely talk shops to the actual workshop*" (Observation of the chairperson of the national MSP).

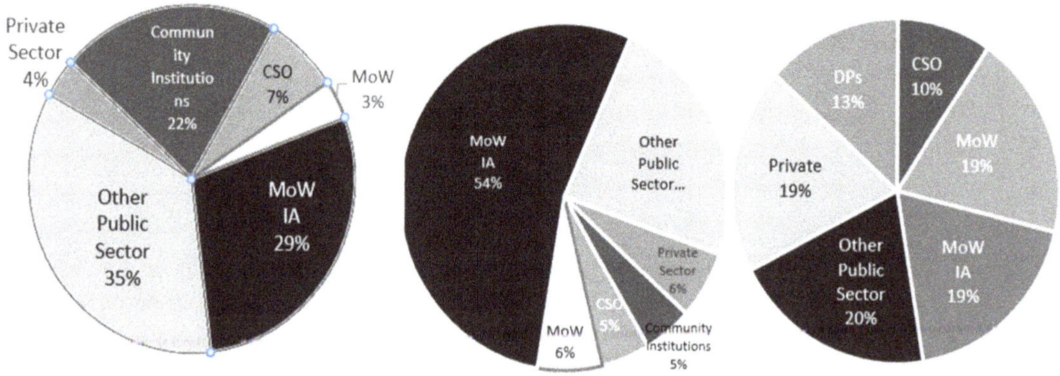

(a) (b) (c)

Figure 5. Percentage balance of sectors in (**a**) Lake Rukwa, (**b**) Lake Nyasa, and (**c**) the National Forum.

3.5. Sectoral Mainstreaming of MSP Undertakings

Engagement of strategic leadership is paramount in securing institutional commitments and conducive grounds for mainstreaming MSP undertakings [38,39]. While acknowledging the need for engaging those who do not hold (government) mandates, WWF stresses the need for engagement of strategic leadership, e.g., in the private sector in determining and committing to a shared water risk [40].

The current assessment considered the level of participation in Lake Rukwa and Lake Nyasa basin forums and compared it with experiences in the uptake of MSP deliberations in other basin forums across Tanzania. We found that there is a considerable constraint on the potential uptake of the MSP deliberations. Arguably, this is because of limited participation of strategic/decision-making levels in the MSPs, as illustrated in Figure 6b, and captured by one director of water resources below. Even in the Lake Rukwa forum, where up to 17 decision-making-level participants were captured, they belonged to the community institutions cluster, i.e., water user associations (WUAs) and/or irrigation associations. As such, these groups may not have sufficient influence on policy compared to the participation of similar groups from the private sector or civil society organizations. Therefore, this potentially translates into limited strategic discussions and securing commitments from the respective partner institutions, which would help sustain the operationalization of these forums. This is related to three other issues: (i) over-reliance on the Ministry of Water or donor financing to operationalize MSPs, (b) lack of sector's own initiatives on WRM in subsequent MSPs, and (c) a general and consistent proposal for the need for fundraising strategy for all platforms, which are not in existence yet. Coupled with the fact that sectors other than the public sector (Section 3.4) are yet to be attracted adequately to these excellent platforms, the business-as-usual is likely to be perpetuated sustainably. The observation is informed by, among others, a lack of initiatives presented by participating sectors showing uptake from previous platform deliberations. The same was included in the challenge tabled by the Director of Water Resources during the opening of the national MSP:

> *"Among the top factors worrying sustainability of these MSP is not only recurrent financing of platforms by donor support but more of sectors picking deliberations, implementing them, and bringing back lessons. This will inform us in finetuning the enabling environment through informed advice to the Minister in charge of the water sector."* (Observation by Director of Water Resources—Ministry of Water).

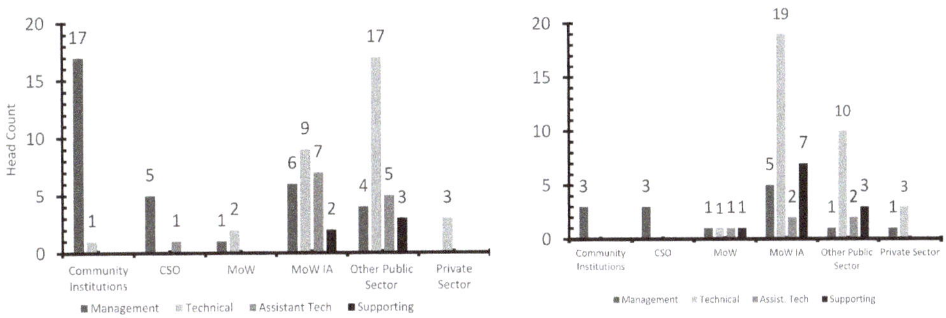

(a) (b)

Figure 6. Participant disaggregation by the level of authority in respective hosting institutions: (**a**) Lake Rukwa and (**b**) Lake Nyasa.

The sentiments above entail the need for the engagement of decision makers from the represented sectors. Outstanding results have been shown by a leap in revenue by the Basin Water Boards when decision makers were engaged. Similarly, an engagement of media houses in the 2021 national MSP has recorded a positive response that needs follow-up to build momentum.

4. Discussions

Scholars have pointed out that gender-based roles frequently put women in direct contact with natural resources such as forests, water, land, and wildlife [41]. Women utilize and conserve these resources to supply basic needs for their families. Kariuki and Birner [42] add that the conservation of natural resources in rural areas cannot be achieved without the involvement and training of women. Therefore, women need not only to be able to fully participate in decision making but also to be enabled to engage in training relating to the management and sustainability of natural resources. The current study has indicated the limited representation of women and movements toward changing the status quo to increase their participation and inclusion of their knowledge, experience, and insights. That said, women's role in NRM is increasingly being recognized, as women have considerable knowledge and experience gained from working closely with their environment. Further, their analytical skills in their community can play a vital role in the sustainable development of water and forest resources. However, both formal and informal organizational rules often exclude women from institutions involved in natural resource management [43]. Structural institutional barriers such as the hierarchical gendered division of labor within water institutions where women's roles are primarily administrative, non-decision-making, non-extension jobs also actively undermine women's participation. Women's participation is usually more successful in initiatives in which coming together creates enhanced resource rights or availability [44]. Although both the water policy of 2002 and the Water Resources Management Act of 2009 mention women and gender in their contents, both documents do not look at the design from a gender perspective. None of them give concrete guidelines or recommendations to make the policies more gender-inclusive. Multi-stakeholder platform coordination will have more informed deliberations if these important players are deliberately facilitated to participate. However, owing to the voluntary nature of participation in MSPs, the inclusion of female members has suffered significantly in the implementation of these platforms.

In relation to the balance of sectors for a robust MSP, the SDG 17.16 and 17.17 stipulate the need for multi-stakeholder partnerships that enhance collaborations between the public sector, private sector, and civil societies [21]. The expectation is that this diverse composition brings about a good blend of mandates, knowledge, experiences, and resources that match the cross-sectoral nature of natural resources challenges well [45]. For instance, the participation of the private sector has been well-captured in the current study as being among the reasons for constructive discussion and trust-building, leading to increased access to private media houses to communicate lessons and higher revenue collections for the Basin Water Boards. A leap from a few hundred to several thousand US$ has been realized in some of the basins such as Lake Victoria and Wami-Ruvu, which forms a critical basis for learning. These benefits have been accrued in circumstances in which private sector participation is less than 10% in basin platforms and less than 20% in the national platform. One can only imagine the increased benefits if participation was well-balanced between the three sectors. This includes areas of resources mobilization, technology transfer, use of wide networks to communicate results, etc. The participation of the private sector is important as it provides mutual benefits in safeguarding its own investments while remaining a good corporate citizen [46,47].

In addition to securing the rightful place of women and girls in MSPs and the need for the right mix of different sector mandates, there is the challenge of linking with the correct authority level for a proper mainstreaming of MSP deliberation. MSPs were established to expand representation and democratize stakeholder participation in water resources management in support of Basin Water Boards (BWBs). Platforms may become an appropriate vehicle to foster cooperative governance between the BWBs, local government, private sector, and other stakeholder interest groups in the interest of integrated water resources management. However, limited participation of strategic level decision-makers from the represented sectors may contribute to limited uptake of MSP deliberations. This is because stakeholders involved in MSPs are numerous with overlapping roles and interests that create competition to establish supremacy and sometimes conflicts [7,48]. The problem is that, although stakeholders are concerned with water quality, quantity, and sustainability, they do not all have the same social position concerning measures proposed or taken to resolve the issues at hand. Moreover, they do not necessarily share the same view of what is desirable or what constitutes the purpose of water resources management [48]. How stakeholders act in relation to the rules and roles that have been taken or assigned to them will determine MSPs' successful implementation and sustainability. This gains more credit in a situation where the water sector is part of broader social, political, and economic development and is influenced by decisions taken by actors outside of the water sector [49]. Drawing lessons from numerous participatory water management initiatives, the authors argue that because of a lack of attention to the complex political contexts in which these initiatives were embedded, the appropriate influence level of participating sectors was not well-represented [50]. These arguments agree with the results of this study, as most participation is at the technical level or below; hence, discussions tend to be largely technical in nature and lack strategic deliberations. For instance, the representing individual frequently lacks the appropriate authority and accountability to make a decision on mainstreaming deliberations from the MSP within their respective sector institution. This can be linked to the observation that a lack of self-championed activities results from missing decision makers. This means that the platform secretariat has to arrange for visits to solicit buy-in from decision makers, increasing the costs of implementing MSPs and undermining ownership.

Consequently, this state of affairs has seen a lack of self-championing of the agreed actions and an over-reliance on donor support to implement platforms. It is argued here that organizers should strive to unpack deliberations for ease of engaging with different levels of authority and seek to act strategically to ensure appropriate decisions and a commitment to MSP. Conroy and Peterson [51] propose a decision model that allows decision makers to develop portfolios of potential management alternatives for their investments, predict risk, estimate consequences, determine weights for objectives, and calculate overall support and trade-offs for each portfolio as well as identify the recommended decision. We argue that top leadership's complete buy-in and commitment to the respective sector is essential for sustainable mainstreaming of platform undertakings of work plans and the budgets of participating sectors.

5. Conclusions

In this study, a critical analysis of the stakeholders' participation and engagement in MSP and their impact on the integrated water resources management is performed. It has been established that none of the platforms achieved the one-third threshold set out in the local and international frameworks on the balance of gender in these representation bodies. In addition, we acknowledge the difficulties in bringing every sector around the table, but the present, skewed balance of sectors in all MSPs may undermine the intended expansion of sector representation in WRM decision-making. In this case, the private sector and civil society organizations are yet to fully participate, support even more, and reap the benefits of these platforms. On top of the limited participation of these sectors, the individuals who participate belong mainly to the technical segment, leaving the decision-making level. The absence of strategic level players impacts the mainstreaming of MSP deliberations in the participating sector and increases the over-reliance on donor support through the Ministry of Water.

The lack of involvement of various stakeholders in multi-stakeholder dialogues may prevent sustainable integrated water resources management at different scales. Furthermore, since dialogues that do not combine the ideas of multiple stakeholders are deficient in articulating the interests of the various stakeholders, the implementation of MSP action plans will be limited in scope. The limitation in MSP scope may create a misunderstanding between what is socioeconomically demanded and what is implemented on the ground at the basin or catchment scale. The policy implication of this study is that in order to have strong and sustainable MSPs for water resources management, both individual and institutional identities need to be well-represented. The role of women cannot be overemphasized in matters of WRM, as is the case for the knowledge, expertise, and resources that the private sector hosts and could bring to play in support of platforms.

Based on this understanding, the following recommendations are proposed:

- A deliberate effort to encourage female participation in the established MSP. The same can benefit from entrusting females with positions of leadership, as is the case for some of the platforms in Tanzania.
- The design of MSP meetings should consider and recognize the time constraints of participating sectors, organizations, and individuals. Moreover, identifying the shared

water risks in priority sites could be an excellent way to entice the participation of this private sector and others that feel a direct impact.

- The participation of decision makers is paramount to self-sustaining MSPs. A strategy for reaching out to top leadership in institutions may help to build interest. In addition, high-level steering committees are worth pursuing. Creating a private-sector-focused group could also help in panning out specific issues of interest and aiming at the participation of the management level, as was tested by 2030 WRG in initial engagement in Tanzania.
- Entrusting leadership roles to non-traditional participating sectors, e.g., the private sector, will increase trust, the sense of responsibility on WRM, and the potential for piggybacking on their networks to mobilize more players.

Author Contributions: Writing—Original draft preparation and writing, O.Z.S.; Writing—Review and editing, M.T. and J.J. All authors have read and agreed to the published version of the manuscript.

Funding: Publication of this paper was financially supported by Water Witness International (www.waterwitness.org) to which the authors are greatly indebted.

Acknowledgments: Editorial: Special thanks to Adam Hejnowicz and Jessica Thorn for critical comments and suggestions. Data acquisition: Most of the data were collected at the respective platforms coordinated by the Ministry of Water (Division of Water Resources), Lake Rukwa Basin Water Board, and Lake Nyasa Basin Water Board, to whom the authors are greatly indebted.

Conflicts of Interest: The authors declare no conflict of interest.

References

1. Al Radif, A. Integrated water resources management (IWRM): An approach to face the challenges of the next century and to avert future crises. *Desalination* **1999**, *124*, 145–153. [CrossRef]
2. Pahl-Wostl, C.; Craps, M.; Dewulf, A.; Mostert, E.; Tabara, D.; Taillieu, T. Social learning and water resources management. *Ecol. Soc.* **2007**, *12*. [CrossRef]
3. Pahl-Wostl, C.; Sendzimir, J.; Jeffrey, P.; Aerts, J.; Berkamp, G.; Cross, K. Managing change toward adaptive water management through social learning. *Ecol. Soc.* **2007**, *12*. [CrossRef]
4. Evers, M.; Höllermann, B.; Almoradie, A.D.S.; Garcia Santos, G.; Taft, L. The pluralistic water research concept: A new human-water system research approach. *Water* **2017**, *9*, 933. [CrossRef]
5. Bammer, G. *Disciplining Interdisciplinarity: Integration and Implementation Sciences for Researching Complex. Real-World Problems*; ANU Press: Canberra, Australia, 2013.
6. Warner, J.F. More sustainable participation? Multi-stakeholder platforms for integrated catchment management. *Water Resour. Dev.* **2006**, *22*, 15–35. [CrossRef]
7. Warner, J. Multi-stakeholder platforms: Integrating society in water resource management? *Ambient. Soc.* **2005**, *8*, 4–28. [CrossRef]
8. Wondolleck, J.M.; Yaffee, S.L. *Making Collaboration Work: Lessons from Innovation in Natural Resource Managment*; Island Press: Washington, DC, USA, 2000.
9. Pahl-Wostl, C. Transitions towards adaptive management of water facing climate and global change. *Water Resour. Manag.* **2007**, *21*, 49–62. [CrossRef]
10. Pahl-Wostl, C.; Gupta, J.; Petry, D. Governance and the global water system: A theoretical exploration. *Glob. Gov. A Rev. Multilater. Int. Organ.* **2008**, *14*, 419–435. [CrossRef]
11. Huitema, D.; Mostert, E.; Egas, W.; Moellenkamp, S.; Pahl-Wostl, C.; Yalcin, R. Adaptive water governance: Assessing the institutional prescriptions of adaptive (co-) management from a governance perspective and defining a research agenda. *Ecol. Soc.* **2009**, *14*. [CrossRef]
12. Engle, N.L.; Johns, O.R.; Lemos, M.C.; Nelson, D.R. Integrated and adaptive management of water resources: Tensions, legacies, and the next best thing. *Ecol. Soc.* **2011**, *16*. [CrossRef]
13. Mena, S.; Palazzo, G. Input and output legitimacy of multi-stakeholder initiatives. *Bus. Ethics Q.* **2012**, *22*, 527–556. [CrossRef]
14. Dentoni, D.; Bitzer, V.; Schouten, G. Harnessing wicked problems in multi-stakeholder partnerships. *J. Bus. Ethics* **2018**, *150*, 333–356. [CrossRef]
15. Walters, C.J.; Holling, C.S. Large-scale management experiments and learning by doing. *Ecology* **1990**, *71*, 2060–2068. [CrossRef]
16. Allen, C.R.; Fontaine, J.J.; Pope, K.L.; Garmestani, A.S. Adaptive management for a turbulent future. *J. Environ. Manag.* **2011**, *92*, 1339–1345. [CrossRef]
17. Megdal, S.B.; Eden, S.; Shamir, E. Water governance, stakeholder engagement, and sustainable water resources management. *Water* **2017**, *9*, 190. [CrossRef]

18. Giordano, M.; Shah, T. From IWRM back to integrated water resources management. *Int. J. Water Resour. Dev.* **2014**, *30*, 364–376. [CrossRef]
19. Agarwal, A.; de los Angeles, M.S.; Bhatia, R.; Chéret, I.; Davila-Poblete, S.; Falkenmark, M.; Gonzalez-Villarreal, F.; Jønch-Clausen, T.; Aït Kadi, M.; Kindler, J.; et al. *Integrated Water Resources Management*; TAC Background Papers, no. 4; Global Water Partnership: Stockholm, Sweden, 2000.
20. Home, D. *Catalyzing Change: A Handbook for Developing Integrated Water Resources Management (IWRM) and Water Efficiency Strategies*; United Nations Department of Economic and Social Affairs: New York, NY, USA, 2004; p. 56.
21. United Nations (UN). *The Seventeen (17) Sustainable Development Goals*; Department of Economic and Social Affairs for Sustainable Development: New York, NY, USA, 2015.
22. Gorre-Dale, E. The Dublin statement on water and sustainable development. *Environ. Conserv.* **1992**, *19*, 181. [CrossRef]
23. Lein, H.; Tagseth, M. Tanzanian water policy reforms—Between principles and practical applications. *Water Policy* **2009**, *11*, 203–220. [CrossRef]
24. United Republic of Tanzania (URT). *The Water Resources Management Act. No 11 of 2009*; Gazette of the United Republic of Tanzania No 20. dated 15th May 2009; Ministry of Water: Dar es Salaam, Tanzania, 2009; Volume 90, pp. 353–424. ISSN 0856-033IX.
25. United Republic of Tanzania (URT). *The National Water Sector Development Strategy 2005–2015*; United Republic of Tanzania: Dar es Salaam, Tanzania, 2006. Available online: https://www.maji.go.tz/pages/strategies (accessed on 30 July 2021).
26. Agyemang, F.S.K.; Silva, E.; Poku-Boansi, M. Understanding the urban spatial structure of Sub-Saharan African cities using the case of urban development patterns of a Ghanaian city-region. *Habitat Int.* **2019**, *85*, 21–33. [CrossRef]
27. United Republic of Tanzania (URT). *Tanzania Water Resources Atlas as of 2018*; United Republic of Tanzania: Dodoma, Tanzania, 2006. Available online: https://www.maji.go.tz/pages/publication (accessed on 30 July 2021).
28. United Republic of Tanzania (URT). *Water Resources Management (Procedure for Nomination of Board Members) (Amendment)*; Ministry of Water: Dar es Salaam, Tanzania, 2020; pp. 1–30.
29. LNBWB. *Report for the Basin Multi Sectoral Forum for Water Resources Management, Tukuyu, Tanzania*; LNBWB: Tukuyu, Tanzania, 2021.
30. LRBWB. *Report for the Basin Multi Sectoral Forum for Water Resources Management, Mbeya, Tanzania*; LNBWB: Tukuyu, Tanzania, 2021.
31. 2030 Water Resources Group (WRG) Tanzania Multi-Stakeholder Platform on Water Resources Management. Available online: https://www.2030wrg.org/tanzania/ (accessed on 30 July 2021).
32. Mwakalila, S. Opportunities and challenges for sustainable water resources management in Tanzania. *Geogr. J.* **2008**, *174*, 154–157. [CrossRef]
33. Mahoo, H.; Simukanga, L.; Kashaga, R.A.L. Water resources management in Tanzania: Identifying research gaps and needs and recommendations for a research agenda. *Tanzania J. Agric. Sci.* **2015**, *14*.
34. Idris, I. Mapping Women's Economic Exclusion in Tanzania. KD4 Helpdesk Report 332. 2018. Available online: https://assets.publishing.service.gov.uk/media/5b432d9e40f0b678bc5d01c1/Barriers_to_womens_economic_inclusion_in_Tanzania.pdf (accessed on 2 December 2019).
35. Morris, M.; Chonya, C.; Koopmanschap, E.; Masanja, M. Sustainable Water Access, Utilization and Management. Report of a Collective Learning Workshop. In Proceedings of the Mafinga Teacher's Resource Center, Mafinga, Tanzania, 25–28 March 2013.
36. Young, M.; Esau, C. CHARTING OUR WATER FUTURE: Economic frameworks to inform decision-making. In *Investing in Water for a Green Economy*; Routledge: London, UK, 2015; pp. 67–79.
37. Jenkins, B.; Gilbert, R.; Nelson, J. *The 2030 Water Resources Group: Collaboration and Country Leadership to Strengthen Water Security 2017*; Harvard Kennedy School: Cambridge, MA, USA, 2017.
38. Agrawala, S.; Carraro, M.; Kingsmill, N.; Lanzi, E.; Mullan, M.; Prudent-Richard, G. *Private Sector Engagement in Adaptation to Climate Change: Approaches to Managing Climate Risks*; OECD Environment Working Papers No. 39; OECD Publishing: Paris, France, 2011.
39. Batchelor, S. Changing the financial landscape of Africa: An unusual story of evidence-informed innovation, intentional policy influence and private sector engagement. *IDS Bull.* **2012**, *43*, 84–90. [CrossRef]
40. WWF. Water Stewardship: Perspectives on Business Risks and Responses to Water Challenges. 2013. Available online: https://wwf.panda.org/?210092/Water-Stewardship--Perspectives-on-business-risk-and-responses--to-water-challenges (accessed on 23 March 2021).
41. Jackson, C. Doing what comes naturally? Women and environment in development. *World Dev.* **1993**, *21*, 1947–1963. [CrossRef]
42. Kariuki, J.; Birner, R. Are market-based conservation schemes gender-blind? A qualitative study of three cases from Kenya. *Soc. Nat. Resour.* **2016**, *29*, 432–447. [CrossRef]
43. Agarwal, B. Environmental action, gender equity and women's participation. *Dev. Chang.* **1997**, *28*, 1–44. [CrossRef]
44. Ibrahim, S. Brief Overview on the Current Situation on Gender and Water Management in Egypt. In Proceedings of the 3rd Arab Water Regional Conference: Research Advancement in Managing Limited Water Resources, Cairo, Egypt, 9–11 December 2006; pp. 9–11.
45. Selsky, J.W.; Parker, B. Cross-sector partnerships to address social issues: Challenges to theory and practice. *J. Manag.* **2005**, *31*, 849–873. [CrossRef]
46. Meißner, N.; Grote, U. Motives, opportunities, and risks for private sector investment in protected areas with international importance: Evidence from German companies. *Environ. Dev. Sustain.* **2017**, *19*, 199–219. [CrossRef]

47. Tai, L. *The Natural Resources Trap: Private Investment without Public Commitment*; MIT Press: Cambridge, MA, USA, 2010.
48. Mdhlovu, P. *Water Governance, Power Relations and Water Conflicts: Exploring the Dynamics of Water Conflict and Its Resolutions between the Mining and Irrigation Sectors in the Olifants Water Management Area (South. Africa)*; University of Pretoria: Pretoria, South Africa, 2018.
49. Moench, M.; Dixit, A.; Janakarajan, S.; Rathore, M.S.; Mudrakartha, S. *Fluid Mosaic: Water Governance in the Context of Variability, Uncertainty and Change; a Synthesis Paper*; Nepal Water Conservation Foundation: Kathmandu, Nepal, 2003.
50. Mirumachi, N.; van Wyk, E. Cooperation at different scales: Challenges for local and international water resource governance in South Africa. *Geogr. J.* **2010**, *176*, 25–38. [CrossRef]
51. Conroy, M.J.; Peterson, J.T. *Decision Making in Natural Resource Management: A Structured, Adaptive Approach*; John Wiley & Sons: Hoboken, NJ, USA, 2013.

sustainability

MDPI

Article

Strengthening Gender Responsiveness of the Green Climate Fund Ecosystem-Based Adaptation Programme in Namibia

Margaret Ndapewa Angula [1,*], Immaculate Mogotsi [2], Selma Lendelvo [3], Karl Mutani Aribeb [4], Aina-Maria Iteta [4] and Jessica P. R. Thorn [5,6]

1 Department of Geography, History and Environmental Studies, Faculty of Humanities and Social Sciences, University of Namibia, P/Bag 13301, Pioneerspark, Windhoek 10023, Namibia
2 Gender Training and Research Programme, Multi-Disciplinary Research Centre, University of Namibia, P/Bag 13301, Pioneerspark, Windhoek 10023, Namibia; imogotsi@unam.na
3 Life Science Division, Multi-Disciplinary Research Centre, University of Namibia, P/Bag 13301, Pioneerspark, Windhoek 10023, Namibia; slendelvo@unam.na
4 Environmental Investment Fund of Namibia, P.O. Box 28157, Auas Valley, Windhoek 10018, Namibia; KAribeb@EIF.ORG.NA (K.M.A.); AIteta@eif.org.na (A.-M.I.)
5 Department of Environment and Geography, York Institute of Tropical Ecosystems, University of York, Wentworth Way, Heslington, York YO10 5NG, UK; jessica.thorn@york.ac.uk
6 African Climate and Development Initiative, Upper Campus, University of Cape Town, Geological Sciences Building Level 6, 13 Library Road, Rondebosch, Cape Town 7700, South Africa
* Correspondence: mangula@unam.na or margaret.angula@gmail.com

Citation: Angula, M.N.; Mogotsi, I.; Lendelvo, S.; Aribeb, K.M.; Iteta, A.-M.; Thorn, J.P.R. Strengthening Gender Responsiveness of the Green Climate Fund Ecosystem-Based Adaptation Programme in Namibia. *Sustainability* **2021**, *13*, 10162. https://doi.org/10.3390/su131810162

Academic Editor: Antonio Boggia

Received: 21 June 2021
Accepted: 3 September 2021
Published: 10 September 2021

Publisher's Note: MDPI stays neutral with regard to jurisdictional claims in published maps and institutional affiliations.

Copyright: © 2021 by the authors. Licensee MDPI, Basel, Switzerland. This article is an open access article distributed under the terms and conditions of the Creative Commons Attribution (CC BY) license (https://creativecommons.org/licenses/by/4.0/).

Abstract: Scholars of gender and climate change argue that gender-blind climate change actions could exacerbate existing inequalities and undermine sustained climate change adaptation actions. For this reason, since 2017, the Green Climate Fund placed gender among its key programming prerequisites, making it the first multilateral climate fund to do so worldwide. However, to date, no lessons to inform planned gender-responsive ecosystem-based interventions in Namibia have been drawn from community-based natural resource management. Thus, this paper aims to share key lessons regarding the way in which gender assessment is useful in enhancing equity in an ecosystem-based adaptation programme for the Green Climate Fund. To this end, we conducted in-depth interviews and group discussions in the 14 rural regions of Namibia with 151 participants from 107 community-based natural resource management organisations (73.5:26.5; male:female ratio). The results identified gender imbalances in leadership and decision-making due to intersecting historic inequalities, ethnicity and geography, as well as other socio-cultural factors in local community-based natural resource management institutions. We also identified income disparities and unequal opportunities to diversify livelihoods, gendered differentiated impacts of climate change and meaningful participation in public forums. Overall, the assessment indicates that considering gender analysis at the initiation of a community-based climate change adaptation project is crucial for achieving resilience to climate change, closing the gender gap, building capacity to increase equity and empowering women in resource-dependent environments in Namibia and Sub-Saharan Africa more broadly.

Keywords: adaptive capacity; climate change adaptation; community-based natural resource management; community-based tourism; gender responsiveness; Green Climate Fund; nature-based solutions; resilience

1. Introduction

The recent special report of the Intergovernmental Panel on Climate Change (IPCC) 1.5 °C confirms that climate change is a major threat to humanity and urgent action is needed [1]. In Sub-Saharan Africa (SSA), increasing temperatures, evapo-transpiration, variable climate and extreme rainfall could impact rural and urban populations severely. This is particularly the case for agricultural and pastoral communities that are highly reliant on natural resources for water-energy-food security in dryland Namibia [2]. While ecosystem-based adaptations are advocated by the Convention of Biological Diversity [3] and its promise to initiate a wider systems transformation is increasingly being recognised by academic and government bodies alike at local, national and international levels [4–6], little has been documented concerning lessons that can be gleaned from experiences of incorporating gender responsiveness in existing community-based natural resource management (CBNRM).

Over the last 15 years, the discourse on gender, climate change adaptation and disaster risk reduction (e.g., [7,8]) indicates that climate change causes significant gender-differentiated vulnerabilities and impacts, for example, due to cultural norms which inhibit adaptation, levels of education and inequitable distribution of roles, resources and power [9–15]. Similarly, in Namibia, early studies confirm that the impacts of climate change on agricultural and ecosystem-based livelihoods are gender-differentiated [16]. Angula (2010) [16] argues that the gender assessment of climate adaptation and mitigation requires a diverse group of competent stakeholders rather than a homogenous group in order to draw from varied experiences and backgrounds to develop solutions in the face of uncertainty [9]. Studies emphasise women's agency and ability to cope with climate change impacts [17] and argue that analyses should go beyond perceiving women as passive victims of climate change [18–20]. More recently, gender responsiveness has emerged as a term which refers to paying attention to the unique needs of females, valuing their perspectives, respecting their experiences, understanding developmental differences between women and men and, ultimately, empowering girls and women [21]. It, furthermore, involves engaging men in climate policies aimed at achieving gender equality and equity [22].

Although there is a growing body of research on adaptation to climate change at the local level, there is still insufficient empirical understanding of gendered-differentiated, adaptive strategies to secure livelihoods [18]. Many publicly financed international projects employ gender-disaggregated data as indicators for achieving gender equality. This is problematic since it assumes that women or men are a homogenous category; consequently, it does not address the underlying causes of gender inequality and does not account for other demographic factors (e.g., culture, age, livelihood and gender) that could make men, the youth, the elderly and others more vulnerable than women [9,13].

We studied a GCF-funded project in Namibia as it was the first multilateral climate fund to place gender among its key programming prerequisites [23,24]. That is, in October 2016, the GCF Board adopted the Gender Policy and Action Plan by decision GCF/B.08/19, which was then updated for 2018–2020 [22]. The Gender Policy and Action Plan is complementary to environmental and social safeguard requirements, and emphasises gender responsiveness, rather than gender sensitivity. In other words, it ensures that remedial actions go beyond raising gender awareness and addressing historical gender biases and inequalities. Ultimately, the gender policy was rooted in its mandate of a paradigm shift towards low-emission and climate-resilient pathways in order to maximise the co-benefits of climate and development action. However, in SSA, the few climate-financed projects that have mainstreamed gender into implementation, monitoring or evaluation, have not undertaken an initial gender assessment to understand the context and to ensure equal gendered participation before commencing [25] (see Appendix A for definitions [21,26]).

Understanding the ways in which unequal gender relationships play out in GCF-funded programmes is particularly important in the context of ecosystem-based adaptation and the co-benefits of ecosystem services for mitigation and adaptation (e.g., carbon sequestration and storage, soil and water regulation, flood attenuation and crop production) [27–29]. One example of ecosystem-based adaptation in Namibia is seen through the CBNRM programme involving rural communities. A gender-responsive approach in CBNRM is critical, because the roles and responsibilities of men and women across Namibia are shaped by socio-cultural norms, traditions and, in part, by their involvement in different kinds of activities regarding livelihood and resource utilisation [9,30]. For instance, women from semi-arid areas where non-timber forest products are abundant are actively involved in forest harvesting. Meanwhile, women in arid areas are actively involved in community tourism. Gender roles, needs and participation in CBNRM are also differentiated. In traditional societies, women are often disinclined to participate in activities that are seen to go against existing traditionally defined roles, most of which can and do present obstacles to participation in climate change adaptation projects. A case study from Kenya illustrates that women's active participation in decision-making and enrolment in activities was hindered because they were represented by their sons [31]. In India, women who were elected to local level institutional governance were represented by their husbands or sons [32]. These examples illustrate that cultural norms and levels of patriarchy limit women's participation in adaptation and developmental activities.

In this paper, we aim to share key lessons regarding the way in which a gender assessment conducted in Namibian rural communities could be useful in developing a gender-responsive, ecosystem-based adaptation project funded by GCF. Our objectives were to:

- assess the gender-differentiated impacts of the effects of climate change on community-based tourism (CBT) in the livelihood-based sector;
- assess the engagement of men and women in the CBNRM sector, their divisions of labour, access, power relations and control of CBT benefits;
- analyse underlying social, economic and political factors that affect the adaptive capacity of men and women and the ways in which they could exacerbate gender inequality; and
- investigate the potential contributions of women and men to ecosystem-based adaptation in order to build resilience to climate change.

Overall, we show that there has been a shift from gender-sensitivity to gender-responsiveness in the ecosystem-based adaptation to climate change. We hope these insights can assist climate-financed projects and programmes to move away from simple gender awareness towards a more comprehensive integration of gender in project activities [22].

Such results have relevant implications for designing programme interventions in meaningful and practical ways, developing national policies, such as the Namibia's Disaster Risk Management Act, National Climate Change Policy and Strategy and Action Plan for Namibia; national targets to attain the Sustainable Development Goals (SDG) 5 (gender equality) and 13 (climate action), the Sendai Framework, the United Nations Framework Convention on Climate Change Gender Policy, the UN's Women Strategic Plan 2018–2021 and the Beijing Declaration and Platform for Action.

2. Materials and Methods

2.1. Study Site Description

The study was carried out in all 13 regions of Namibia (see Figure 1). Namibia was selected because rural communities are among the most vulnerable to the impacts of climate change [33]. Furthermore, the GCF funded an ecosystem-based adaptation programme that focused on sustainable harvesting, consumption, equitable access and benefit-sharing of these resources [9]. We focused on CBNRM and community-based tourism through the conservation of biodiversity, since these sectors were identified at the national level as having the greatest potential to diversify livelihoods, generate wider economic and developmental gains and address the adaptation deficit at the local level in rural Namibia [34,35]. Furthermore, tourism and non-timber forest products, as well as non-consumptive benefits of wildlife management offered through CBNRM, are reliant on water, biodiversity and landscapes that are strongly affected by climate change [36]. Nevertheless, the CBNRM sector is characterised by rigid gender roles, resulting in inequitable benefit sharing.

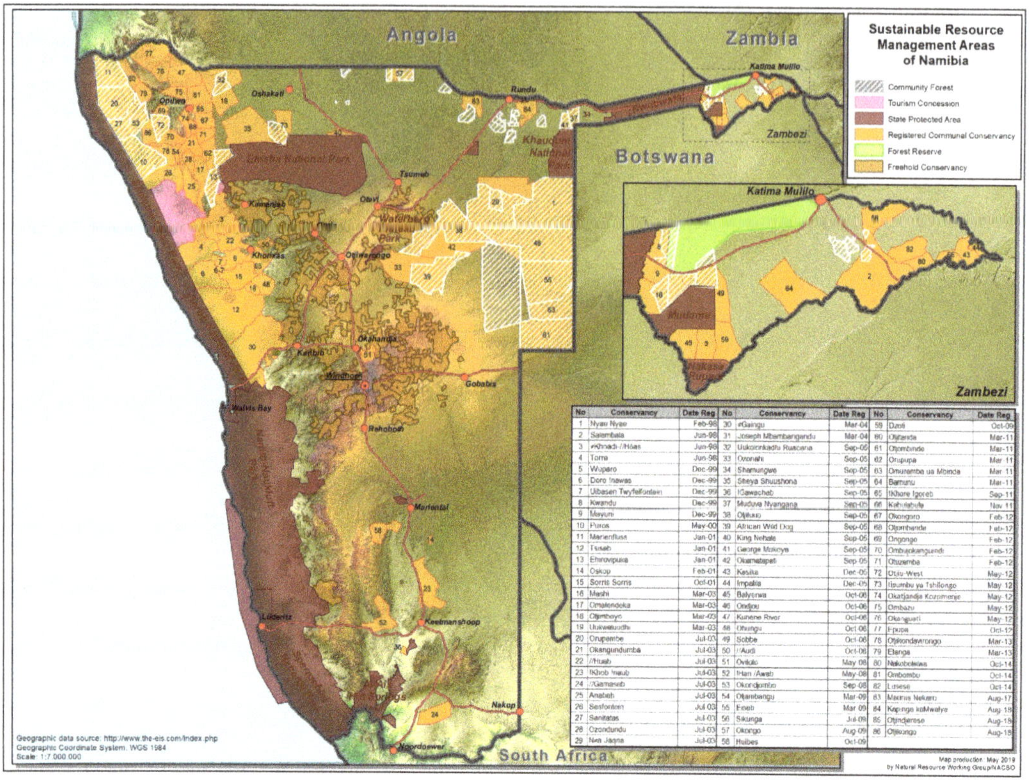

Figure 1. A map of Namibia illustrating the distribution of conservancies and community forests which represent the main CBNRM institutions in the country (Source: [37]).

2.2. *Data Collection*

2.2.1. Focus-Group-Discussion Workshops

Between March 2017 and December 2017, we conducted regional (sub-national level) focus-group-discussion workshops to ensure the inclusion of people's views at all levels. Consultative workshops offered a platform to elicit diverse worldviews through multi-layered reflections, develop a collective understanding that promoted an open and detailed discussion among participants and generated forms of data distinct from interviews, interactions and observation [38]. Due to the vastness of the country, regions were clustered into five groups. Five two-day focus-group workshops involved 151 participants. Participants represented key stakeholders knowledgeable in CBNRM, representatives of conservancies, community forestry committees, traditional authorities, the tourism sector, women's movements, staff members from the Ministry of Agriculture Water and Land Reform, as well as the Ministry of Environment, Forestry and Tourism, who worked or lived in the study areas (see Table 1).

Consultative discussions covered the following topics: aspects that shaped the nature and implementation of CBNRM initiatives; the ways in which climate change carries gender-differentiated implications concerning roles, needs, rights, priorities, access to and control over resources and decision-making processes; socio-economic, cultural and institutional gaps that prevent men and women from responding and adapting equitably to the impacts of climate change; the influence of ethnicity, income and class on socio-economic relationships and gendered adaptive capacity; ways in which the GCF projects could either reinforce or reduce the barriers to adaptation caused by gender inequalities; visions of effective and sustainable solutions.

Table 1. Namibian regional consultative workshops and the number and gender of participants.

Cluster	Region	Workshop Location	Number of Participants		
			Total	Male	Female
South	Hardap, //Karas	Mariental	17	12	5
Central	Omaheke, Otjozondjupa, Erongo	Otjiwarongo	13	10	3
Western	Kunene	Opuwo	38	30	8
North-Central	Oshikoto, Ohangwena, Oshana, Omusati	Ondangwa	26	14	12
North-East	Kavango West, Kavango East, Zambezi	Rundu	57	45	12
	Total		**151**	**111 (73.5%)**	**40 (26.5%)**

Cumulatively, the workshop displayed male dominance (74%), although smaller group discussions included female representatives. Attempts were made to have an equal gender balance in participation, and the invitations addressed to the CBNRM management committees explicitly indicated a request for a balanced gender representation. Most workshops were male-dominated, except in the workshop in the north–central regions. The low representation of women could be attributed to the following social factors: women's low representation in community-based institutions, women not being encouraged to attend meetings in the presence of men (especially in the case of the OvaHimba in the Kunene region) or mobility constraints related to the gendered division of labour (e.g., household chores and childcare).

2.2.2. In-Depth Interviews with Informants

In July 2017, we conducted 15 in-depth informant interviews with a subset of participants who attended the workshops. Key informants were selected by UNAM researchers, Environmental Investment Fund (EIF) staff members and other consultants that were involved in developing the GCF ecosystem-based adaptation programme for Namibia. Key informants were purposively sampled by targeting individuals who impacted or were impacted by CBNRM institutions in Namibia. Stakeholders included inter alia non-governmental organisations working with conservancies, community-based organisations leaders, chairpersons of conservancies and community forests, community members, Ministry of Environment, Forestry and Tourism staff members, as well as Traditional Authority secretaries and leaders. These in-depth interviews covered the same questions that also formed part of focus-group-workshop discussions (see Appendix B for a list of questions). Each interview lasted 60–80 min. Both the workshops and interviews were conducted mainly in English and the dominant local language in each area, via translators.

2.3. Analysis

To understand the existing inequalities in the distribution of responsibilities and power in conservancies and community forests in Namibia regarding climate adaptation, three frameworks were utilised to structure the data collection (i.e., informing the content of the interview schedule) and analyse the outputs of the interviews and focus groups. The Harvard Analytical Framework was employed to frame questions and analyse data related to roles and responsibilities, allocation of resources and productive and socially reproductive work. The Social Relations Approach Framework was useful in assessing and analysing existing inequalities in the distribution of resources, responsibilities and power [39]. Additionally, the IPCC vulnerability framework was applied to identify the impact (exposure and sensitivity) and adaptive capacity of communities to climate change [9]. Interview and discussion transcripts were analysed by means of thematic analysis, employing a predetermined set of deductive codes grouped into the following main themes: ecological and economic (livelihoods); employment, education and skills (to access opportunities and income); cultural and traditional practices (hindering or promoting agency); institutional and governance (decision-making and participation).

3. Results

Overall, patriarchal norms continue to limit the equitable access to, control of and benefit from natural resources and community-based tourism, as well as other ecosystem services. In this context, patriarchy is understood as a socio-political system that is embedded in cultural norms and practices that favour males as the dominant figure in society [40]. As a result, the limitations regarding the potential of ecosystem services in the adaptation of climate change are not only biophysically related, but are also socially induced.

3.1. *Gendered Division of Labour in CBNRM Institutions*

In Namibia, gendered norms can exclude women from diversifying income sources in ecosystem-based adaptation projects as a means to secure livelihood, and influence the way in which women and men may employ some adaptation strategies over others. This was evidenced in the finding that employment from tourism and natural resource livelihoods were demarcated by a gender division of labour [41,42]. For example, the results show that overall, women were mainly involved in activities around cultural tourism, such as being cultural dance performers, working in crafts, cooking traditional dishes and fulfilling hospitality roles such as launderers, cleaners, waitresses and receptionists in tourism accommodation. On the other hand, men controlled higher-paid activities, such as game driving, trophy hunting, tour guiding, hiking guiding, fish harvesting and timber harvesting. In the north-eastern, western and north-central regions, stakeholders aired their views that "women cannot be game trackers and skinners because there is blood, and women are not comfortable to be near blood" (Rundu workshop participants) or "women are defenceless, they are not brave enough and will run away from wild animals, therefore, cannot be appointed to be game guards" (Opuwo and Ondangwa workshop participants). This suggests that men were still reluctant to accept women's participation in traditionally male roles as they were perceived to be physically weak or as not having the character to hunt, given the danger associated with trophy hunting. Such male-dominated perceptions inhibit women's participation in what is considered "men's work", while some women endorse it out of fear of cultural sanction.

Beyond roles, the temporal character of labour also differs. Men's ecosystem-based tourism activities tend to be seasonal during peak tourism seasons, with short-term contracts in lodges, campsites and information offices. Men also acquire work during the construction phases of tourism establishments (e.g., drilling boreholes). Meanwhile, women's activities are associated with more sustained permanent employment [43,44]. The number of working hours also differ. In this way, often the participation of women in CBNRM activities beyond the household and village is limited or absent.

According to respondents, the gender division of labour is more rigid in rural communities, where women are expected to stay at home, look after children, the elderly and the sick, as well as clean, cook or collect water [44]. Meanwhile, men work more often in urban areas. According to [45], most conservancy staff members across Namibia are men (76%) and the proportion of women who were elected as treasurers stands at 43%, while only 13% are elected as chairpersons.

Sometimes, where a vacuum has been created because of male out-migration to urban areas, opportunities emerge for women to take up roles that were traditionally assigned to men. For example, this was found to be the case among the Nama living in the southern region, where in recent years women have begun serving as advisors to the Traditional Authority, a role that was traditionally reserved for men. Despite some women's involvement in such male traditional roles, we found that very few men ventured into traditional female roles, which may be because these roles have fewer to no financial gains.

Given the opportunities to address gender imbalances while enhancing their adaptive capacity through programmes that fund ecosystem-based adaptation, workshop participants were asked to prioritise livelihood diversification that would earn an income to reduce food insecurities in conservancies and community forestry reserves. The ecosystem-based adaptation and alternative livelihood activities that were prioritised included:

- Non-consumptive tourism activities, such as game viewing, driving and hiking;
- non-timber forest product activities, such as harvesting medicinal plants, basket weaving and beekeeping;
- horticultural production in water-abundant areas, including hydroponics and fog capture (in north-eastern and western Namibia), to contribute to food security; and
- cultural tourism involving "living museums", where people visit and stay in cultural villages, in addition to "landscape tourism" while protecting communities from cultural romanticism and assimilation (e.g., Damara and OvaHimba communities in the Kunene region).

3.2. Gender Imbalances in Leadership and Decision-Making

Given that most ecosystem-based adaptation projects recognise the importance of including traditional authorities in project planning and implementation, it is essential to acknowledge and actively counter the way in which such structures can hinder equitable participation in decision-making over community natural resources.

We found that, in general, in conservancies and community forestry committees, women are not equal partners in resource management. This can be explained in part because CBNRM management committees in Namibia are constituted of community members and advisors or councillors from the Traditional Authority [46]. These advisors from the Traditional Authorities are typically men.

On average, of 35% of women were conservancy committee members, while the majority were males. The proportion of women in CBNRM management committees varied by region, with some conservancies having no women in their management committees and others comprising more than the required 50% female representation. In particular, the north-central and southern regions have been more successful than other regions in narrowing the gender gap. For instance, in the north-central regions, 60% of the conservancy committee were women. In Erongo, Otjozondjupa and Omaheke (central regions), there was a 50% female representation in CBNRM committees. This was more than in the western and north-eastern regions, which are generally the strongest in observing traditional values.

In terms of the representation of women in leadership positions in CBNRM committees, females holding leadership roles ranged from 0% (in Ehirovipuka Conservancy) to 67% (in Otjimboyo Conservancy). Notably, there was a female professional hunter in one community in the Otjozondjupa region and 3 of the 7 game guards were female in the same region, while in southern Namibia, there was equal representation in the management committee, but the executive committee positions were dominated by men. For instance, in the Nico-Noord conservancy in the Hardap region, 3 of the 6 executive members were female and 2 were additional members without portfolios and one was a treasurer.

In the north-central, central, and southern regions women were well represented, sometimes serving as advisors in the Traditional Authority; however, this does not automatically translate into women influencing decision-making. At the same time, in other regions where women were reluctant to take up leadership positions, they tended to show a high level of participation in several voluntary community initiatives.

These findings illustrate that adaptive capacity is gender-differentiated and that these contextual nuances need to be understood before any ecosystem-based intervention. It is likely that women who can re-negotiate their roles in decision-making and develop a range of proactive ecosystem-based management strategies will decrease their risk of exposure.

3.3. Gendered-Differentiated Impacts of Climate Change

Drought, floods and high temperatures were the three main climate hazards reported to increase exposure and vulnerability. The nature of women's and men's economic opportunities within community-based tourism is affected differently by climate change through reduced earnings. Climate risks have the potential to lead to job losses and reduce household income, thereby contributing to the migration of skilled staff members to other areas in search of alternative employment. Climate change has a direct impact on the landscape of an area, resulting in the loss of wildlife species, vegetation and soil, which can also reduce the performance of the tourism industry and, thus, earnings. Climate change can also lead to human–wildlife conflicts due to the growing scarcity of resources. Women become more vulnerable owing to job losses in the tourism sector, caused by limited employment choices at the local level, as well as the fact that women are generally less mobile when seeking employment elsewhere and women generally receive lower wages compared to their male counterparts [43]. When comparing regions, we found that women in western and northeastern Namibia tended to be more vulnerable due to their limited capacity to contribute to making timely decisions at the household and community levels, their low-income earning potential from tourism and their lesser access to information.

3.4. Differentiated Gendered Meaningful Participation in Public Forums

Understanding how existing structures in Namibia hinder or support gender-equitable and inclusive stakeholder engagement and consultations throughout the design and implementation of the ecosystem-based project is central to the success of any climate-financed programme. It was found that persistent patriarchal norms inhibited women's meaningful participation in the decision-making processes of CBNRM institutions.

Namibia's patriarchal governance structure goes back centuries [47]. Historically, in many regions, tradition and religion dictated gendered relationships and entrenched male domination in the structure and leadership of social organisations. In the last century, in Omusati, Oshana, Oshikoto, Ohangwena (north-central regions), Hardap and ||Karas (southern regions), cultural norms shifted somewhat. That is, the influence from Lutheran and Catholic missionaries relaxed, to a degree, some of the rigid gendered roles and allowed women to attain education and literacy. With Independence in 1990, Namibia saw the introduction of gender equality laws in the Constitution [48].

Despite this progress, we found evidence that patriarchy still continues to affect the meaningful participation of women in community consultations and meetings related to CBNRM. Across all workshops, participation was male-dominated (73.5%). For instance, in workshops in the Kavango West, Kavango East and Zambezi (north-eastern regions), fewer women attended (21%). In these areas, women's participation in local-level governance was often passive and limited to meeting attendance, with little or no input in discussions. Yet, male participants were unaware that structural norms had the potential to inhibit female participation, stating: "No cultural limitations or reasons are discriminating against our women to get work in the tourism sector. They just need to be empowered" (male participant from Rundu workshop). Similarly, in the western and north-eastern regions, cultural norms inhibited women from contributing to discussions in public or in the presence of men. Although the southern regions showed a more gender-balanced representation in their committees, only one female CBNRM committee member of 17 participants attended the workshop. These results highlight the ways in which women's voices and perspectives could be silenced, or left out from prioritising livelihood diversification projects, such as GCF ecosystem-based adaptation projects.

Yet, the level of dominance in patriarchal systems varied across ethnic groups and regions. For instance, in the north-central regions, more women attended (46%) and spoke in community meetings. It emerged from this workshop that "women are more trustworthy to occupy the treasurer position in the committees, unlike men who are seen to be likely to mismanage funds" (male participant from Ondangwa workshop). On the other hand, in the western region, cultural perceptions emerged as one of the biggest obstacles in obtaining gender equality among the OvaHimba community. Men who did not participate in male-dominated activities were ostracised, as were women who spoke openly in meetings [49].

4. Discussion

Based on a national study conducted in the 14 regions of Namibia, this study contributes to the empirical literature on gender responsiveness to financing in an ecosystem-based adaptation climate. We argue that a gender-responsive approach in any climate change programme or policy is essential from the outset. We therefore suggest three key implications for future climate financing for ecosystem-based adaptation in Namibia: align with existing institutions; build the capacity to support meaningful participation and representation in decision-making processes; engage both men and women of all ages and positions of leadership for men to play other roles in order to challenge cultural norms and to work actively to address gendered divisions of labour.

4.1. Align Ecosystem-Based Adaptation Governance with Existing CBRNM Institutional Set-Up

One of the key lessons our assessment offers is that aligning climate adaptation governance with the CBNRM institutional set-up offers an opportunity to ensure equal representation and participation in decision-making and leadership. Namibia currently does not have institutions with a mandate to discuss climate change at regional-level (sub-national levels) and the way this links with other cross-cutting policy targets. Therefore, we recommend that an appropriate platform should be identified to oversee ecosystem-based adaptation at the community and regional level, rather than the set ting up of new committees. Employing an existing institutional set-up both at regional (sub-national) and community levels presents a potentially efficient and cost-effective opportunity to integrate climate change adaptation for GCF-accredited entities, such as the Environmental Investment Fund, while achieving the decentralisation of efforts. To this end, some programmes ((i) CDKN (Climate and Development Knowledge Network) Knowledge Brokering Project Namibia and (ii) IDRC funded CLARE (Uptake of Climate Adaptation research results in Africa) Namibia project) are emerging to build capacity and raise an awareness of the ways in which climate change commitments fit in with their existing institutional targets (e.g., rural development); however, these need to be scaled up. Where they are most prominent in Namibia, differentiated levels of patriarchy that influence equitable participation, gender imbalances in leadership and decision-making need to be addressed among local-level institutions. Knowing the level of dominance and the dynamics of patriarchy and its influence on the governance of local institutions is crucial to ensure that not only women, but all participants, distinguished by multiple forms of social differentiation (e.g., ethnicity, age, education, social capital), are meaningfully involved [11].

4.2. Levels of Participation and Decision-Making Agency

Increased participation opportunities for women in CBNRM can enhance direct, tangible and intangible benefits [44,50]. Despite the silence on gender representation in the National Policy on CBNRM in Conservancies and Community Forests' management committees [46], there exists the political commitment from the government to embrace gender equality across all sectors in Namibia. For instance, recent legal reforms in this policy require 50% gender representation in positions of leadership in the governing body of conservancies and community forest reserves that are gazetted as CBNRM institutions [45]. This serves as an example to be replicated in other governance and community-based management structures working on ecosystem-based adaptation programming.

In Namibia, through ecosystem-based adaptation programmes, women can participate as equal partners to men. Where capacity to participate is lacking in terms of the skills and awareness required, the Gender Action Plan for the GCF ecosystem-based adaptation project specifies the need for men and women to be granted equal opportunities to participate in, and benefit from the fund through the progressive and efficient mainstreaming of gender dimensions, while avoiding, minimising or mitigating the gender-related adverse impact of subprojects [51].

Different approaches to mainstreaming participation can enhance and reinforce one another [44]. It is indicated that effective representation could be enhanced among women if their satisfaction levels with conservancy benefits were high. Participation also increases where household benefits are tangible and members are satisfied with conservancy efforts. Moreover, the equitable and active participation of women in local institutions is enhanced through increasing the meeting attendance by females or women being voted into leadership positions. This allows women to be part of a collective voice, leading to the strengthening of common identities and local democracy [9].

Furthermore, local ecosystem-based adaptation policies and planning should create conditions that foster autonomous adaptation at the household level, and provide public support for planned adaptation when autonomous adaptation is insufficient.

Our findings resonate with other literature that shows that the gendered nature of everyday realities and experiences of women and men tend to be overlooked when it comes to developing and strengthening the adaptive capacities of local communities [13]. Arora-Jonsson [52] warns that this oversight could lead to the incorrect formulating of gender issues in policy development. There is a tendency to portray women as vulnerable, weak, poor and socially isolated, rather than seeing them as negotiating and dealing regularly with different kinds of change in their lives [53], particularly in ecosystem-based adaptations. Ramchurjee [54] alluded to the entry points for women's employment and opportunities for creating self-employment in small- and medium-sized, income-generating activities, thus creating paths towards the elimination of poverty for women and local communities.

4.3. Diversification of Livelihoods Should Account for Gendered Divisions of Labour

Climate financing should consider the way in which ecosystem-based adaptation and livelihood diversification options intersect with gendered divisions of labour and other forms of differentiation. In Namibia, we found that, because of the high economic value associated with male roles, there is a tendency for the development and policy interventions to encourage women to venture into what is perceived as traditionally male roles as a means to equalise income levels with those of men. However, such interventions can have negative impacts by overburdening women if their traditional roles remain unchanged. Thus, we contend that climate-financed interventions should strive to engage both men and women of all ages and positions of leadership for men to play other roles, equalise income disparities, raise awareness of the value of so-called "women's work" and ensure more support in order for women to perform their reproductive roles (e.g., paid maternity leave, childcare) [55]. Furthermore, cultural barriers that hinder men and women to venture into non-conventional gender roles must be addressed.

Another lesson for the GCF ecosystem-based adaptation programming is that local leadership structures should capitalise on skills development among both women and men. Similar to [9,49,54], we found that women's interests were represented less in negotiations of private ventures (e.g., trophy hunting). This appears to be, in part, due to a lack of negotiation and legal skills among community members, and this leads to a conflict over the control of the funds generated. To counter this, local leadership, government, industry, NGOs and international agencies can support training and extension programmes to influence adaptation processes positively. An example of this includes the initiatives of the Namibian Association of CBNRM Support Organizations (NACSO), which train women in public speaking, harvesting and entrepreneurial skills, and has led to women occupying leadership positions; however, this needs to be scaled up and out to include technologies that increase yields, produce goods and reduce environmental degradation [55,56].

Livelihood diversification ecosystem services that supplement traditional agricultural livelihoods and have the potential to withstand climate shocks should be prioritised. This should be accompanied by strengthening the value chains to enhance the marketing of natural products which improve returns for women and the community. An example of this is the establishment of organisations by communities to harvest, market and sell the devil's claw and other natural products [55]. Despite the existence of natural product markets, such initiatives have not involved many communities in Namibia.

A key lesson for climate financing is to engage in initial discussions on the potential value of all activities for ecosystem-based adaptation, irrespective of gender norms and cultural relations [57]. This should be followed by prioritising interventions to ensure that they address local needs and avoid reinforcing existing gendered responsibilities (e.g., males are inclined to participate in the construction of community-based tourism and wildlife infrastructure or eco-tourism activities, such as commercial hunting). Any ecosystem-based adaptation initiative in the CBNRM requires consultations that are sensitive and are aimed at addressing patriarchy and existing gendered cultural, age and other inequities [9,18].

5. Conclusions

This study provides important insights from community-based tourism, community forestry and wildlife management initiatives in Namibia that can inform the future mainstreaming of gender equity into interventions regarding ecosystem-based adaptation. We show that gender roles and cultural factors intersect to hinder the participation of women in discussions and the implementation of programmes. Patriarchy affects the participation of women in leadership, decision-making and livelihood diversification. The gendered division of labour differentiates income earning potential and participation in biodiversity-related activities, such as nature-based tourism. Thus, given the intersection of heterogeneous gender, but also ethnicity, education, historical and socio-cultural factors, context is important. Understanding such variations is critical when designing climate-financed programmes for ecosystem-based adaptation.

A gender analysis at the outset of a community-based, climate change adaptation project is essential to identify the hindering and facilitating factors to the implementation of effective climate adaptation and to put in place mitigation measures to reduce existing gender inequalities. Gender-responsive actions for local-level ecosystem-based adaptation projects should include capacity-building for women, engaging with men to address gender stereotypes towards women's participation and encouraging the active inclusion and participation of the youth and marginal communities. Interventions should be designed in association with gender performance indicators and sex-disaggregated targets to the monitoring and evaluation of initiatives [9].

Although there is no silver bullet regarding the way in which a GCF project can achieve gender equality, we hope that these insights contribute to a more comprehensive assessment of gender dynamics before designing any ecosystem-based adaptation intervention in Namibia and sub-Saharan Africa more broadly. Programming should be adaptive, agile and sensitive to the socio-cultural context and must not be applied homogenously either across Namibia or other nations. Systemic shifts will require the time and commitment of many actors across scales and decades.

Further research is needed to investigate the ways in which cultural norms and patriarchy could be addressed in the context of being explicitly acknowledged in the implementation of ecosystem-based adaptation projects and programmes to ensure that the rollout of these initiatives are gender-responsive. In the same vein, it is critical to explore ways in which women can contribute and collaborate in those processes, helping to create positive change in local policies and practices [6]. Future research could assess ecosystem-based adaptation interventions retrospectively, and evaluate over time whether climate-financed interventions have achieved the desired outcomes.

Author Contributions: Conceptualisation, M.N.A., S.L., I.M., J.T.R.T.; methodology, M.N.A., S.L., I.M.; investigation, S.L., I.M., A.-M.I.; formal analysis M.N.A., S.L., I.M.; writing—original draft preparation, M.N.A., S.L., I.M., K.M.A., J.P.R.T. and A.-M.I.; writing—review and editing M.N.A., S.L., I.M. and J.P.R.T.; supervision, I.M. and K.M.A.; project management, A.-M.I., I.M. and S.L.; funding acquisition, K.M.A. and J.P.R.T. All authors have read and agreed to the published version of the manuscript.

Funding: This research was funded by the Environmental Investment Fund of Namibia through GCF (project number EDA FP024) and UK Research and Innovation's Global Challenges Research Fund (UKRI GCRF) through the Development Corridors Partnership project (project number: ES/P011500/1).

Institutional Review Board Statement: The study was conducted according to the guidelines of the Declaration of Helsinki and was approved by the Institutional Review Board (or Ethics Committee) of the Environmental Investment Fund of Namibia.

Informed Consent Statement: Informed consent was obtained from all participants involved in the study. All participants agreed and consented to participate in this study when they confirmed the invitations.

Data Availability Statement: Not applicable.

Acknowledgments: We acknowledge all participants who were interviewed for data collection. We are grateful for the participation of the University of Namibia, Ministry of Environment, Forestry and Tourism, Regional Councils and Traditional Authorities. We further acknowledge the African Women in Climate Change Science Fellowship of the African Institute of Mathematical Sciences and the Next Einstein Forum and the Climate Research for Development (CR4D) Postdoctoral Fellowship CR4D-19-21 implemented by the African Academy of Sciences (AAS) in partnership with the United Kingdom's Department for International Development (DfID) Weather and Climate Information Services for Africa (WISER) programme and the African Climate Policy Center (ACPC) of the United Nations Economic Commission for Africa (UNECA) for supporting J.P.R.T on this work. Statements made and views expressed in this work are solely the responsibility of the authors.

Conflicts of Interest: The authors declare no conflict of interest.

Appendix A. Definition of Gender Concepts

Gender analysis: A critical examination of the way in which differences in gender roles, activities, needs, opportunities and rights/entitlements affect men, women, girls and boys in certain situations or contexts. Gender analysis examines the relationships between females and males and their access to and control of resources and the constraints they face relative to each other.

Gender awareness: Being conscious of the fact that men and women have different roles, responsibilities and needs.

Gender biased: Making decisions based on gender that result in favouring one gender over the other which often results in contexts that are favouring men and/or boys over women and/or girls.

Gender blindness: The failure to recognise that the roles and responsibilities of men/boys and women/girls are given to them in specific social, cultural, economic and political contexts and backgrounds. Projects, programmes, policies and attitudes which are gender blind do not consider these different roles and diverse needs, maintain the status quo and will not help to transform the unequal structure of gender relations.

Gender mainstreaming: Mainstreaming can be defined as re-organising, improving, developing and evaluating policy-making processes to incorporate a gender perspective in all policies at all levels and all stages.

Gender-responsive programming and policies: Intentionally employing gender considerations to affect the design, implementation and results of programmes and policies. Gender-responsive programmes and policies reflect girls and women's realities and needs in components, such as site selection, project staff, content and monitoring. Gender-responsiveness means paying attention to the unique needs of females, valuing their perspectives, respecting their experiences, understanding developmental differences between girls and boys, women and men and, ultimately, empowering girls and women.

Gender-sensitive: Programmes, projects and policies that are aware of, and address, gender differences.

Patriarchy is defined as the justification to marginalise women in education, the economy, labour market, politics, business, family, domestic matters and inheritance rights.

Appendix B. Gender Assessment Questions

Key informants' questions (in-depth interviews one-on-one)

1. What are the key vulnerabilities to drought, flooding, high temperature, shifting rainy season, low crop yields, reduced livestock yield that are facing men and women in selected conservancies?
2. How are the communities responding in these areas? Are men and women responding differently?
3. What are the external factors that help or hinder the community to respond? Which ones are institutional? Which ones are cultural and gendered?
4. What capacities are lacking in these communities that make men and women more vulnerable? Probe: which ones are specific to women and which ones are specific to men? Youth/marginalised communities such as the San, Himba or Zemba?
5. Who is more vulnerable to climate change impacts?

Group discussions in workshops:

Vulnerability: Exposure and sensitivity

1. What are the CBNRM communities exposed to? These are biophysical impacts such as changes in temperature (high temperature and how it affect livestock, crops and wildlife), the same with reduced rainfall, high rainfall, and these include drought and flooding. Probing questions and what we were paying attention to when asking the exposure question included: what are they mostly exposed to? What women say, what men say? Who is most exposed to which—men or women?
2. What are the key vulnerabilities facing your communities (i) related to climatic factors such as drought, flood, high temperature, water scarcity etc.; (ii) related to non-climatic factors such as unemployment, livestock theft, human–wildlife conflict, HIV/AIDS and other health issues, etc.; (iii) related to cultural norms and values such as gender stereotypes, discrimination of marginalised, cultural beliefs hindering better responses etc.; (iv) related to governance such as traditional regulations, institutional support and lack of support, social politics, etc.?
3. Which livelihood is most sensitive to impacts of climate change? What is the gendered level of dependency on natural resources and other climate-dependent sectors by community members?

Adaptive capacity

1. When you are faced with climate change-related impacts such as floods, drought, high temperature and associated water scarcity, how do you cope?
2. Who make decisions regarding farming preparedness and response when faced with climate change impacts? How do cultural gender relations affect this?
3. Who responds in the household and community regarding food security in the household? livestock? water?
4. How do you benefit or how useful is CBNRM (conservancies and community forests) contributions towards enabling your capacity to respond to climatic impacts and non-climatic impacts?
5. Who make decisions and controls the harvesting of natural resources and access to it in your community?
6. What type of employment and income-generating activities are offered in your conservancy or community forest? Do men and women participate? which ones are dominated by men and which ones by women?

Questions relevant for the EDA (Empower to adapt) project activities

1. Which activities and interventions do you suggest should be included in the GCF-funded EDA (Empower to adapt) project? Which income-generating activities need to be strengthened? Which income activities need to be introduced? Which capacity building programmes? Which skills development?
2. What challenges do you currently face in your conservancy or community forest? Financial and institutional? Management committees and administration?
3. When traditional authority emerged in discussions, follow-up questions concentrated on how it acts as a barrier or an enabler to CBNRM programmes and how potentially it could act as a barrier or enabler to the GCF funding of the EDA project.

References

1. Allen, M.R.; de Coninck, H.; Dube, O.P.; Hoegh-Guldberg, O.; Jacob, D.; Jiang, K.; Zhou, G. *Technical Summary in Global Warming of 1.5 °C above Pre-Industrial Levels and Related Global Greenhouse Gas Emission Pathways, in the Context of Strengthening the Global Response to the Threat of Climate Change, Sustainable Development, and Efforts to Eradicate Poverty*; World Meteorological Organisation: Geneva, Switzerland, 2018.
2. Angula, M.N.; Kaundjua, M.B. The changing climate and human vulnerability in north-central Namibia. *JAMBA* **2016**, *8*, 1–7. [CrossRef]
3. Convention on Biological Diversity. CBD—Convention of Biological Diversity. In *Connecting Biodiversity and Climate Change Mitigation and Adaptation: Report of the Second Ad Hoc Technical Expert Group on Biodiversity and Climate Change*; CBD Technical Series 41; Convention on Biological Diversity: Montreal, QC, Canada, 2009.
4. IPCC. *Global Warming of 1.5 °C. An IPCC Special Report on the Impacts of Global Warming of 1.5 °C above Pre-Industrial Levels and Related Global Greenhouse Gas Emission Pathways, in the Context of Strengthening the Global Response to the Threat of Climate Change*; Masson-Delmotte, V., Zhai, P., Pörtner, H.-O., Roberts, D., Skea, J., Shukla, P.R., Pirani, A., Moufouma-Okia, W., Waterfield, T., Eds.; World Meteorological Organisation: Geneva, Switzerland, 2017.
5. Sitas, N.; Prozesky, H.; Esler, K.; Reyers, B. Exploring the gap between ecosystem service research and management in development planning. *Sustainability* **2014**, *6*, 3802–3824. [CrossRef]
6. Wamsler, C.; Pauleit, S. Making headway in climate policy mainstreaming and ecosystem-based adaptation: Two pioneering countries, different pathways, one goal. *Clim. Chang.* **2016**, *137*, 71–87. [CrossRef]
7. Dankelman, I. Climate-change: Learning from gender analysis and women's experiences of organising for sustainable development. *Gend. Dev.* **2010**, *2*, 21–29. [CrossRef]
8. MacGregor, S. Gender and climate change: From impacts to discourses. *J. Indian Ocean. Reg.* **2010**, *6*, 223–238. [CrossRef]
9. Environmental Investment Fund of Namibia. *Gender Assessment Report for the Ecosystem-Based Adaptation (EDA) Project*; Environmental Investment Fund: Windhoek, Namibia, 2017.
10. Babugura, A. *Gender and Climate Change: South Africa Case Study*; Heinrich Boell Foundation-Southern Africa: Cape Town, South Africa, 2010.
11. Otzelberger, A. *Gender-Responsive Strategies on Climate Change: Recent Progress and Ways Forward for Donors*; University of Sussex, BRIDGE/IDS: Falmer, UK, 2011.
12. Goh, A.H.X. *A Literature Review of the Gender-Differentiated Impacts of Climate Change on Women's and Men's Assets and Well-Being in Developing Countries*; CAPRi working papers 106; International Food Policy Research Institute (IFPRI): Washington, DC, USA, 2012.
13. Alston, M. Gender mainstreaming and climate change. *Women Studies Int. Forum* **2014**, *47*, 287–294. [CrossRef]
14. Moosa, C.S.; Tuana, N. Mapping a research agenda concerning gender and climate change: A review of the literature. *Hypatia* **2014**, *29*, 676–694. [CrossRef]
15. Angula, M.N.; Menjono, E. Gender, culture and climate change in rural Namibia. *J. Stud. Humanit. Soc. Sci.* **2014**, *3*, 225–238.
16. Angula, M. *Gender and Climate Change: Namibia Case Study*; Heinrich Boell Stiftung—Southern Africa: Cape Town, South Africa, 2010.
17. Williams, M.J.; Lentisco, A.; Badayos-Jover, M.B.; Pedroza-Gutierrez, C.; Giri, K.; Siar, S.; Gopal, N.; Shanthi, B.; Ferrer, A.J.G.; Sumagaysay, M.B.; et al. Gender as the missing link for improved climate change adaptation in fisheries and aquaculture. In Proceedings of the Fish Adapt: The Global Conference on Climate Change Adaptation for Fisheries and Aquaculture, Bangkok, Thailand, 8–10 August 2016.
18. Vincent, K.; Mkwambisi, D. *Gender, Agriculture and Climate Change in Malawi*; UMFULA: Lilongwe, Malawi, 2017.
19. Rao, N.; Lawson, E.T.; Raditloaneng, W.N.; Solomon, D.; Angula, M.N. Gendered vulnerabilities to climate change: Insights from the semi-arid regions of Africa and Asia. *Clim. Dev.* **2019**, *11*, 14–26. [CrossRef]
20. Rao, N.; Singh, C.; Solomon, D.; Camfield, L.; Sidiki, R.; Angula, M.; Poonacha, P.; Sidibe, A.; Lawson, E.T. Managing risks, changing aspirations and household dynamics: Implications for wellbeing and adaptation in semi-arid Africa and India. *World Dev.* **2020**, *125*, 104667. [CrossRef]
21. UNICEF Regional office for South Asia. *Gender Equality: Glossary of Terms and Concepts*; UNICEF: Kathmandu, Nepal, 2017.

22. Green Climate Fund. Gender Policy and Action Plan. 2018. Available online: https://www.greenclimate.fund/documents/20182/24946/GCF_B.08_19_-Gender_Policy_and_Action_Plan.pdfGCF/B.08/19 (accessed on 8 October 2019).
23. Green Climate Fund. Gender. Available online: https://www.greenclimate.fund/projects/gender (accessed on 10 June 2021).
24. Terry, G. No climate justice without gender justice: An overview of the issues. *Gend. Dev.* **2009**, *17*, 5–18. [CrossRef]
25. Schalatek, L. *From Innovative Mandate to Meaningful Implementation: Ensuring Gender-Responsive Green Climate Fund (GCF) Projects and Programs*; Heinrich Boell Foundation, North America: Washington, DC, USA, 2015.
26. Salaam, T. A brief analysis on the situation of women in Nigeria today. In *Democratic Socialist Movement*; Greenwood Publishing Group: Westport, CT, USA, 2003.
27. Munang, R.; Thiaw, I.; Alverson, K.; Liu, J.; Han, Z. The role of ecosystem services in climate change adaptation and disaster risk reduction. *Curr. Opin. Environ. Sustain.* **2013**, *5*, 47–52. [CrossRef]
28. Huq, N.; Bruns, A.; Ribbe, L.; Huq, S. Mainstreaming ecosystem services based climate change adaptation (EbA) in Bangladesh: Status, challenges and opportunities. *Sustainability* **2017**, *9*, 926. [CrossRef]
29. Locatelli, B. Ecosystem Services and Climate Change. 2016. Available online: http://hal.cirad.fr/cirad-01264738 (accessed on 26 August 2021).
30. Angula, M.N. A Gendered and Intersectional Analysis for Understanding Vulnerability to the Changing Climate within Socially Diverse Communities in Semi-Arid Regions, North-Central Namibia. Ph.D. Thesis, University of Cape Town, Cape Town, South Africa, 2017, (unpublished work).
31. Kariuki, J.; Birner, R. Are market-based conservation schemes gender-blind? A qualitative study of three cases from Kenya. *Soc. Nat. Sci.* **2015**, *29*, 432–447. [CrossRef]
32. ASSAR. Challenging assumptions about gender and climate adaptation. It's not always what, or who, you think. In *CARIAA-ASSAR Infographics*; The University of Cape Town: Cape Town, South Africa, 2018.
33. ASSAR. What does global warming of 1.5 and higher means for Namibia? In *CARIAA-ASSAR Policy Brief*; University of Cape Town: Cape Town, South Africa, 2018.
34. Republic of Namibia. *National Policy on Climate Change for Namibia—2011*; Ministry of Environment: Windhoek, Namibia, 2011.
35. Republic of Namibia. *National Climate Change Strategy & Action Plan*; Ministry of Environment and Tourism: Windhoek, Namibia, 2013.
36. Republic of Namibia. *Third National Communication to the United Nations Framework Convention on Climate Change*; Ministry of Environment and Tourism: Windhoek, Namibia, 2015.
37. Namibia Association of CBNRM Support Organisations (NACSO). *Sustainable Resource Management Areas of Namibia*; Namibia Association of CBNRM Support Organisations: Windhoek, Namibia, 2019. Available online: http://www.nacso.org.na/resources/map (accessed on 22 June 2021).
38. Orngreen, R.; Levinsen, K.T. Workshops as a research methodology. *Electron. J. e-Learn.* **2017**, *15*, 70–81.
39. Vincent, K.; Wanjiru, L.; Aubry, A.; Mershon, A.; Nyandiga, C.; Cull, T.; Banda, K. *Gender, Climate Change and Community-Based Adaptation: A Guidebook for Designing and Implementing Gender-Sensitive Community-Based Adaptation Programmes and Projects*; United Nations Development Programme: New York, NY, USA, 2010.
40. Thakadu, O.T. Communicating in the public sphere: Effects of patriarchy of knowledge sharing among community-based organizations leaders in Botswana. *Environ. Dev. Sustain.* **2017**, *20*, 2225–2242. [CrossRef]
41. Baum, T. *International Perspectives on Women and Work in Hotels, Catering, and Tourism*; Bureau for Gender Equality Working Paper (1/2013); International Labour Office: Geneva, Switzerland, 2013.
42. Otobe, N. Gender dimensions of the world of work in a globalised economy. In *Harvesting Feminist Knowledge for Public Policy: Rebuilding Progress*; Jain, D., Elson, D., Eds.; Sage: New Delhi, India, 2011; pp. 150–169.
43. Khatiwada, L.K.; Silva, J.A. Mitigating Gender Inequality in Rural Regions: The Effects of Tourism Employment in Namibia. *Int. J. Tour. Res.* **2015**, *17*, 442–450. [CrossRef]
44. Lendelvo, S.; Munyebvu, F.; Suich, H. Linking women's participation and benefits within the Namibian community-based natural resources management program. *J. Sustain. Dev.* **2012**, *5*, 27–39. [CrossRef]
45. Ministry of Environment, Tourism and Forestry/Namibian Association of CBNRM Support Organisations. *The State of Community Conservation in Namibia (Annual Report 2019)*; MEFT/NACSO: Windhoek, Namibia, 2021.
46. Republic of Namibia. *Guidelines for the Management of Conservancies and Standard Operating Procedures*; Ministry of Environment and Tourism: Windhoek, Namibia, 2013.
47. Ambunda, L.; De Klerk, S. Women and Custom in Namibia—A Research Overview. In *Women and Custom in Namibia. Cultural Practice versus Gender Equality?* Ruppel, O.C., Ed.; Macmillan Namibia: Windhoek, Namibia, 2008; pp. 43–82.
48. Lafont, S. *Belief and Attitudes towards Gender, Sexuality and Traditions amongst Namibian Youth*; Monograph No.5. Legal Assistance Centre: Windhoek, Namibia, 2010.
49. Lendelvo, S.; Angula, M.N.; Mogotsi, I.; Aribeb, K. Towards the reduction of vulnerabilities and risks of climate change in the community-based tourism, Namibia. In *Natural Hazards—Risk Assessment and Vulnerability Reductions*; do Carmo, J.S., Ed.; IntechOpen: London, UK, 2018. [CrossRef]
50. Mogotsi, I.; Lendelvo, S.; Angula, M.; Nakanyala, J. Forest resources management and utilization through a gendered lens in Namibia. *Environ. Nat. Resour. Res.* **2016**, *6*, 79–90.

51. Green Climate Fund. *Mainstreaming Gender in Green Climate Fund Projects. A Practical Manual to Support the Integration of Gender Equality in Climate Change Interventions and Climate Finance*; Green Climate Fund: Yoensu-Gu, Korea, 2017.
52. Arora-Jonsson, S. Forty years of gender research and environmental policy: Where do we stand? *Women Stud. Int. Forum* **2014**, *47*, 295–308. [CrossRef]
53. Okali, C.; Naess, L.O. Making Sense of Gender, Climate Change and Agriculture in sub-Saharan Africa: Creating Gender-Responsive Climate Adaptation Policy. In *Future Agricultures: Working Paper 057*; University of Sussex: Brighton, UK, 2013. Available online: www.future-agricultures.org (accessed on 22 June 2021).
54. Ramchurjee, N. *"Tourism" a Vehicle for Women's Empowerment: Prospect and Challenges*; University of Mysore: Manasagangotri, India, 2011.
55. Call, M.; Sellers, S. How does gendered vulnerability shape the adoption and impact of sustainable livelihood interventions in an era of global climate change? *Environ. Res. Lett.* **2019**, *14*, 1–15. [CrossRef]
56. Rivera, P.F.; Bardales Espinoza, W.A.; Ochoa, W. Escenarios futuros de cambio climático para Guatemala. In *Primer Reporte de Evaluación del Conocimiento Sobre Cambio Climático en Guatemala*; Castellanos, E.J., Paiz-Estévez, A., Escribá, J., Rosales-Alconero, M., Santizo, A., Eds.; Editorial Universitaria UVG: Guatemala City, Guatemala, 2019; pp. 40–61.
57. Ihalainen, M.; Basnett, B.S.; Larson, A.; Duchelle, A.; Thu Thuy, P.T.; Djoudi, H. *What Should Be Included in the Green Climate Fund's New Gender Policy and Action Plan? Lessons from CIFOR's Research and Analyses*; CIFOR: Bogor, Indonesia, 2017. [CrossRef]

sustainability

MDPI

Article

Aligning Resilience and Wellbeing Outcomes for Locally-Led Adaptation in Tanzania

Emilie Beauchamp [1,*], Nigel C. Sainsbury [2], Sam Greene [1] and Tomas Chaigneau [2]

[1] International Institute for Environment and Development, London WC1V 7DN, UK; sam.greene@iied.org
[2] Environment and Sustainability Institute, University of Exeter, Penryn TR10 9FE, UK; nigel_sainsbury@sfu.ca (N.C.S.); t.w.b.chaigneau@exeter.ac.uk (T.C.)
* Correspondence: emilie.beauchamp@iied.org

Abstract: Interventions to address climate adaptation have been on the rise over the past decade. Intervention programmes aim to build the resilience of local communities to climate shocks, and ultimately their wellbeing by helping them to better prepare, adapt and recover. Resilience, similar to human wellbeing, is a multidimensional construct grounded in local realities and lived experiences. Yet current evaluation frameworks used in resilience programming rarely consider what resilience means in local contexts prior to implementation. This means policy designs risk failing to improve resilience of communities and creating unintended negative consequences for communities' well-being. Better processes and indicators for assessing resilience are needed. This paper explores the interplay between local predictors of resilience and wellbeing to assess the validity of self-assessed indicators as part of frameworks to measure resilience. We draw from research on the Devolved Climate Finance (DCF) mechanism implemented between 2014 and 2018 in Tanzania. We find that different factors explain resilience when compared to wellbeing; while resilience is primarily influenced by relationships, wellbeing is correlated with livelihoods. This shows that incentives to improve resilience differ from those of wellbeing. Climate and development practitioners must adopt locally grounded framings for resilience and wellbeing to ensure interventions track appropriate indicators, towards positive outcomes.

Keywords: climate adaptation; social resilience; human wellbeing; Devolved Climate Finance; resilient development; Tanzania; monitoring and evaluation

Citation: Beauchamp, E.; Sainsbury, N.C.; Greene, S.; Chaigneau, T. Aligning Resilience and Wellbeing Outcomes for Locally-Led Adaptation in Tanzania. *Sustainability* **2021**, *13*, 8976. https://doi.org/10.3390/su13168976

Academic Editors: Adam P. Hejnowicz and Jessica P. R. Thorn

Received: 14 June 2021
Accepted: 28 July 2021
Published: 11 August 2021

Publisher's Note: MDPI stays neutral with regard to jurisdictional claims in published maps and institutional affiliations.

Copyright: © 2021 by the authors. Licensee MDPI, Basel, Switzerland. This article is an open access article distributed under the terms and conditions of the Creative Commons Attribution (CC BY) license (https://creativecommons.org/licenses/by/4.0/).

1. Introduction

An increasing awareness of the negative impacts of climate change on communities, including from the rising frequency of climate shocks and stressors, is spurring organisations to support adaption to these changes to enhance human welfare and wellbeing. Climate change however brings challenges that are wide-reaching, with the only expectation for the future being that it will be nothing similar to the present or past [1]. It is critical that governments, funding agencies and interventions integrate climate risks into plans and policies to progress sustainable development in the face of the climate crisis [2]. The concept of resilience is now widely used in policy and practice yet its meaning and measurement remain contested [3–5].

Resilience is seen as a pivotal concept in helping communities deal with climate related stressors and shocks [6]. Several definitions of resilience co-exist, spanning disciplines from engineering, ecology, psychology and social sciences [7,8]. Given the socio-ecological context of climate adaptation interventions, we here refer to climate resilience as the capacity of individuals, communities or systems to adapt, reorganize and evolve into more desirable configurations in face of climate shocks, leaving them better prepared for future climate change impacts [9–11]. Several factors affect resilience of individuals and of systems: resilience varies across scales and contexts, depending on the local manifestations of climate change (exposure), the degree to which people depend on affected resources (sensitivity) and on their capacity to adapt to or take advantage of the changes they experience (adaptive capacity) [12]. Global accords such as the United Nation's Paris Agreement and 2030 Agenda for Sustainable Development [13] see resilience to climate variability and change as a key development priority to improve wellbeing [14,15]. This rise in policy interest has inevitably led to calls for identifying robust ways of measuring resilience across scales, and validation of the linkages between resilience and wellbeing outcomes [16,17]. Yet, how well local conceptualizations of both concepts compare and align locally is rarely explored (Chaigneau et al. In Review). This positive and synergistic relationship between wellbeing and resilience indicators remains a powerful yet unevidenced conceptual assumption with important implications for the design of development and climate intervention programmes—and the wellbeing of communities across the globe.

Linking Resilience and Wellbeing

Since the rise of resilience as a key concept, the term has been taken up by different policy actors and communities and has been interpreted, presented and applied in different ways in the humanitarian and development sphere [18]. The uptake of the term is similar to the one observed for the concept of human wellbeing as an alternative to income indicators of development progress [19].

The adoption of resilience to understand community response to climate shocks is reflected in the multitude of frameworks which seek to measure multiple dimensions of resilience through practical assessments [4,20]. There is however, no emerging consensus on what metrics or methods to use—neither for resilience, nor wellbeing [17]. Further, many of these frameworks measure more tangible factors, such as assets, livelihood strategies, financial or social capitals. These factors do not capture everything that encompasses what resilience is, or what influences when and how resilience capacities help buffer the negative effects of shocks and stressors [18]. A similar pattern is found for wellbeing, for which emphasis is often placed on objective and easily quantifiable indicators such as wealth and income over relational or subjective domains of wellbeing [21,22]. Increasingly, there are calls to assess how resilience and wellbeing are lived and experienced rather than through observable variables [23]. Subjective measures of resilience [24] and wellbeing [25,26] can enable an investigation of the key issues that people highlight when explaining their vulnerability to, and capabilities to deal with, climate change (rather than their proxy [1]), and how this relates to their wellbeing.

The complexities inherent within multi-dimensional concepts such as a resilience and wellbeing, together with the need for collecting and communicating evidence on the climate resilience of communities, raises questions about how best to measure, monitor and assess resilience and its consequence for wellbeing. Numerous measurement frameworks for climate and development programmes have emerged over the past decade to operationalise the concepts, and aggregate results within and between intervention programmes [17]. However, as resilience and wellbeing are not directly observable, they must be inferred from the measurement of items that can be observed [27]. As such, both are latent variables and, with a broad range of definitions in existence, quantitative measurement of resilience and its relationship to wellbeing therefore poses a significant challenge [28].

Complex sets of quantitative indicators are often used for monitoring and evaluating projects that seek to build resilience as a means to improve long-term wellbeing. The guidance to measure the number of people whose resilience has been improved as a result of intervention programmes often necessitates multiple time consuming and expensive steps. These steps often entail resilience metrics based on measured resilience at a household or individual scale, using a set of externally generated project-specific indicators (e.g., DFID 2014 in Brown [1]. There are a number of well-documented drawbacks of such an approach. Firstly, resilience capacities that experts choose to measure may not relate to lived experiences of resilience across different contexts. Secondly, variables identified often prove difficult to measure objectively. Finally, resilience indicators often require variables to be combined into a single value that can cause difficulties with relative weighting and addressing interactions between different resilience domains [28].

Rather than investing greater effort and more resources in developing definitive measurements of resilience and wellbeing, Levine [3] argues that organisations should focus on becoming better at quantifying things that really matter to communities. Importantly, it is the understanding of the determinants of resilience and wellbeing in different contexts that is more important than attempting to quantify the outputs of resilience-focused interventions [29].

Within resilience frameworks proposed by development actors, resilience is often framed normatively as an intermediate outcome that is conceptually linked to the long-term impact of improved wellbeing variables, such as food security or poverty reduction [30–32]. From a programming perspective, there is a hypothesized link between increasing resilience—or having better adaptive capacities—and improving wellbeing. For example, this was the case of the 3As framework of the global climate programme Building Resilience and Adaptation to Climate Extremes and Disasters (BRACED), which was funded from 2015 to 2019 by the UK Department for International Development [33,34]. Indeed, the preferred way to assess the value of a resilience measure is its ability to predict a proxy outcome of interest, usually relating to food security, nutrition status or other objective measures of wellbeing [35].

Wellbeing and resilience are often assumed to be related, reinforcing or synergistic [36,37]. However, very little of the policy literature refers to the possibility of trade-offs inherent in advancing resilience as a means to improve wellbeing [38]. When multiple domains of resilience are considered, tensions between these—and with the multidimensional nature of wellbeing—may reveal each other [39]. For example, in the context of acute poverty, trade-off between provision of basic needs and fundamental rights such as food security with system resilience have been found in small-scale fisheries in West Africa [40]. Greenpeace have also highlighted that projects focused on achieving wider wellbeing goals through increased food production have left farmers less resilient as they are increasingly dependent on external inputs and resources that are too costly for farmers [41]. More generally, Medecins Sans Frontieres have also stated that building resilience can be at odds with approaches to crises that seek to enhance wellbeing [42]. They state that building resilience can become an excuse for inaction and consequently can ignore wellbeing impacts from resilience interventions.

Despite the popularity of development interventions and academic research involving resilience and wellbeing separately, there has been limited research to date exploring the linkages between the two concepts. The interplay resilience and wellbeing has only recently started to be explored conceptually in social-ecology and development [23]. Yet empirical evidence of how resilience and wellbeing outcomes pan out in reality remains scant, due in part to lack of reliable local data from either local government and development projects [17,43].

This paper addresses this gap by exploring the alignment and trade-offs between two local indicators of resilience and wellbeing, as part of a devolved climate and development project in Tanzania. We expect that predictors of local resilience and wellbeing will broadly align together, as per the normative development frameworks seeking increasing wellbeing from increased resilience. We use this analysis to assess the implications for framing and designing climate resilience programmes for sustainable development outcomes. This research draws on secondary data from monitoring and evaluation (M&E) exercises applied during the pilot of the Devolved Climate Finance (DCF) mechanism in Tanzania—more specifically its 2017 household baseline survey—by the International Institute for Environment and Development. DCF is a research action and advocacy project supporting local people to access to locally controlled adaptation funds in three districts in Northern Tanzania.

2. Materials and Methods

2.1. Study Site

The DCF mechanism aims to support locally led adaptation to climate change by enabling communities to access funding for locally prioritised public good investments [44]. To date, it has been piloted across four countries: Kenya, Tanzania, Mali and Senegal [45–48]. The DCF mechanism consists of four dimensions: (1) the creation of local adaptation committees that identify and implement resilience investments based on inclusive community consultations and pre-defined fund criteria; (2) the use of participatory resilience assessments and climate information-informed local decision-making tools, through which communities identify climate stresses, opportunities and resilience-building priorities; (3) the development of local climate adaptation funds managed by local governments to finance locally-prioritised public good investments in resilience; and (4) local monitoring to assess effectiveness of resilience investments, support iterative learning and inform future planning [44].

This article focuses on the implementation of the DCF mechanism in three northern districts in the Arusha Region of Tanzania (Figure 1) between 2014 and 2018. The project was funded by the UK government's Department for International Development's UK Aid (now the Foreign, Commonwealth and Development Office), which provided a US$4.8 million (£3.8 million) grant under its 'Assisting public institutions and markets to become resilient to effects of climate change in Tanzania' (AIM 4 Resilience) programme. The project aimed to strengthen the government of Tanzania's capacity to channel climate finance to the local level, at scale, for adaptation, by piloting the DCF mechanism in the three districts of Monduli, Longido and Ngorongoro. The mechanism was overseen by a consortium led by the President's Office—Regional Administration and Local Government and provided a system for accessing climate finance and channeling it, through existing government systems, towards a sub-national planning process that enables locally planned and delivered local investments for resilience. The DCF mechanism upholds the principle that local communities have existing strategies for managing climate variability and are best placed to identify investments that will support their adaptive strategies. In practice, most local investments aim to reinforce livelihood systems and productive assets, as part of a range of climate adaptation strategies. Through the mechanism, £1 million was directed towards 35 public good investments identified by community-led committees.

The three districts of Monduli, Longido and Ngorongoro comprise varied agro-ecological systems and a great variety of access to public services or transport infrastructures. Longido, Ngorongoro and parts of Monduli are dryland ecosystems characterised by variability and unpredictability of rainfall across time and space and high levels of biodiversity. Other parts of Monduli include the East African Rift Valley with milder temperatures and a bi-modal rainfall pattern.

As in many dryland regions, communities rely on flexible and varied livelihood strategies and associated cultures. Economies are dominated by produce from agro-pastoral activities that use flexibility, negotiated seasonal migration and customary resource management systems to maximise productivity during periods of rainfall and minimise loss during dry spells. Most pastoralists in the region identify as Maasai. Non-Maasai, and particularly those living in the valley regions of Monduli, engage in cultivation of crops including Maize, Millet and other vegetables, often for export—although Maasai farmers, particularly wealthy ones, also often cultivate crops. In normal times the local economy is bolstered by tourism generated by the rich biodiversity and famed wildlife, as well as artisanal mining and growing urban centres.

The DCF project operated across 10 administrative divisions in the district authorities of Monduli, Longido and Ngorongoro. Divisions consist of 3–5 wards, with wards in the three districts having an average population of 12,000. A history of inappropriate development interventions, marginalisation and policies that undermine local livelihood strategies including seasonal mobility, combined with climate shocks (droughts, flooding) and slow-onset threats (temperature change, changing rainfall patterns, windstorms) have culminated in widespread poverty and vulnerability to climate change. Longido is particularly drought prone and 55% of its population falls below the Tanzanian basic needs poverty line [49].

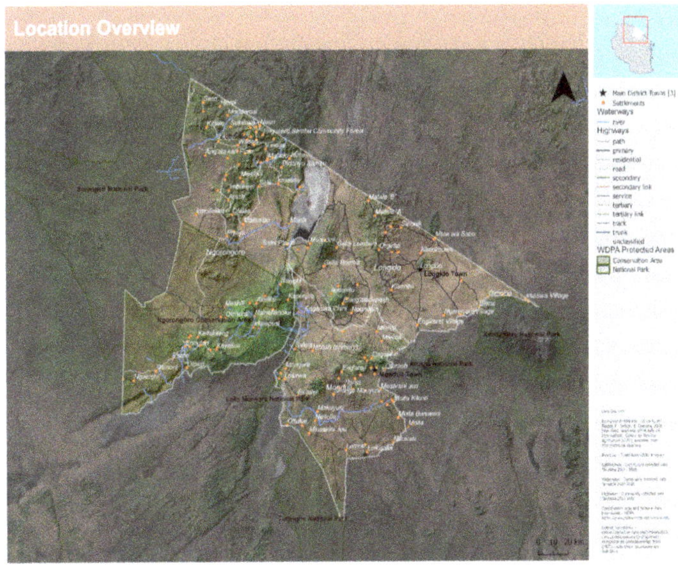

Figure 1. A project map Monduli, Longido and Ngorongoro in Northern Tanzania [50].

2.2. Methods

This analysis uses secondary data collected in the frame of its project M&E exercises. The Tanzania DCF project team developed a household survey to measure resilience and identify changes over time in household resilience guided by the TAMD framework, and other M&E exercises in Mali and Senegal [17]. The survey design aimed at recording changes in core outcomes of household wellbeing and resilience, along with influencing factors in villages across years. This article is based on the baseline data collected for a longitudinal analysis. However, due to cuts in the programme funding, there was no possibility to collect data later or after the intervention. Given this programmatic M&E purpose and the desire to test simple indicators that local actors could use, the DCF team framed resilience in the context of climatic shocks and tested indicators of subjective resilience and community wellbeing ranking. The household survey was designed and administered by LTS International, an M&E consulting firm working with DCF project partners to track indicators identified by communities as valuable for increased resilience in planning meetings for DCF investments. The survey was reviewed and edited by a community expert from Northern Tanzania for context relevance. This article uses anonymized secondary data for this analysis.

Data collection took place across 19 villages in April and May 2017, six in Longido, six in Monduli and seven Ngorongoro. Villages for data collection were selected by grouping those likely to benefit from DCF interventions by thematic areas including water access, livestock health, agriculture or market access. An equal number of villages were selected across Monduli, Longido and Ngorongoro. Villages were relevant, a selection was made to ensure villages with characteristics representative of the overall community, reflecting dominant livelihoods in the region.

Data collection for community wellbeing rankings were developed using participatory discussions with mixed and women-only groups of 8–12 people from sampled villages. Participants described characteristics of four or five categories of wellbeing, which differed for men and women. They then ranked a representative group of 150 households in their villages against each category. Through the process, they further refined the categories Defining characteristics included size of family, ability to obtain food for the household number of livestock and land ownership for households cultivating crops. Data was collected from 572 households across 17 villages in in April and May 2017. Enumerators were identified by project partners as well by local government officers from sampled villages. Local language enumerators able to speak Ma, the Maasai language, were necessary to ensure participants could offer meaningful responses, which were translated during the interview into KiSwahili and English for recording. Training was carried out with enumerators by the M&E consultant one week before data collection. Stratified random sampling was used to identify households for interviews. Households were selected using the wellbeing categories from wellbeing ranking exercises to ensure proportional representation of households from each category.

2.3. Variables and Statistical Analysis

This analysis is concerned with two primary response variables: community ranking of household wellbeing, and household subjective resilience. Subjective indicators were included as part of other proxy variables to better understand determinants of perceived resilience and wellbeing. Subjective approaches to measurement are argued to add value to more traditional, objective measures for three reasons. Firstly, they suggest that they can improve our understanding of the drivers of latent concepts such as resilience and wellbeing. It offers a valuable opportunity to capture the perspectives of those who know most about their own resilience and wellbeing, and the factors that contribute to it: the people themselves [49]. Secondly, they can reduce the questionnaire burden on respondents. This is especially relevant where the main goal of a questionnaire is to investigate the level, rather than the drivers, of resilience and wellbeing [28]. Finally, they can offer more valid cross-cultural comparisons, as they measure an individual's perception of whether their overall resilience capacities are sufficient to maintain and/or improve wellbeing within the context of shocks and stressors that they currently experience and are likely to experience in future.

Subjective wellbeing was first locally defined before being assessed through a local ranking exercise. Wellbeing—or "Maisha Bora" in KiSwahili, "Ingishu Sidai" in Maa, is broadly understood as having a "good life" and doing well economically. For some, it is closely linked to available wealth. The ranking exercises used the locally defined concept of wellbeing to identify categories of wellbeing within a village through a series of focus groups (see Table 1). Enumerators asked participants to state whether a household had higher or lower wellbeing when compared to the average in the village. They then asked why the household ranked so, through which locally assessed definitions of wellbeing were developed.

Depending on the village, the wellbeing ranking process identified four or five categories of wellbeing, with women picking out different characteristics to mixed groups. The precise nature of the categories differs from village to village, and the variable cycles of drought and rainfall necessarily introduce fluidity between them: several years of good rains can see some households gain greater wellbeing and move upwards through categories—but drought, disease or other factors can have significant negative impacts.

Subjective resilience was assessed in the household survey on a scale of 1 to 10, as per the respondent's perceptions. We asked, "On a scale of 1–10, 1 being the lowest, 10 being the highest, please rate how resilient you feel to the impacts of climate change?".

Explanatory variables were selected using theoretical justification, an assessment of key factors of resilience, and mixed factor analysis to check for multi-collinearity. Table 2 shows the justification, format and transformation for variables used in the statistical analysis.

Table 1. Overview of typical characteristics identified under different levels of wellbeing through the wellbeing ranking exercise. Mixed and women focus groups are shown, aggregated across villages.

Category	Women's Groups	Mixed Group
High *(Irkasis)*	Married to a man with livestock 3 meals a day Can buy food when needed Access to health services Own medium sized businesses Able to dress well Farm ownership (non-Maasai households) Children educated to college level Small poultry farms Access to credit Water storage and rainwater harvesting	Ownership of cattle (150+) Ownership of sheep and goats (400+) Children educated to college level Access to health services Able to hire labour and vehicles to support livelihood Land ownership (for cultivating households) −10 acres + Rainwater storage facilities Able to support others in the community Stone houses with iron roofs
Upper Middle *(Irkasis Lepolos)*	Own small businesses Children go to school with support from small businesses Some access to credit	Ownership of cattle, <150 cattle Ownership of sheep and goats <400 Children educated to form 4 level Access to health services Some vehicle access Own secondary businesses Access to tractors (farming only) Land ownership 5–7 acres Donkey access
Lower Middle *(Irmentati)*	Dependant on others, particularly Irkarsis Participate in women's self-help groups Dependent on family/clan for support Work as casual labourers or petty traders Often widowed or without husbands Children educated up to form 6 Some credit access through village communitybanks Often single parents or absentee husbands	Work as casual labourers Ownership of cattle − <10 cattle Ownership of sheep and goats <30 Dependent on state support for health services Reduced food access
Low *(Ilaisanik)*	Reduced nutrition intake Unreliable businesses—*"pata potea"* selling firewood, vegetables, eggs <2 cattle <2acre of land	Work as casual labourers Reliant on others in the community for support No cattle ownership Sheep/Goat ownership <10 Reduced Food Access
Lowest *(Ndorobo)*	Often elderly Few possessions or income generating activities Dependent on others for support: permitted to grow beans in between Maize	Destitute No livestock owned Dependent on others in the community for survival

Separate statistical models were estimated to test the effect of livelihood explanatory variables on subjective resilience and wellbeing. A proportional odds cumulative logit generalised linear mixed effects (POCL) model analystical framework was selected, reflecting the ordinal categorical nature of the response variables. The proportional odds cumulative logit model is suitable in this case as it reflects the appropriate amount of information contained within the ordinal response variable, in that the variable represents an underlying continuous interval/ratio variable and so neither loses the ordinal information nor assumes a ratio or interval value that is not present. Further, the use of the POCL model is also appropriate because we assume that the coefficients across levels of the dependent variable are constant (proportional odds). The POCL models were fitted using the Ordinal package [51] in R Programming with the village treated as a random effect to account for spatial differences in the response variables. A detailed description of the POCL model can be found in Supplementary Materials (Tables S1 and S2). Model validation tests (optimization convergence, parameter accuracy and likelihood function performance), as required by for ordinal logistic regression, were performed using the Ordinal package [51] and were satisfied.

An Information Theoretic (IT) approach to model selection [52,53] was adopted to reflect the absence of a single truth for relationships between resilience, wellbeing and the predictor variables we have selected. The IT approach recognises that the data was not collected in an experimental setting and that an average of possible models will provide the best insight into the workings of the system. The 'MuMIn' package [54] was used to create a selection of candidate models from the global model up to two points lower than the best model AIC. The global model consisted of the variables summarised in Table 2.

Table 2. Hypothesis and table of variables used for the statistical analysis.

Variable	Direction of Effect	Justification	Format and Transformation
Response Variables			
Subjective resilience	N/A	Response variable	10 level response scale consolidated to 1–4 scale to homogenise distribution (1–2 = 1, 3–4 = 2, 5–6 = 3, 7–10 = 4)
Wellbeing	N/A	Response variable	Six level response scale consolidated to 1–4 scale to homogenise distribution (5–6 = 4)
Explanatory Variables			
Gender of head of household (HHH)	Negative for female HHH	Reduces adaptive capacity for women through agency and assets domains	None
Major livelihood	Negative for pastoralists	Affects adaptive capacity through assets, flexibility and agency domains	None
Livelihood diversity	Positive	Enhances adaptive capacity through flexibility domain	Converted to total count of number of major livelihood and complementary incomes sources for each respondent.
Livelihood group membership	Positive	Enhances adaptive capacity through organization and learning domain	None
Receipt of climate information	Positive	Enhances adaptive capacity through learning domain	None
Water access for domestic use in dry season	Positive	Enhances capacity through assets domain	None
Livestock ease of pasture access all season score	Positive	Enhances capacity through assets and flexibility domain	None
Livestock market access	Positive	Enhances capacity through assets and flexibility domain	None
Storage of harvest	Positive	Enhances capacity through assets and flexibility domain	4 level categorical responses consolidate to two ("No" = "no", "yes" = "yes", "Yes, in my house" = "yes", "Yes, in a communal storage facility" = "yes")
Access to agricultural extension services	Positive	Enhances capacity through learning domain	None
Use of irrigation	Positive	Enhances capacity through assets domain	None

3. Results

After respondents were removed due to data absence, 392 (resilience) and 410 (wellbeing) responses were used in the models. Of the sample used in the wellbeing model, 48% heads of household were female and 52% male. Major livelihoods of the household were 45% agro-pastoralist, 40% pastoralist, 10% farmer, 2% pastoralist-business, 2% business and 1% agro-pastoralist-business. Below we report the results from the regressions on self-assessed resilience (Table 3) and wellbeing ranking (Table 4).

3.1. Resilience

The most important predictor variables for self-reported resilience in the average model were the major livelihood of the household, membership of a livelihood group and gender of head of household. However, only the coefficient estimates for gender of head of household and livelihood group membership had confidence intervals that did not cross zero, providing high confidence in the direction of effect. Female head of households were found to have a lower subjective resilience than men. Respondents that were members of a livelihood group had higher subjective resilience than those who were not. In order of importance in the average model (from more to less), storage of harvest, use of irrigation, livelihood diversity, climate information and ease of access to livestock markets were also found to play a role in subjective resilience, although the confidence intervals for their coefficient estimates cross zero indicating low confidence in the direction of effect. Details of explanatory variable importance in the average model can be found in the Supplementary Materials.

Table 3. Results from average model (conditional average) for resilience. Conventional symmetric so-called Wald confidence intervals were run for the parameters and are presented alongside the estimate and standard error, all on the log odds scale. The importance of each variable to the average model, and the number of models in which the variables appear are shown in Table S1 in Supplementary Materials.

Parameter	Categorical Base Level	Estimate	Std Error	CI (2.5%)	CI (97.5%)
1|2	N/A	−0.81573	0.41317	−1.626	−0.006
2|3	N/A	0.72689	0.41197	−0.081	1.534
3|4	N/A	2.85900	0.43901	1.999	3.719
Major livelihood (Agro-pastoralist)	Pastoralist	0.71067	0.38221	−0.038	1.460
Major livelihood (Agro-pastoralist-business)		0.99860	1.02518	−1.011	3.008
Major livelihood (Business)		1.20880	0.80441	−0.368	2.785
Major livelihood (Farmer)		−0.57144	0.53840	−1.627	0.484
Major livelihood (Pastoralist-business)		−0.64994	0.79651	−2.211	0.911
Gender (Female)	Male	−0.58767	0.19740	−0.975	−0.201
Livelihood group membership (Yes)	No	0.46499	0.22861	0.017	0.913
Storage of harvest (Yes)	No	0.51670	0.32002	−0.111	1.144
Use of irrigation (Yes)	No	0.57968	0.45588	−0.314	1.473
Livelihood diversity (Number of livelihoods)	N/A	0.22720	0.18227	−0.130	0.584
Receipt of climate information (Yes)	No	0.19065	0.21595	−0.233	0.614
Livestock market access (Somewhat difficult)	Always accessible	0.56667	0.34236	−0.104	1.238
Livestock market access (A little difficult)		0.36569	0.35761	−0.335	1.067
Livestock market access (Very difficult)		0.03825	0.38372	−0.714	0.790

Table 4. Results from average model (conditional average) for wellbeing. Conventional symmetric so-called Wald confidence intervals were run for the parameters and are presented alongside the estimate and standard error, all on the log odds scale. The importance of each variable to average model, and the number of models in which the variables appear are also shown. See Table S2 in Supplementary Materials for additional details.

Parameter	Categorical Base Level	Estimate	Std Error	CI (2.5%)	CI (97.5%)
1|2	N/A	−1.51831	0.31805	−2.142	−0.895
2|3	N/A	−0.22500	0.30626	−0.825	0.375
3|4	N/A	1.49210	0.31336	0.878	2.106
Major livelihood (Agro-pastoralist)	Pastoralist	−0.23339	0.29832	−0.818	0.351
Major livelihood (Agro-pastoralist-business)		−0.67400	0.86520	−2.370	1.022
Major livelihood (Business)		−0.50057	0.77468	−2.019	1.018
Major livelihood (Farmer)		1.32243	0.44961	0.441	2.204
Major livelihood (Pastoralist-business)		−0.55534	0.63701	−1.804	0.693
Gender (Female)	Male	0.32378	0.18885	−0.046	0.694
Livelihood group membership (Yes)	No	−0.28282	0.21242	−0.699	0.134
Domestic water access (Somewhat difficult)	Very difficult	−0.33826	0.36932	−1.062	0.386
Domestic water access (Moderately easy)	Very difficult	−0.14374	0.32353	−0.778	0.490
Domestic water access (Very easy)	Very difficult	0.53277	0.42822	−0.307	1.372
Livelihood diversity (Number of livelihoods)	N/A	−0.10755	0.17719	−0.455	0.240
Use of irrigation (Yes)	No	−0.25168	0.40900	−1.05	0.550
Ease of access to pasture in dry season (Difficult)	Easy	−0.12472	0.23288	−0.581	0.332
Receipt of climate information (Yes)	No	−0.08802	0.20004	−0.480	0.304

3.2. Wellbeing

Major livelihood was found to be the most important predictor of household wellbeing, mirroring importance for explaining self-reported resilience. Farming households were shown to have higher wellbeing than pastoralist households (coefficient estimate confidence intervals did not cross zero). Gender was also important in predicting wellbeing, although the direction of effect had greater uncertainty (confidence intervals cross zero). The same uncertainty affects the remaining predictors of wellbeing, which in order of importance in the average model (high to low) were domestic water access, livelihood diversity, use of irrigation, ease of access to pasture in the dry season and receipt of climate information. Details of explanatory variable importance in the average model can be found in the Supplementary Materials.

3.3. Resilience-Wellbeing Equivalence

Of the nine explanatory variables in the average models for either resilience or wellbeing, six featured in both average models (major livelihood, gender of head of household, livelihood grop membership, livelihood diversity, crop irrigation and receit of climate information). Two predictor variables (resilience model) and one (wellbeing model) had confidence intervals that did not cross zero (high confidence in direction of effect) (Table 5): difference between agriculture and pastoralist major livelihood on wellbeing; gender of head of household on resilience; and livelihood group membership on resilience. As such, high confidence effects of explanatory variables on resilience or wellbeing were not equivalent, which prevented comparison of the direction of effect of variables on resilience and wellbeing. Three explanatory variables only appeared in one of the average models: domestic water access in dry season and ease of access to pasture in dry season only affected wellbeing; and ease of access to livestock market only affected wellbeing.

Table 5. Comparison between the effects of aspects of adaptive capacity on resilience and subjective wellbeing (* important in system but 95% confidence intervals of coefficient estimates cross zero; ** important in system and 95% confidence intervals of 95% confidence intervals do not cross zero; N/A means that variable does not feature in average model are not important in system).

Variable	Resilience	Wellbeing	Comparison
Major livelihood	*	**	=
Gender of head of household (female)	**	*	=
Livelihood group membership	**	*	=
Domestic water access in dry season	N/A	*	≠
Livelihood diversity	*	*	=
Crop irrigation	*	*	=
Ease of access to pasture in dry season	N/A	*	≠
Receipt of climate information	*	*	=
Ease of access to livestock market	*	N/A	≠

4. Discussion

We analyzed household survey data from the baseline of the climate and development project to assess if, and how, factors affecting local concepts of resilience and wellbeing aligned. We found that while six predictors featured in both average models, high confidence predictor variables were not shared between models. The results show that while both resilience and wellbeing can conceptually feature similar sets of proxy predictors, meaningful predictors that have high influencing effects are not aligned. This falls against our initial assumption that predictors of wellbeing would match those of resilience. The relative importance of some variables was high in both models; however, their lack of high confidence effect means this cannot be used as a basis for replicable and transferable results.

Only the difference between agriculture and pastoralist major livelihood influenced wellbeing outcomes with high confidence, whereas both gender of the household head and membership of a livelihood group did so for resilience outcomes. This falls in line with conceptualizations that livelihoods and wellbeing are closely related [55–57], not simply in terms of income but also identity. Resilience results also point to the importance of power, agency and relationships as key determining factors [12]. While predictors of resilience and wellbeing are not shared, they nonetheless point to strong links between both concepts and local power dynamics. The observed cleavages between farmers and pastoralists, and between men and women head of households mean climate and development interventions must sit on robust interpretation and navigation of local power dynamics and intersectionality. Strategies benefitting one or some social groups may be detrimental to others.

Understanding and respecting local social norms and power dynamics require taking appropriate time and resources at project inception, something current project life cycles rarely allow. This research encourages practitioners and donors to secure enough resources at the expense of making entire programmes miss expected yet misunderstood outcomes. Bottom-up work that puts local people and rules at the center of interventions are required not only to produce positive outcomes, but also to start addressing the decolonization of aid.

This means practitioners must explore carefully what are the hypothesized pathways of change between resilience outcomes and wellbeing impacts, at a disaggregated scale. This is essential to avoid unintended negative consequences to local wellbeing by programming activities that could lead to maladaptation and ill being in the long-term. Developing with local communities involved a shared understanding of the underlying mechanisms that underline correlations is critical to reach positive effects. Our analysis shows that resilience-seeking interventions may need to programme different sets of activities, such as the creation of cooperative, and gender-focused work, whereas wellbeing-seeking interventions must ensure livelihood benefits are reached. These are far from incompatible, yet this shows how several completing strategies must be used to achieve both sustained resilience and wellbeing—rather than focusing on few select activities.

Our analysis shows that subjective indicators of wellbeing and resilience are a useful part of indicators used to understand impacts on these latent variables. Wellbeing and resilience are fluctuating states, and more importantly can be a state of mind. Ensuring that researchers and practitioners understand and monitor how different observable factors affect subjective perceptions and lived experiences are key in designing programmes that truly build resilient and sustainable capabilities in local communities, regardless of the nature of the shocks and changes occurring. Using subjective indicators can further inform whether building resilience to climate shocks should be considered separate to building resilience to other shocks and stressors for intervention programming. Ultimately, feeling resilient and well are key desirable outcomes that should drive sustainable development.

This research provides empirical evidence that go against hypothesized theories of change, resilience does not straightforwardly contribute to wellbeing [39]. We show that the misalignment between their predictors suggests the relationship between the multiple dimensions of resilience and wellbeing are not linear. However, this research has not allowed commenting on whether the conceptualizing resilience as a short or medium-term outcome and wellbeing as a long-term impact appropriately reflects local realities. While this research applies practically to Tanzania, other regional research has shown that localized resilience and wellbeing predictors do not align. Practitioners must therefore carefully monitor and track changes across several indicators for potential trade-offs.

Further empirical research on local conceptualisations of wellbeing and resilience, both qualitative or quantitative, is needed to verify how predictors change through time. Building collections of robust local case studies would promote building regional, and ultimately global, syntheses and meta-analyses that could provide a robust evidence base for practitioners and policy makers to understand and plan to improve resilience and wellbeing appropriately. This would also enable building conceptual framework that truly reflect local and lived realities. Similarly, power dynamics analysis should feature as an ongoing exercise throughout projects to ensure unequal dynamics are not reproduced or entrenched.

5. Conclusions

Both resilience and wellbeing are multidimensional concept, that cannot be tracked and assessed without a series of both qualitative and quantitative indicators that can reflect their latent dimensions. These indicators must be locally contextualized and interrogated prior to development interventions, at the risk of perpetuating incorrect assumptions about how resilience and wellbeing interact in the local communities involved. Community power dynamics, social identities must become integral parts of programming rather than quick inception assessments to ensure the trade-offs between—and within—resilience and well-being dimensions, do not lead to maladaptation and illbeing.

Supplementary Materials: The following are available online at https://www.mdpi.com/article/10.3390/su13168976/s1, Table S1, Table S2, Model Specification.

Author Contributions: Conceptualization, E.B. and T.C.; methodology, E.B. and N.C.S.; software, E.B. and N.C.S.; validation, E.B. and S.G.; formal analysis, E.B. and N.C.S.; investigation, S.G.; resources, E.B.; data curation, N.C.S.; writing—original draft preparation, E.B., N.C.S., S.G. and T.C.; writing—review and editing, E.B.; visualization, E.B. and N.C.S.; supervision, E.B.; project administration, E.B.; funding acquisition, E.B. and T.C. All authors have read and agreed to the published version of the manuscript.

Funding: This research was made possible by funding from Economic and Social Research Council Impact Acceleration Award Programme, grant reference 2705.001 and the support of the International Institute for Environment and Development.

Institutional Review Board Statement: Not applicable, secondary data. The data collected was designed and conducted by LTS International, following the International Institute for Environment and Development Ethics guidelines. This article uses a secondary anonymized dataset for analysis.

Informed Consent Statement: Informed consent was obtained from all subjects involved in the study.

Data Availability Statement: The data presented in this study are available on request from the corresponding author. The data are not publicly available due to shared ownership between project partners.

Acknowledgments: The authors wish to acknowledge contributions from all members of the consortium of the DCF AIM4Resilience project in Tanzania: Hakikazi Catalyst; the District Governments of Monduli, Longido, Ngorongoro; the Regional Administration of Arusha; LTS International (Now part of NIRAS International); the President's Office, Regional Administration and Local Government; the Tanzanian Institute for Rural Development and Planning; the Local Government Training Institute, Tanzania; and the Divisional Adaptation Planning Committees of the DCF programme. Special thanks to Alais Morindat, Ally Msangi, Kaiza Victor, Joseph Rutabingwa, Flora Assey. We are also grateful for the participation and contributions of community members interviewed as part of the research.

Conflicts of Interest: The authors declare no conflict of interest.

References

1. Jones, L.; Tanner, T. 'Subjective resilience': Using perceptions to quantify household resilience to climate extremes and disasters. *Reg. Environ. Chang.* **2017**, *17*, 229–243. [CrossRef]
2. Brown, K. *Resilience, Development and Global Change*; Routledge: Abingdon, UK, 2016.
3. Brooks, N.; Anderson, S.; Aragon, I.; Smith, B.; Kajumba, T.; Beauchamp, E.; D'Errico, S.; Rai, N. *Framing and Tracking 21st Century Climate Adaptation*; IIED: London, UK, 2019.
4. Levine, S. *Assessing Resilience: Why Quantification Misses the Point*; Overseas Development Institute: London, UK, 2014; p. 26.
5. Schipper, E.L.F.; Langston, L. *A Comparative Overview of Resilience Measurement Frameworks Analysing Indicators and Approaches*; Working Paper 422; Overseas Development Institute: London, UK, 2015; p. 30. [CrossRef]
6. Sharifi, A. A critical review of selected tools for assessing community resilience. *Ecol. Indic.* **2016**, *69*, 629–647. [CrossRef]
7. Eakin, H.C.; Lemos, M.C.; Nelson, D.R. Differentiating capacities as a means to sustainable climate change adaptation. *Glob. Environ. Chang.* **2014**, *27*, 1–8. [CrossRef]
8. Alexander, D.E. Resilience and disaster risk reduction: An etymological journey. *Nat. Hazards Earth Syst. Sci.* **2013**, *13*, 2707–2716. [CrossRef]
9. Searle, B.A.; Pykett, J.; Alfaro-Simmonds, M.J. Introduction to wellbeing research. In *A Modern Guide to Wellbeing Research*; Edward Elgar Publishing: Cheltenham, UK, 2021.
10. Adger, W.N. Social and ecological resilience: Are they related? *Prog. Hum. Geogr.* **2000**, *24*, 347–364. [CrossRef]
11. Adger, W.N. Vulnerability. *Glob. Environ. Chang.* **2006**, *16*, 268–281. [CrossRef]
12. Folke, C. Resilience: The emergence of a perspective for social–ecological systems analyses. *Glob. Environ. Chang.* **2006**, *16*, 253–267. [CrossRef]
13. Cinner, J.E.; Adger, W.N.; Allison, E.H.; Barnes, M.L.; Brown, K.; Cohen, P.J.; Gelcich, S.; Hicks, C.C.; Hughes, T.P.; Lau, J.; et al. Building adaptive capacity to climate change in tropical coastal communities. *Nat. Clim. Chang.* **2018**, *8*, 117–123. [CrossRef]
14. United Nations United Nations Framework Convention on Climate Change, Paris, France. 2015. Available online: https://unfccc.int/sites/default/files/english_paris_agreement.pdf (accessed on 1 June 2021).
15. Béné, C.; Wood, R.G.; Newsham, A.; Davies, M. Resilience: New Utopia or New Tyranny? Reflection about the Potentials and Limits of the Concept of Resilience in Relation to Vulnerability Reduction Programmes. *IDS Work. Pap.* **2012**, *2012*, 1–61. [CrossRef]
16. Szaboova, L.; Brown, K.; Chaigneau, T.; Coulthard, S.; Daw, T.M.; James, T. Resilience and Wellbeing for Sustainability. In *Ecosystem Services and Poverty Alleviation*; Routledge: London, UK, 2018; pp. 273–287.
17. Jones, L.; Harvey, B.; Cochrane, L.; Cantin, B.; Conway, D.; Cornforth, R.J.; De Souza, K.; Kirbyshire, A. Designing the next generation of climate adaptation research for development. *Reg. Environ. Chang.* **2018**, *18*, 297–304. [CrossRef]
18. Beauchamp, E.; Abdella, J.; Fisher, S.; McPeak, J.; Patnaik, H.; Koulibaly, P.; Cissé, D.; Touré, M.; Bocoum, A.; Ndao, M.; et al. Resilience from the ground up: How are local resilience perceptions and global frameworks aligned? *Disasters* **2019**, *43*, S295–S317. [CrossRef]
19. Béné, C.; Frankenberger, T.; Griffin, T.; Langworthy, M.; Mueller, M.; Martin, S. 'Perception matters': New insights into the subjective dimension of resilience in the context of humanitarian and food security crises. *Prog. Dev. Stud.* **2019**, *19*, 186–210. [CrossRef]
20. Gough, I.; McGregor, J. *Wellbeing in Developing Countries: From Theory to Research*; Cambridge University Press: Cambridge, UK, 2007.

21. Ifejika Speranza, C.; Wiesmann, U.; Rist, S. An indicator framework for assessing livelihood resilience in the context of social-ecological dynamics. *Glob. Environ. Chang.* **2014**, *28*, 109–119. [CrossRef]
22. Reyes-García, V.; Babigumira, R.; Pyhälä, A.; Wunder, S.; Zorondo-Rodríguez, F.; Angelsen, A. Subjective Wellbeing and Income: Empirical Patterns in the Rural Developing World. *J. Happiness Stud.* **2015**, *17*, 773–791. [CrossRef]
23. White, S.C. Bringing Wellbeing into Development Practice. In *Wellbeing in Developing Countries Research Group Working Paper*; University of Bath/Wellbeing in Developing Countries Research Group: Bath, UK, 2009; p. 50.
24. Brown, K. Resilience, Development and Global Change. In *Resilience, Development and Global Change*, 1st ed.; Routledge: London, UK, 2015; pp. 1–208. [CrossRef]
25. Jones, L.; Samman, E.; Vinck, P. Subjective measures of household resilience to climate variability and change. *Ecol. Soc.* **2018**, *23*, 23. [CrossRef]
26. Clark, D. *Adaptation, Poverty and Development: The Dynamics of Subjective Well-Being*; Springer: Berlin, Germany, 2012.
27. White, S.C. Analysing wellbeing: A framework for development practice. *Dev. Pract.* **2010**, *20*, 158–172. [CrossRef]
28. Joseph, J.; McGregor, J. *Wellbeing, Resilience and Sustainability*; Springer: Cham, Switzerland, 2019.
29. Clare, A.; Graber, R.; Jones, L.; Conway, D. Subjective measures of climate resilience: What is the added value for policy and programming? *Glob. Environ. Chang.* **2017**, *46*, 17–22. [CrossRef]
30. Brown, K. Global environmental change: A social turn for resilience? *Prog. Hum. Geogr.* **2014**, *38*, 107–117. [CrossRef]
31. Marshall, N.; Marshall, P.; Tamelander, J.; Obura, D.; Malleret-King, D.; Cinner, J.E. *A Framework for Social Adaptation to Climate Change: Sustaining Tropical Coastal Communities and Industries*; IUCN: Gland, Switzerland, 2010.
32. Bennett, N.J.; Dearden, P. Why local people do not support conservation: Community perceptions of marine protected area livelihood impacts, governance and management in Thailand. *Mar. Policy* **2014**, *44*, 107–116. [CrossRef]
33. Béné, C.; Frankenberger, T.; Nelson, S. Design, Monitoring and Evaluation of Resilience Interventions: Conceptual and Empirical Considerations. *Inst. Dev. Stud.* **2015**, *2015*, 26.
34. DFID. *Methodology for Reporting Against KPI4—Number of People Whose Resilience Has Been Improved as a Result of Project Support*; DFID: London, UK, 2014.
35. Bahadur, A.; Peters, K.; Wilkinson, E.; Pichon, F.; Gray, K.; Tanner, T. *The 3As: Tracking Resilience across BRACED*; ODI: London, UK, 2015.
36. Constas, M.; Frankenberger, T.; Hoddinott, J. *Resilience Measurement Principles: Toward an Agenda for Measurement Design*; Food Security Information Network: Rome, Italy, 2014.
37. Frankenberger, T.R.; Constas, M.A.; Nelson, S.; Starr, L. *Nongovernmental Organizations Approaches to Resilience Programming*; International Food Policy Research Institute: Washington, DC, USA, 2014.
38. Smith, L.C.; Frankenberger, T.R. Does Resilience Capacity Reduce the Negative Impact of Shocks on Household Food Security? Evidence from the 2014 Floods in Northern Bangladesh. *World Dev.* **2018**, *102*, 358–376. [CrossRef]
39. Robards, M.D.; Schoon, M.L.; Meek, C.L.; Engle, N.L. The importance of social drivers in the resilient provision of ecosystem services. *Glob. Environ. Chang.* **2011**, *21*, 522–529. [CrossRef]
40. Chaigneau, T.; Camfield, L.; Coutlhard, S.; Jones, L.; Matthews, N.; Mcquistan, C.; Szaboova, L.; Daw, T.; James, T.; Reyers, B.; et al. Reconciling wellbeing and resilience for sustainable development. *Nat. Sustain.*. in press.
41. Béné, C.; Evans, L.; Mills, D.; Ovie, S.; Raji, A.; Tafida, A.; Kodio, A.; Sinaba, F.; Morand, P.; Lemoalle, J.; et al. Testing resilience thinking in a poverty context: Experience from the Niger River basin. *Glob. Environ. Chang.* **2011**, *21*, 1173–1184. [CrossRef]
42. Thomson, K.F.; Kruszewska, I.; Tirado, R. *Building Environmental Resilience: A Snapshot of Farmers Adapting to Climate Change in Kenya*; Greenpeace Research Laboratories: Technical Report: 04-2015; University of Exeter: Exeter, UK, 2015.
43. Whittall, J.; Philipis, M.; Hofman, M. Opinion and Debate: Building Resilience by Deconstructing Humanitarian Aid. Available online: https://www.msf-me.org/article/opinion-and-debate-building-resilience-deconstructing-humanitarian-aid (accessed on 11 June 2021).
44. Beauchamp, E.; Hirons, M.; Brown, K.; Milner-Gulland, E.J. Twenty priorities for future social-ecological research on climate resilience. *Environ. Res. Lett.* **2020**, *15*, 105006. [CrossRef]
45. DCF Alliance. *The Devolved Climate Finance Mechanism: Principles, Implementation and Lessons from four Semi-Arid Countries*; Project Report; IIED: London, UK, 2019.
46. Crick, F.; Hesse, C.; Orindi, V.; Bonaya, M.; Kiiru, J. *Delivering Climate Finance at the Local Level to Support Adaptation: Experiences of County Climate Change Funds in Kenya*; Working Paper; IIED: London, UK, 2019.
47. Greene, S. *Lessons from Piloting the Decentralised Climate Finance Programme, Tanzania*; IIED: London, UK, 2019.
48. NEF; IIED; IED. Afrique Consortium Knowledge on Adaptation and the Reduction of Climate Risks and Disasters in Mali and Senegal; 2014. Available online: https://www.neareast.org/download/materials_center/Report_Reduction-of-Climate-Risks-Mali-and-Senegal.pdf (accessed on 11 June 2021).
49. World Bank Group. *Tanzania Mainland Poverty Assessment: Tanzania's Path to Poverty Reduction and Pro-Poor Growth Part 1*; World Bank: Washington, DC, USA, 2019.
50. DCF Alliance. Scaling up Devolved District Climate Finance in Tanzania: Community-Based Mapping of Pastoralist Resources and Their Attributes. Unpublished—Internal report.
51. Christensen, R.H. Ordinal—Regression Models for Ordinal Data. 2019, R Package Version 2019.12-10. Available online: https://CRAN.R-project.org/package=ordinal (accessed on 1 March 2021).

52. Anderson, D. *Model Based Inference in the Life Sciences: A Primer on Evidence*; Springer: New York, NY, USA, 2008.
53. Burnham, K.P.; Anderson, D.R. *Model Selection and Multimodelinference: A Practical Information-Theoretic Approach*, 2nd ed.; Springer: New York, NY, USA, 2002.
54. Barton, K. MuMIn: Multi-Model Inference 2020. Available online: https://cran.r-project.org/web/packages/MuMIn/index.html (accessed on 10 August 2021).
55. Deaton, A. Income, Aging, Health and Wellbeing around the World: Evidence from the Gallup World Poll. *J. Econ. Perspect.* **2008**, *22*, 53–72. [CrossRef]
56. Sina, D.; Chang-Richards, A.Y.; Wilkinson, S.; Potangaroa, R. A conceptual framework for measuring livelihood resilience: Relocation experience from Aceh, Indonesia. *World Dev.* **2019**, *117*, 253–265. [CrossRef]
57. DeJaeghere, J.; Morris, E.; Bamattre, R. Moving beyond employment and earnings: Reframing how youth livelihoods and wellbeing are evaluated in East Africa. *J. Youth Stud.* **2020**, *23*, 667–685. [CrossRef]

sustainability MDPI

Article

Discourse and Practice of REDD+ in Ghana and the Expansion of State Power

Shannon Johnson

Rights and Resources Initiative, Washington, DC 20009, USA; sjohnson@rightsandresources.org

Abstract: Green market mechanisms, as part of the architecture of climate finance, have become key components of international environmental frameworks. One of the most widely known mechanisms for climate change mitigation has been the creation of Reducing Emissions from Deforestation and Forest Degradation (REDD+). The purpose of this paper is to trace global discourses and narratives throughout REDD+ official documents and compare them to the implementation on the ground to determine the extent that REDD+ meets its stated objectives in the Ghanaian context. Then, given the gaps in discourse and practice, this paper aims to define the inexplicit consequences, or rather instrumental effects, of REDD+. Discourse analysis of official REDD+ documents and land policies combined with qualitative interviews and focus groups to determine the linkages between discourse and practice of REDD+ and the impacts of these gaps. While critical civic environmentalism, highlighting environmental justice as a core principle, was somewhat incorporated into official discourse from the international to the national level, the depoliticization of the discourse and the apolitical nature of interventions make these justice concerns negligible and create gaps in discourse and practice. These gaps create disjointed, infeasible policies that establish REDD+ as a fad to bring in financial resources that expand state control of forest resources under the veil of social-ecological responsibility. As a result, state power expands into rural areas, allowing for greater control over land and forests at the expense of local communities.

Keywords: ecological modernization; civic environmentalism; policy implementation; carbon market; state expansion; discourse analysis; neoliberal environmentalism; Ghana; REDD+

Citation: Johnson, S. Discourse and Practice of REDD+ in Ghana and the Expansion of State Power. *Sustainability* **2021**, *13*, 11358. https://doi.org/10.3390/su132011358

Academic Editors: Adam P. Hejnowicz and Jessica P. R. Thorn

Received: 23 July 2021
Accepted: 22 September 2021
Published: 14 October 2021

Publisher's Note: MDPI stays neutral with regard to jurisdictional claims in published maps and institutional affiliations.

Copyright: © 2021 by the author. Licensee MDPI, Basel, Switzerland. This article is an open access article distributed under the terms and conditions of the Creative Commons Attribution (CC BY) license (https://creativecommons.org/licenses/by/4.0/).

1. Introduction

The last twenty years has witnessed an increasing utilization of so-called market-based and financial approaches (e.g., certification, green bonds, payments for ecosystem services, permits, and trading schemes) within the environmental policy and regulatory landscape to address large-scale environmental grand challenges such as biodiversity loss and climate change [1–4]. The turn towards the use and development of market and financial mechanisms and instruments as ways to reduce or reverse environmental harms is often viewed as a political extension of neoliberalism [5–7]. From this vantage, so-called "neoliberal environmentalism" aims to internalize environmental externalities such that environmental problems are solved through privatization, commercialization, and commodification of natural resources and ecosystems [8], and increased private and public–private partnerships instead of state governance [9].

The burgeoning neoliberal environmental agenda can been seen in the rise of climate and green financing—especially in the wake of the Paris Climate Agreement (2015) and 2030 Agenda for Sustainable Development (2015)—which comprises a whole suite of policy mixes, activities and investment funds, targeted at local to global scales, and built around state and private-sector opportunities for low-carbon transitions, greenhouse gas emissions reductions, infrastructure development and ecosystem protection and restoration [10–13]. A notable example of climate financing, which has been the subject of widespread scholarly attention over the last decade is Reducing Emissions of Deforestation and Forest

Degradation (REDD+), developed as part of the UN-REDD program under the auspices of the United Nations Framework Convention on Climate Change (UNFCCC). Designed to address the linkage between environmental degradation, greenhouse gas emissions and climate mitigation in so-called developing countries, the REDD+ narrative advances the idea of bringing together a diversity of stakeholders to create "a financial value for the carbon stored in forests by offering incentives for developing countries to reduce emissions from forested lands and invest in low-carbon paths to sustainable development" [14]. However, since the beginning, REDD+ has been highly contested for the way it frames both the issue and the solution for deforestation [15–17].

Crucially, REDD+ policies are not neutral, but instead, are constituted and representative of dominant perspectives and knowledge regimes, and products of competing discourses [18,19]. Official REDD+ discourse, at international, national, and sub-national levels, emerges from competing narratives and power relations shaped to both appease and deliver a consensus compromise for all parties of the UNFCCC Conference of the Parties (COP) [20–22]. This is demonstrated by two distinct narratives. Supporters of REDD+ argue for the merits of green capitalism, particularly that the market is a solution to incorporate environmental externalities while promoting sustainable development [23]. The key "green development" assumptions of REDD+ are that forest carbon can be accurately measured; it will be valued higher than other forest commodities; and it will result in equitable poverty alleviation [24]. In contrast, political ecology critics question the "win–win" narrative of green capitalism, arguing that there are inherent trade-offs to valuing forests for primarily for their carbon. They argue that conservation is inherently linked to political, economic, technical, ecological, and social issues that REDD+ does not address [23].

REDD+ created several programmatic elements to address critical concerns such as social and environmental safeguards to protect local interests and co-benefits to forest conservation, collectively known as the "Cancun Safeguards" [25]. REDD+ safeguarding is acknowledged to be a complex issue [25], which partly explains differences in national-level progress (Jagger et al., 2014). At the same time, REDD+ does often have significant and detrimental impacts on local communities, frequently undermining livelihoods, institutions and raising social and environmental justice concerns [26], through for example restricting local forest practices [27].

In West Africa, there is a growing body of research examining stakeholder participation in REDD+ projects [28], and the impacts of REDD+ on local communities such as smallholder famers [29], justice and governance [30], equity [31], and the politics of design and implementation [32]. Across all these areas, challenges and problems have been identified, with some (e.g., [32]) highlighting the national political framing of REDD+ as being particularly instrumental in determining the effectiveness of implementation.

It is against this backdrop that this paper explores the depoliticization, or the removal of issues from political contention [33], of REDD+ discourse in Ghana, pushing the envelope to further extend the understanding of how the political construction of discourse shapes national REDD+ policy processes and impacts. Specifically, this paper combines a critical political ecology approach with poststructuralist discourse theory to examine (i) key legal and REDD+ policy documents, and (ii) national- and district-level policy actors directly or indirectly involved in REDD+ implementation, to critically assess the relationship between REDD+ discourse and practice.

Importantly, what the case of REDD+ in Ghana shows is that official discourse ignored political and power dimensions and so rendered the intervention a purely technocratic process shaped by bureaucratic "governmentality", in which societies, economies, and government bureaucracies respond in reflexive, straight-forward ways to policies and plans. This "depoliticization" is reflected in official REDD+ discourse, which effectively enables the Ghanaian state to divert responsibility for policies that underly deforestation, maintain decision-making power, and enact contradictory national policies in sustainable forestry. At the same time, financial resources are also directed towards the State for the infeasible actualization of the forest carbon commodity, leaving critical concerns unaddressed in practice. As a result, the Ghanaian government reaps financial gains without addressing the primary causes of deforestation under the veil of social inclusion. Social and environmental safeguards, then, legitimize expanding state power in practice. This power is cyclical in nature, which expands with each new, depoliticized fad in forest governance. The instrumental effects, or consequence, of an expanding state without local community rights over resources are ultimately criminalization and continued deforestation and forest degradation.

2. Methodological Framework and Approach

2.1. Theoretical Approaches: Critical Political Ecology and Poststructuralist Discourse

Critical political ecology is place based and incorporates both discourse analysis and qualitative research. Blaikie and Brookfield define political ecology as combining the concerns of ecology and political economy—including the shifting dialectic between society and land-based resources [34]. Forsyth expands this original definition to "critical" political ecology to include a "politicized understanding of environmental explanation beyond the epistemology offered by the critique of capitalism" ([35], p. 7). Critical political ecology seeks to integrate structural and poststructuralist attention on state and industry, in particular, closely examining how interactions between these actors results in the co-construction of environmental discourses [35].

Poststructuralist discourse theory acknowledges that discourse is a socially constructed phenomenon. It goes further, however, in arguing that this social construction is not simply about subject and object but also encompasses the fluctuating dialectic relationship between them. It is that fluctuation from which meaning emerges, and thus the theory accounts for possible changes in meanings, advancing that meaning (and meaning making by extension) is never fixed and is always to an extent unstable [36]. When discourse is more repeated, the theory posits, it becomes more sedimented and eventually naturalized [24].

This is particularly relevant for new concepts such as carbon storage, carbon market, and restoration that might be common terms and concepts within international policy but relatively alien at the national and especially local levels. In this way, when applied in the context of REDD+ policy, poststructuralist discourse can help associate a certain type of knowledge (international) with power and legitimacy, while devaluing discourse around comprehensive forest management discourse from local communities.

2.2. Discourse Analysis

Discourse analysis seeks to understand the debates, narratives, and storylines surrounding a policy. In the case of forest management, discourse determines the meanings that society constructs around forests, which has multi-faceted meanings that make up a field of discursivity [24]. For discourse analysis, 10 official REDD+ policy documents relevant at both the international and national levels were selected, alongside 8 legal documents relating to ownership of land and timber in the Ghanaian content. The documents were coded for key themes including participation, illegalities, land and timber rights, deforestation driver, results-based funding, carbon payments, and social and environmental safeguards. The discourse analysis links the framing of each programmatic aspect of REDD+ with a distinct global narrative, and these narratives were compared to Ghanaian laws and policies relating to land and forests.

2.3. Key Informant Interviews and Focus Groups

To explore how REDD+ policy discourse relates to REDD+ practice at the national and sub-national levels, during February and March 2019, a series of stakeholder interviews were undertaken through a combination of semi-structured interviews (n = 20) and focus groups (n = 2). A diverse set of stakeholders was identified, spanning policy, practice, academic and public and private sectors, largely drawn from snowball sampling. In total,

20 individuals were interviewed—four officials from the Forestry Commission's Climate Change Unit (Ghana's REDD+ entity); two from the Forestry Commission's district office at Sefwi Wiawso; nine members of forestry civil society; one professor of natural resources; one forestry development consultant; one USAID senior advisor; and two professionals of partner organizations working on Ghana's REDD+ strategy. The focus groups were among various staff members in the Climate Change Unit and with community members from Sefwi Wiawso.

All interviews were conducted in English and followed best practice guidelines as stipulated and approved by the Ethics committee at Geneva Graduate Institute. Crucially, all participants consented to be interviewed. All interviews were fully anonymized.

2.4. Relating REDD+ Discourse and Practice

To compare discourse and practice, we follow Ferguson's goal "to connect observed discourse regularities to non-discursive practices and institutions" [37]. This is based on Foucault's contention that discourse is practice and has real effects [38]. Foucault first theorized the "conceptual apparatus" to describe how problems in society were constructed to justify the need for a solution [38]. Similarly, Ferguson described the "development apparatus" as a conceptual apparatus that creates a problematic designed to justify an international intervention [37]. Additionally, objects of analysis are formed partly by discourse that describes them and do not exist naturally [39].

A key construct in these problematics is the principle of governmentality in which societies, economies, and government bureaucracies respond in reflexive, straight-forward ways to policies and plans. The notion of "governmentality" and its discourse suggest that state control is determinant ([37], p. 64). The state apparatus is deemed neutral and appears as an apolitical machine for implementing development projects as it represents "the people," often described as a homogenous group. Governmentality exhibits the extreme state-centeredness of development discourse because it assumes that there is no other way to improve [37]. This narrative is relevant to REDD+ because coordination and funding is only directed to the state rather than communities, CSOs, and other stakeholders. From the beginning of the UN-REDD program, the UN Framework Convention on Climate Change (UNFCCC) set up the state as the primary planner and implementer of the programs on a voluntary basis. Thus, States are not only seen as the most legitimate actor to implement REDD+ and improve rural community livelihoods, but the only way to intervene from an international perspective.

However, like other institutions, government is always an exercise of power [37]. The state has two functions: formally, to deliver social services, and informally, to use power to benefit some over others [37]. As a result, political turmoil and politics are absent in discourse, where "the people" are undifferentiated. Political and structural causes of poverty are systematically erased and replaced with technical ones, and the "modern" capitalist nature of society is systematically understated/concealed [37]. Simultaneously, international development discourse sets up "an object, out there, not part of the study but external to it" and itself as "an expertise and intelligence" that are entirely separate [39].

Neoliberal environmentalism, in a manner similar to the development industry, relies on technical "experts" to function and legitimize apolitical interventions [40]. Environmental neoliberal interventions are often depicted as common sense, objective or neutral through a process of depoliticization, or "to remove issues from political contention" [33], as opposed to value-laden and normative, political, issues due to considerations of equity and justice [40]. "Expert knowledge," then, becomes a way to empower market actors and others while marginalizing locals and context-specific concerns [41]. These conceptual frameworks are heavily used throughout the analysis of this study.

3. Results and Discussion

The analysis for this paper draws heavily from examples in international development, particularly the work of Ferguson who examined the implications of apolitical interventions in Lesotho. As in his work, this research recognizes that structures are multi-layered, polyvalent, and often contradictory that often overtake intentional practices. In this way, intentional plans are never fully realized as "*outcomes of planned social interventions can end up coming together into powerful constellations of control that were never intended [...] but more effective for being subject less*" ([37], p. 19). Outcomes that seem as side effects of unsuccessful attempts become "instrumental effects" that are instrumental in some other aim [38]. Ferguson demonstrates that development projects "*squash political challenges to the system by enhancing powers of repression, and reposing political questions of land, resources, and jobs as technical*" ([37], p. 270). The instrumental effect of these programs was to strengthen the power of the state as a powerful source of funds and site of patronage [39].

Ferguson contends that state expansion, as an instrumental effect, is not intentional nor centralized [37]. The state does not "rationalize and centralize" power relations, rather it "*grabs and loops around existing power relations to cinch them together like a knot*" ([37], p. 274). Bureaucratic state power, then, is a mode of power that relies on state institutions but exceeds them. REDD+ operates in a similar way, depoliticization discourse and the notion of governmentality squash political challenges to the program and create disjointed policies at the national level which establishes REDD+ as an unsustainable conservation fad. The fad, then, allows for the expansion of the state.

3.1. Formation of REDD+ Discourse: State-Centered Technical Interventions

3.1.1. Competing REDD+ Narratives

Critical political ecology needs to consider the framing of discourse to legitimize intervention. Discourse is a "shared way of apprehending the world" that relies on assumptions, judgements, and contentions ([42], p. 8). It is critical to examine discourse because those who subscribe to it interpret information in a certain narrative or account [42]. In this section, the dominant and critical discourses and corresponding storylines that have been drawn out of policy discourse, and how REDD+ has evolved as a result are traced. These narratives frame REDD+ interventions and have become integrated together at the international and national level.

Ecological modernization is the dominant discourse that frames REDD+ thinking and policy practice in negotiations, drawing support from UNEP, FAO, and the World Bank in their operationalization of REDD+ [43], as the program primarily commodifies forest–carbon and encourages restructuring forest–people relations towards market-oriented public–private partnerships [19]. Ecological modernization (EM) scholars advance that EM provides a coherent framework that incorporates market and monetary processes and redefines the role of the state and markets in environmental reform beyond narrow neoliberal confined [44]. However, critics continue to emphasize that EM has a technocratic outlook, maintains the continuity of the capitalist order, which is fundamentally at odds with environmental prosperity, and prefers light touch green reformist approaches [44]. Principally, ecological modernization discourse still posits a largely market-focused orientation (even if it advocates for different models and approaches for connecting diverse actors, sectors, and institutions) as the primary solution for environmental externalities. It cements the role of the market and technological innovation to be able to reverse the negative impact economic development has had on the environment (Nielsen 2014).

On the other hand, civic environmentalism, while a heterogeneous term, brings concerns of environmental justice, ecological sustainability, equity, local knowledge systems and the inclusion of local stakeholder participation. This discourse is critical of EM's mainstream discourse and its reliance on markets and technical experts to solve environmental problems, and instead emphasizes relations of power between actors [23]. Civic environmentalism questions the "win–win–win" storyline of ecological modernization, arguing that community-based conservation inherently involves trade-offs and highlights the disjuncture between ideals of poverty alleviation and actual practice [19]. Furthermore, some civic environmentalists argue that community-based conservation is just a tool for the expansion of neoliberalism to further capital accumulation into rural areas [19].

Civic environmentalism values social and environmental safeguards more than the commodification of forest carbon fluxes [23]. EM discourse, conversely, suggests that the reason deforestation occurs is primarily a consequence of market failure within the forestry sector and the undervaluing (in economic terms) of carbon, which disincentives local communities from favoring forest conservation. The solution, therefore, is to create a carbon market and "teach" local communities the economic value they can receive from retaining carbon in standing forests. The solution is technical and apolitical and ignores issues of power. In contrast, civic environmentalism centers power and non-Western knowledge. It places politics between the North and South and their power relations as a key driver, for which local knowledge and holistic valuation could be a solution. Table 1 below showcases common storylines associated with both civic environmentalism and ecological modernism in REDD+ discourse.

Table 1. Storylines for Ecological Modernization and Civic Environmentalism (based off Nielson 2014).

Ecological Modernization	Civic Environmentalism
Cost-efficiency: reducing deforestation is most cost efficient mitigation strategy for climate change	Beyond markets: emphasis should be on equity and legitimacy rather than effectiveness and efficiency
Win–win–win: REDD+ helps reduce emissions, improve forest conservation, and reduces poverty	Beyond markets: REDD+ involves trade-offs between economic growth and sustainable forest management
Market rationale: market is key to internalize environmental costs due to innovation in private sector	Local, not global: local knowledge not adequately used during policy process
Carbon accounting: forests are subject to management and control through technical advances	Biodiversity: problematizes valuing forests purely for carbon
Technocratic rationale: societies can manage environmental cycles, so carbon becomes governmental/political domain	North–South divide: countries in South lose control of forestland, allowing North to defer responsibility for mitigation

3.1.2. REDD+ Discourse

Before REDD+ emerged in international law, the link between forests and carbon were situated within scientific inquiry. In 2005, at the Coalition for Rainforest Alliance in Papua New Guinea, an official document claimed deforestation as "*the single largest source category of emissions in the developing world*" and that "*without a more complete market valuation, standing forests cannot overcome the economic opportunity costs associated with their conservation*" [45]. Low- and middle-income countries with high concentrations of forest advocated for a program that would compensate for their conservation while mitigating carbon and ensure common but differentiated responsibilities (CBDR). It was called "RED." RED continued to evolve to include reducing forest degradation (REDD) and role of conservation, sustainable management of forests, and enhancement of forest carbon stocks (REDD+). Table 2 outlines key COP provisions related to both ecological modernization and civic environmentalism narratives.

Table 2. Narratives in official REDD+ discourse.

Conference	Elements Derived from Ecological Modernization	Elements Derived from Civic Environmentalism
2005 Coalition for Rainforest Alliance in PNG	Market valuation Economic opportunity costs	–
COP 13 (2007)	Technology transfer Mobilization of resources	Sustainable forest management
COP 16 (2010)	Framework for conservation of forest carbon stocks Enhancement of forest carbon stocks	Environmental and social safeguards to protect or avoid risks while promoting benefits Alternative policy approaches for integral and sustainable management of forests
COP 17 (2011)	Financing options for results-based actions	Transparency and effectiveness of safeguards
COP 18		Methodological issues of non-carbon benefits
COP 19 (2012)	Measuring, reporting, verification (MRV) National forest monitoring systems Encouraging public–private partnerships	Governance and measures to ensure social and environmental benefits and coordination of support and institutional arrangements

As seen in Table 2, the core problematization of power for civic environmentalism discourse (local, not global storyline and North–South divide) is replaced by technocratic interventions when the narratives enter into the international framework. Where equity and legitimacy were key aspects of the beyond markets storyline for civic environmentalism discourse, technical social safeguards take their place. The result is a REDD+ program that "administrates" and "governs" an intervention rather than addressing concerns of equity, legitimacy, and local knowledge. In this way, the REDD+ official discourse mirrors Ferguson's development discourse by rendering deforestation apolitical.

These COP decisions also solidified the need for technical experts, particularly in forest carbon stock and monitoring, evaluating, and verification. The emphasis on results-based funding also ensures a slow process, requiring national-level frameworks well established after high-level dialogue, sizeable initial investment, and forest carbon as central to programmatic success.

3.1.3. REDD+ in Ghana

Ghana, the first REDD+ program in West Africa, began to develop a national REDD+ strategy in 2008 and submitted its final REDD+ strategy in 2016. Ghana's REDD+ strategy has been designed to directly meet requirements of the Warsaw Framework on REDD+ and other COP decisions. It integrates key frameworks, particularly from the Warsaw Framework, encouraging public–private partnerships, information sharing, results-based payments and actions, technical expertise and assessments, carbon market, safeguards, forest monitoring systems/MRV, and non-carbon benefits [46]. Ghana's vision for REDD+ is "to significantly reduce emissions from deforestation and forest degradation over the next twenty years, whilst at the same time addressing threats that undermine ecosystem services and environmental integrity in order to maximize co-benefits from forests" [46].

Discursively, ecological modernization and civic environmentalism are integrated at the international and national level to a certain extent. Without focusing on the power and agency of various stakeholders, and pursuing apolitical interventions, depoliticized critical discourse obscures its own aim. Competing narratives result in integrated discourse and finally practice, all the while losing an emphasis on power that rooted the first discussions. REDD+ in Ghana depoliticizes and deemphasizes civic environmentalism in the problematization of deforestation and the implementation of social and environmental safeguards. However, these components still legitimize a market-based forest carbon mechanism that all actors on the ground acknowledge is not politically or economically feasible.

3.2. Depoliticizing Drivers of Deforestation

The goal in REDD+ discourse is to make Ghana a promising candidate for technical forestry intervention. To frame a "forestry apparatus" for a REDD+ intervention informed under UNFCCC, the national government needs to be a neutral instrument for a solution rather than a part of the problem [37,39]. As Hajer notes, environmental conflict is not due to a specific predefined problem between actors, but rather "a complex and continuous struggle over the definition and the meaning of the environmental problem itself" ([47], p. 5). The production of knowledge and discourse is ultimately a display of power, of complex social and political influences. Ghana's REDD+ strategy, likewise, attempts to depoliticize the problem of deforestation, particularly by side-lining market and tenure factors which would implicate the state.

Ghana's REDD+ strategy focuses on eliminating illegal activities that contribute to deforestation rather than addressing the market and policy factors that create illegal markets. Ghana's REDD+ strategy claims that it is "well-anchored within a favorable policy environment due to the National Climate Change Policy (2013) and the National Forest and Wildlife Policy (2012), which is foundational to REDD+ implementation" [46]. However, the legal framework has yet to recognize individual and collective management rights nor sufficiently regulate the timber market. The REDD+ strategy document also lists the indirect drivers of deforestation, including population growth and development, global markets, weak law enforcement, tree tenure and low stampage prices ([46], p. 31). However, the main drivers of deforestation are the "primary" criteria for assessing strategy options, which gives more weight to them over structural and political causes of deforestation ([46], p. 32). While these underlying drivers are minimally acknowledged, they are consistently minimized.

3.2.1. Land and Tree Tenure as Deforestation Driver

Throughout Ghana's REDD+ strategy, legal reforms in the land and tree sectors are constantly de-emphasized, thereby largely placing the fault for deforestation on local communities for illegal harvesting and agricultural practices. For example, REDD+ gives three primary interventions to counter deforestation in Ghana, ranked by urgency. The first two are targeted towards the practices of local communities, including improving land use and addressing wood harvesting and agricultural practices. Policy and legislative reforms are last on the list. In REDD+ strategy documents, Ghana was given a "further development required" in legal reform because policies have not been translated into legally binding laws—particularly tree and carbon tenure [48]. Yet, the "next steps" for the progression of REDD+ fail to mention legal reform as a prerequisite for REDD+ readiness. This section demonstrates how land and tree tenure are underlying causes to deforestation and forest degradation, as communities lack ownership and management rights to their natural resources.

There are three types of land in Ghana: natural parks, on-reserve, and off-reserve. REDD+ largely targets off-reserve communal land to diminish deforestation and forest degradation. Community off-reserve land are often vested in chiefs, and off-reserve exchanges of land require authorization from them, accounting for control of approximately 80% of the land in Ghana Prior to British colonial rule, much of the land in Ghana was communally owned without official title/deeds. Then, in the late 18th-century colonial era, customary law accredited people with legal capacity to carry out valid alienation of group-held property [49]. Chiefs or heads of families were usually the accredited individuals. As communal property became seen as chief's property over the years, industry developed on the land in the form of permanent cultivation, mining, and timber. Due to these first efforts to cement private land tenure, chiefs gained communal lands and partnerships with key industrial, colonial actors, leaving local farmers with little opportunity to accumulate wealth.

The 1983 International Monetary Fund and World Bank structural adjustment programs further exacerbated the accumulation of land management by shifting property from communal to individual ownership [49]. The privatization of land, structured around neoliberal rational of markets and private actors, gave most land management power to chiefs as individual owners. Land tenure was inundated by insecurity of title, especially as the population began migrating from the countryside to the city and increased land transactions [49]. Land insecurity is a primary driver for deforestation and forest degradation in these regions, but it is largely because communities are not aware that the chiefs should manage the land for them, and they have customary rights over these areas. Instead, they are largely seen as the chief's private property, thereby perpetuating alternative livelihoods into forest areas. This could improve with Ghana's Land Act (Act 1036) enacted in 2020 which requires chiefs to be transparent, fair, and impartial in making decisions over land. However, it still does not give customary land tenure, including management and use rights, back to communities. Land tenure insecurity, especially the accumulation of land by chiefs to use for private gain, is an underlying cause of deforestation and forest degradation that REDD+ does not address.

The lack of secure tree tenure for local farmers also fuels deforestation and forest degradation in Ghana. There are two legal categories for trees in Ghana: planted and naturally occurring Tree tenure includes right to own or inherit, right to plant, right to use, and right of disposal. The state hold, or appear to hold, ownership and management rights due to lack of clarity in the law, misinformation, and lack of communal governance. Ghana's 1962 Concession Act vests timber resources and naturally occurring timber trees in the President of the Republic of Ghana on behalf of the People of Ghana [50], although forest-owning communities assumed their ownership rights had been lost to the State [51]. The 1992 Constitution vests naturally occurring timber in the president but in trust for the stools (chiefs) concerned, which is to be managed by the Forestry Commission in recognition of pre-existing customary rights [52].

While farmers own and manage planted trees, they have no ownership or management rights of naturally generated trees, even on their private property. Naturally occurring trees still require the same amount of management as planted trees. It is illegal for farmers to sell timber from naturally occurring trees from their land, instead they can only use it for subsistence purposes. Yet, timber companies can access trees on their land (Director, PAB Development Consults). Hence, farmers have an incentive to cut trees down, burn them, and/or sell them to illegal chain saw operators, which fuels deforestation and perpetuates the illegal timber market. It is important to note that illegal chain-saw operators buy timber at a lower rate than the legal market, meaning that farmers lose that potential income by selling it illegally.

To tackle illegal logging, legal reform is necessary to give individual farmers ownership and management rights of naturally occurring trees on their farms as well as community's explicit ownership and management rights of forest reserves. The 2012 Forest and Wildlife Policy was formed during the REDD+ negotiations and attempted to incorporate individual and collective benefits for tree tenure. It says "to enact legislation that will allow communities/individuals to benefit from trees on their land. Benefits accruing from resources individually and collectively" [53]. While benefitting from forest resources individually and collectively is an important component of reducing deforestation, it does not give individuals or communities explicit ownership or management rights of forests. Furthermore, the 2012 policy contradicts the 1962 Concession Act. Meaning, the 1962 Concession Act would need to be repealed in order for the 2012 Forest and Wildlife Policy to have legislative backing and come into effect. The Tree Tenure Reform, submitted in 2016, is the latest attempt to ensure "ownership of naturally occurring timber trees off reserve are vested in the communities and stools concerned," but there is little hope for it to be adopted [51].

Politics and power are central to why these reforms have yet to take place. The Director of PAB Development Consultants, who drafted the Tree Tenure Reform for the Ministry of Land and Natural Resources, notes that he has been involved with tree tenure reform for 25 years, and "there have been so many policy reform discussions in the sector that have not seen the light of day. I'm not very optimistic that any major reforms are going to take place because there's too much vested interest in the sector" (Director, PAB Development Consultants). Yet, Ghana's REDD+ strategy frames legal reform as probable ([46], p. 62). For communities to benefit from forest resources, the State and chiefs would need to give up some of their power. The State would have to implement legal reform, mandating naturally-occurring trees to individuals and ownership and management rights to communities for off-reserve community forests. Further, politics and timber are interconnected, as the private sector funds political campaigns (Director PAB Consulting). Chiefs, who benefit from the misinformation and management rights, are well respected and powerful "so no government wants to have conflict with chiefs." Ghana's REDD+ strategy depoliticizes legal reforms in the land and tree tenure sector, which allows the state and chiefs to maintain ambiguous control of these resources and leaves local communities without sustained access to forest resources that they need for their livelihoods.

3.2.2. Timber Market as Driver of Deforestation

Ghana's REDD+ strategy defines the problem of deforestation as local, rather than driven by domestic timber market factors (quotas, ease of access, etc.) set by the national government. Yet, the domestic market for timber in Ghana makes it difficult and expensive for Ghanaians to access legal timber markets. Present laws are not workable—usually the nearest sawmill that sells domestic timber is distantly located, and the timber market is not decentralized to district and local levels. Additionally, Ghana set the quota for export timber at 70% export, leaving 30% for domestic consumption. Local demand exceeds 30% of timber produced, which creates the conditions for illegal timber markets. The Programs Officer for EcoCare Ghana estimates that 7/10 domestic lumber comes from illegal sources. Even though timber companies pay the same stampage fees for domestic and international markets, they make a higher profit selling on the international market and the state can charge higher taxes. So, the state has an incentive to maintain the high export quota, regardless of the demand for timber in the domestic market. Without addressing the accessibility for domestic timber and reforming export/import quotas in Ghana, the illegal timber market will continue.

In order to curb deforestation and forest degradation, land and tree tenure need to be equitably secured by communities, including ownership, management, and use rights. Additionally, the domestic timber market needs to have a higher quota to meet demand. The necessary reforms, which would diminish state financial resources, contradict the "win–win–win" storyline of environmental modernism, and instead demonstrates the trade-offs in national economic growth and sustainable forest management, outlined by the beyond markets storyline of civic environmentalism discourse which is not present in official REDD+ discourse. The objectives of the program do not target these structural causes of deforestation due to their political nature, which would cut into the financial resources of the state. The timber market and tree tenure are not sufficiently addressed in the REDD+ strategy because the state has an incentive to maintain its power and resources, thereby contradicting the premise of a neutral state.

3.3. *Depoliticized Social and Environmental Safeguards*

Civic environmentalism discourse has largely led to the inclusion of social and environmental safeguards in neoliberal interventions, with explicit links to transparent governance and inclusionary participation [19]. Environmental and social safeguards have been set up to ensure that Ghana's REDD+ strategy "does no harm" to local farmers and populations, following the World Bank's requirements that were developed from UNFCCC guidelines ([46], p. 66). One social safeguard that is based on the "beyond market" storyline that emphasizes equity and legitimacy is the "full and effective participation of all UN-REDD Program stakeholders—partner countries, donors, indigenous peoples, civil society organizations, participating UN agencies—while ensuring streamlined decision-making processes and clear lines of accountability" [54].

However, the deemphasis on power relations renders this social safeguard impotent. Further, the differing political interests of stakeholders are not accounted for in negotiations. REDD+ in practice in Ghana questions whether civic environmentalism discourse can be effectively integrated into programs largely concerned with technical social interventions that ignore the political roots of inequalities between stakeholders. Once again, this depoliticization of participation allows powerful actors, such as the state, to ultimately remain in control of the decision-making process to maintain their own power and resources.

3.3.1. "Full and Effective" Participation of Communities

Ghana has included an approach "that prioritizes the need for an accountable and participatory process with effective participation of women and local communities" ([46], p. 66). First, participation does not guarantee governance or decision-making power, it only allows the discussion between various stakeholders. The official stakeholders are civil society, the private sector, government, research and academia, and development partners, whilst communities are noticeably absent as stakeholders ([46], p. 71). Nuances of social systems, politics, and history are not considered to ensure "full and effective participation." Instead, these processes are co-opted to legitimize that focus on technical rather than systematic interventions. REDD+ safeguards in Ghana overlook the political and competing interests of actors. The lack of financial resources and time also limit REDD+ from gaining "full and effective" participation from communities (REDD+ consultant).

Communities have limited, if any, participation in high-level decision making. REDD+ in Ghana has progressed fast at the national level, but communities lack an understanding of what policies were decided (IUCN officer). While a participatory forestry governance mechanism in Ghana was set up by civil society to bring local representatives to engage at the national level—the National Forestry Forum (NFF)—it lacked the funding to continue (Programs Officer for EcoCare Ghana, Programs Officer for Forest Watch Ghana). NFF has not taken place in over two years. Without the NFF, the Climate Change Unit has difficulty deciding which communities to include in consultation, so community representatives end up being random or politically motivated rather than representative (Safeguards Working Group member). As a result, communities are not represented in Ghana's Multi-Stakeholder Implementing Communities (MSIC), the highest level in policy development.

When local representatives are included in the discussion, they are not "effective" collaborators because of unequal technical knowledge of REDD+, and thus lack the capacity to contribute effectively to the Working Groups, particularly in the MRV working group (Programs Officer for EcoCare, Country Coordinator for IUCN, National Deputy Director for ROCHA). This affirms Nielsen's limit of civic environmentalism that assumes a participatory process based on different power dynamics is even possible [23]. By depoliticizing the participation of stakeholders, REDD+ Ghana allows for powerful, decision-making actors to maintain power.

What happens in most cases is that communities are represented instead, in national-level committees and sub-working groups, by traditional authorities and civil society organizations. It is assumed that the interests of these stakeholders align with those of communities and hence the community voice will be well represented, yet that it is not necessarily the case [55]. Civil society organizations may be pro-industry, who would seek large areas of land for REDD+ projects, or pro-community, who would protect community rights to land (Program Officer, Forest Watch Ghana). Additionally, civil society representatives recognize that they cannot represent community interests (Program Officer, EcoCare). Additionally, chiefs are often invited to represent community interests in REDD+ negotiations as powerful actors. However, as outlined earlier, chiefs often have different interests from the communities largely due to a history of dispossession of communal land. Selecting chiefs as community representatives loses a sense of transparency and accountability in the negotiation process (Officer, Casa Watch Ghana).

Similarly, actors at the community level are taken as a homogeneous group, when, in practice, there are divisions of interest. In the cacao farms in South western Ghana, tenant farmers are some of the most vulnerable in Ghanaian society, yet they are not represented in community REDD+ processes because they do not have customary titles (Officer, Casa Watch Ghana). In national-level policy meetings, issues with tenant farmers are not represented, especially if it is a traditional authority who is summoned (Officer, Casa Watch Ghana). These traditional authorities have been consulted in every aspect of REDD+, as a result they automatically receive 2% of benefits from carbon market.

The State ultimately makes decisions about REDD+ as the most powerful actor, especially because UN-REDD program funds flow directly through the state without any funding mechanism for other stakeholders. Ghana's REDD+ strategy acknowledges that the national government will "play a lead role" during policy and legislation discussion, but the discourse assumes that the State equally values perspectives from other stakeholders. The State engages with other stakeholders in discursive participation but ultimately decides based on their own interest (Country Coordinator for IUCN). This demonstrates the limit of the notion of governmentality and the "full and effective" participation safeguard, which does not transfer ownership or management of the program to local communities who are the primary beneficiaries of the program. Instead, it assumes that the State "represents the people" in an apolitical manner.

Ghana's REDD+ strategy overlooks historical and political power in favor of a policy that paints them all as equally effective actors with common interest, just as discourse in development projects conceptualizes "the people" as an agglomeration of individuals, thereby reducing political and structural causes of poverty and layers of society [37]. Ciplet and Timmons point out that environmental neoliberalism has increased over time in international environmental governance of climate change, thereby negating precautionary and equity-based concerns: *"The neoliberal pursuit of transparency is not preoccupied with other political goals such as democracy, empowerment of diverse stakeholders, and improved governance, but is instead used to preempt stronger, compliance forms of regulatory action (Roberts 1998; Haufler 2010), reinforce neoliberal norms of individual responsibility (Mason 2008), and elevate concerns of powerful actors over others under a veil of neutrality"* ([40], p. 151). Furthermore, Scheba and Schebab describe this veil of neutrality as "inclusionary" discourse and practices in REDD+ to legitimize new markets without regard to power imbalances and structural inequalities [19]. These authors therefore "question the emancipatory power of 'inclusive' practices that actively script-out larger historical context and insufficiently recognize it as an uneven playing field" [19]. In this way, REDD+ democratic engagement of participation paradoxically deepens the processes of depoliticization by overlooking "the underlying political economic dimensions of poverty and exclusion" ([19], p. 528)

3.3.2. Consistent Actions with National Forest Programs

The first social and environmental safeguard of REDD+ is "consistent actions with national forest programs." It is key to note that this discourse is already depoliticized because it does not say "consistent actions with national programs," implicating cohesive programming with extractive industries and other competing interests. Regardless, all interviewees agreed that REDD+ must not be implemented as a separate stand-alone program, but rather integrated into the national forestry governance landscape. By establishing REDD+ within the Forestry Commission, as opposed to setting up a new system, there is some sustainability and continuity (Director of Climate Change Unit). Additionally, the safeguards feedback and grievance mechanism, which will receive and address any project-related complaints, can be used to address forestry complaints beyond REDD+ (Program Director, Tropenbus Ghana).

However, integration of programs is not sufficiently combined due to competing interests. REDD+ has brought funding to the Climate Change Unit, but other departments within the Forestry Commission that are essential for sustainable forest management are "starving of resources, such as the wildlife department" (Deputy Director for ROCHA). Additionally, the Forestry Commission has not adequately integrated two separate international initiatives—the REDD+ secretariat under the Forestry Services Division and Forest

Law Enforcement Governance and Trade (FLEGT) under Timber Validation Department. The Programs Officer for EcoCare Ghana sees this conflict problematic particularly at the community level, where farmers will choose between both programs to participate, whether to choose timber or carbon which will always result in timber.

Additionally, although 47 district-level offices are key components of REDD+ implementation, information about REDD+ is not well dispersed throughout the district levels, and funding to implement education programs to the communities does not trickle down to the regional or district level. While the Forestry Commission has the institutional capacity to decentralize REDD+, it still has not happened (Officer, Care International). After 10+ years of negotiation, REDD+ still operates from a national level.

The lack of integration at the state level is the result of competing interests within government agencies, which reflects the notion of government as an exercise of power [37]. Contradictory and incoherent policies and programs allow for the State to accomplish its second, informal goal—to use power to benefit some over others [37]. Additionally, international fads in forestry create transient government priorities to acquire funding. The depoliticization between forestry sectors and state agencies create disjointed policies in REDD+ practice. The shifting international and government priorities is another reason that local communities should have governance and management rights over their forests. Legal reform, as already shown, is highly unlikely.

3.4. *Forest Carbon Market Acquires State Funding without Benefitting Communities*

3.4.1. Commodification of Carbon

Neoliberal environmental conservation models move environment and social relationships into the realm of commodities, connect to the carbon accounting storyline of ecological modernism [56]. REDD+ sets the market for a newly establish commodity, forest carbon [24]. The justification for forest carbon as a means to diminish deforestation and forest degradation is deeply rooted in ecological modernization, especially the market rationale and carbon accounting storylines. The measurement of carbon forest is a prerequisite for commodifying avoided deforestation, which is estimated based on the forest's biomass [24]. A forest monitoring system was established that combines remote sensing technology with ground-based sampling to estimate carbon emissions reductions and removals. Reference levels create a baseline to measure emissions reductions compared to "business as usual" projections, which include deforestation, degradation, and carbon stock enhancement ([46], p. 78). Emission Reductions Payments are calculated by social performance indicators and emission reduction indicators ([57], p. 27).

REDD+ uses an already established metric, the ton of carbon dioxide equivalents (tCO2e for short). The performance of REDD+, which determines future payment, is based on emission reductions, a calculation of tCO2e. These calculations are the basis for defining successful outcomes and progress. Ghana's National Climate Change Policy Master Plan defined progress in terms of carbon sinks and carbon stocks ([46], p. 84). However, academics have contested the measurement of forest carbon through baselines estimates, thereby questioning the legitimacy of its commodification [58]. Given the complexity and creation of a commodity that does not physically exist, there is high uncertainty in the carbon market. Marx defines a commodity as a physical thing that satisfies human needs in exchange for something else, as something that has use value, exchange value, natural material, and labor [56]. Yet, tCO2e is not a physical thing nor is it a natural material. It only has an exchange value, so the baseline can be inflated during calculation. This section demonstrates how uncertainty, low price, and high cost of accounting make economic benefits of forest conservation negligible for local communities, while bolstering the State's budget.

3.4.2. Carbon Payments and Results-Based Funding

To achieve 100% performance for REDD+ Ghana, there must be 10 million tCO2e emission reduction worth US$50 million for the period 2018–2024. In this scenario, US$2 million will cover fixed costs, US$1.44 million as a performance buffer, US$33.465 million for communities and authorities, and US$13.095 million for the government ([57], p. 28). The funding system is results-based, strictly on reduction in emissions as "only verified reductions in deforestation and degradation will trigger carbon payments from the FCPR-CF to be stared between identified beneficiaries" ([57], p. 23). The results-based financing, which requires strong reporting, is posited as "one of the main differences between traditional development project models and REDD+" ([46], p. 72).

The carbon market and carbon accounting are prioritized by REDD+ because of "results based funding," which also means that most of the REDD+ readiness funding is diverting for Monitoring, Verification, and Reporting (MRV). Several interviewees name MRV as the greatest challenge to REDD+ in Ghana, particularly because it requires regulatory data collection every five years. This means that external funding is necessary to keep up with MRV, especially as the government is not likely to pay for it if they do not have a financial interest (Director of PAB Development Consultants).

Given the operational costs of MRV, there is little funding left for other aspects of REDD+, such as participatory governance. MRV is by far the most expensive aspect of the REDD+ program, accounting for US$1,140,000 out of US$1,990,812, or 57.26% of the fixed costs of the program from 2019 to 2024. The next most expensive item is consultants that account for US$180,000, or 9.04% of the overall fixed costs ([59], p. 25). Consultants, or technical experts, are also largely required for the MRV process. The financial resources are spent on travel to international conferences (Deputy Director, ROCHA).

Even with the expense of measuring forest carbon, communities do not see the benefit of "selling" forest carbon due to the complicated accounting and low price of carbon. All interviewees noted that individual monetary carbon payments would be ideal as it would incentivize farmers to keep trees on their off-reserve farm. However, individual carbon payments are not enough to incentivize local farmers to not cut down trees as timber (Ghana's REDD+ strategy staff). Given the choice between competing commodities (timber, coco) directly, farmers would not save trees for carbon. Yet, the cost of MRV and carbon accounting make individual payments for carbon impossible. To commodify carbon for individual benefits, it would require assigning carbon benefits to specific trees (Program Officer Tropenbus). The accumulation of carbon is not the same for every tree in the same landscape, which makes it more complicated than the timber analogy (Professor of Forestry, Kwame Nkrumah University). Therefore, the amount of work required to generate biomass carbon stock of each tree in the landscape would surpass the money coming in terms of operational costs (Professor of Forestry, Kwame Nkrumah University). The low price of carbon also makes it so that households will likely not gain a significant amount of financial incentive for participating in REDD+ (Country Coordinator for IUCN).

Due to the low price of carbon on the carbon market and complexity of carbon accounting for individuals, Ghana's REDD+ strategy decided on collective instead of individual benefits in which communities have a fund to use for community projects. The community benefit-sharing system sets up community development programs for services that the state should provide. As a Program Officer for Forest Watch Ghana notes, "they're giving fertilizers anyways. They're building hospitals anyways. But if you look at REDD+ carbon scheme, they will say they will use carbon money. But it's something they're supposed to get in the first place". So, these services are promised to be paid for by the state and the carbon payments, meaning that the state can pocket the excess funds. Government can use the carbon money to finance its own national budget (Program Officer, Forest Watch Ghana). In this way, REDD+ provides another avenue for the state to gain financial resources.

3.5. Co-Benefits: Where Discourse and Practice Align Temporarily

Emission Reductions Payments are only based on tCO2e emissions that do not incorporate non-carbon benefits. Due to commodification of carbon, every other value and articulation of forests lose meaning. In other words, forests become "carbonified" [24].

Corbera notes that forest carbon could render invisible interconnectedness of ecosystem elements, an expressed concern by several CSO representations [60]. Yet, REDD+ is only viable with policies to address deforestation without undermining livelihoods [60]. Of course, forests have a wider spectrum of values to local communities beyond their carbon storage capacity. They are necessary for livelihood strategies, household firewood, source of wood, a site of biodiversity, or as a recreational space. The biodiversity storyline of the civic environmentalism discourse exemplifies this concern well, noting that forests should not be valued solely for their carbon.

Due to the limitations of the REDD+ interventions that do not provide economic incentive to halt deforestation, Ghana's REDD+ strategy has adapted to focus on how else they can incentivize local farmers and communities to conserve forests outside of the international framework: ecosystem services and co-benefits to improve livelihoods. The Climate Change Unit credits the enforcement of safeguards measures to ensure biodiversity and ecosystem services "are given adequate attention" ([61], p. 76). The Climate Change Unit focuses on essential commodities that add direct value like coco to generate individual income for farmers. The co-benefit strategy allows for both financial and non-financial incentives for forest conservation. Given that farmers would not choose the carbon commodity over timber, cocoa, rubber, or any other commodity which has a stable, higher price and accessible market, Ghana's REDD+ strategy seeks to minimize the importance of carbon payments and highlight instead co-benefits (Director, Climate Change Unit). GCFRP noted that the "priority non-carbon benefits have the potential to carry the program, even if performance is low, and are meant to secure engagement in the program and success over the long-term (past the ERPA time-frame) ([61], p. 19)". All interviewees expressed that this was not only the best option, but the only feasible way to implement REDD+.

The World Bank formally accepted Ghana's Emission Reduction Program for the Cocoa Forest Mosaic Landscape in 2014, which seeks to reduce emissions driven by cocoa farming, other agriculture, illegal logging, and illegal mining [46]. Most coco is grown through smallholder farmers with an average of 1-hectare plot of land, so these farmers would directly benefit from increased yield (Head of Climate Change Unit). The program, then, is two-fold: restoring forests but also increasing yields. Ghana's REDD+ strategy Cocoa Initiative is the most advanced REDD+ program which has set up co-benefits, attracting the attention of both the private sector and farmers. Optimal yields come from incorporating trees into the cocoa landscape for shade, which will produce higher coco yield. Then, the farmers would also be less likely to encroach on forest reserves. The incorporation of trees also qualifies as a climate-smart condition for which buyers would be willing to pay premium prices.

Similarly, the GCF's REDD+ Shea Program in the Northern landscape will focus on increased income by targeting the supply chain (Global Shea Alliance Development officer). With access to storage infrastructure, the Shea Program will generate revenue 30–50% more than the normal market price, which will become apparent in the 1st and 2nd shea season (Global Shea Alliance Development officer). Community members have noticed the depletion of these ecosystem services over time, which motivates them to protect biodiverse, sustainable forestry (Several interviews). While these ecosystem services are undervalued in international discourse around REDD+, Ghana's adaptive approach incorporates them.

However, the co-benefits system may not be enough to deter deforestation in Ghana. In the short term, REDD+ could be successful only because it will increase their yields, but these co-benefits are not a long-term sustainable forestry management plan (Care International officer). Land use will likely change in the long term in Ghana. For example, cocoa yields are predicted to drop again due to climate change and timber industry will increase to take its place, making Ghana's REDD+ strategy less of an incentive (Care International officer). Additionally, the co-benefits strategy overlooks that local communities and Indigenous Peoples likely already understand the co-benefits and livelihoods opportunities that the forests offer. REDD+ only connects them to the private sector for export-oriented goods rather than their customary livelihood practices.

Additionally, the State can easily alter its current strategy of co-benefits. Competing state interests could use REDD+ to create environmental enclosures through territorialization for carbon conservation [60]. As Asiyanbi et al. note, rendering carbon visible renders others aspects of forests invisible, which could lead to aggressive government actions to pursue commodities (timber, carbon, cocoa) that provide the most financial gain to the state and politicians [32]. The co-benefits component of REDD+ may be advantageous right now for the private sector, government, and communities, but that may not always be the case. Without secure tenure and decision-making power for local communities, the State could easily shift its focus to more profitable land uses for them. Without a framing to ensure that local communities still have access to their land even if political and economic interests shift, co-benefits in REDD+ provide only temporary poverty alleviation dependent on international markets.

3.6. Tracing Discourse and Practice

Civic environmentalism was the root of official REDD+ discourse; however, the discourse renders the interventions purely technical and apolitical. In practice, the social and environmental safeguards allow the state to avoid responsibility for policies that encourage deforestation, ensure decision-making power, and acquire financial resources under the veil of social-ecological responsibility. The following two tables (Tables 3 and 4) summarize civic environmentalism and ecological modernization through debate narratives, official discourse, and the resulting practice that were outlined in the above sections.

As civic environmentalism moves from a debate narrative to REDD+ practice in Ghana, it becomes weaker. The debate narratives focus on power. It calls out the varying political power and responsibility in relation to the Global North–Global South divide as well as the local–global knowledge paradigm. The emphasis on equity and legitimacy, also, assumes that the status quo in forestry conservation is neither. It recognizes that some actors will need to "lose" some power in the event of a trade-off between economic growth and sustainable forest management.

Table 3. Civic environmentalism from discourse to practice.

Debate Narratives	Official Discourse	In Practice
Beyond markets: emphasis should be on equity and legitimacy rather than effectiveness and efficiency	Social safeguards: "full and effective" participation	"Full and effective" participation is not adequately funded nor is there adequate, informed, equal representation from all stakeholders
Beyond markets: REDD+ involves trade-offs between economic growth and sustainable forest management	–	–
Local, not global: local knowledge not adequately used during policy process	–	–
Biodiversity: problematizes valuing forests purely for carbon	Environmental safeguards: Sustainable forestry management Non-carbon benefits "Consistent actions with national forest programs"	Receives no "results-based" funding; incentivized through agricultural programs and private-sector partnership (co-benefits); inconsistent with other forestry programs
North–South divide: countries in South lose control of forestland, allowing North to defer responsibility for mitigation	–	–

Table 4. Ecological Modernization: from discourse to practice.

Ecological Modernization	Official Discourse	In Practice
Cost-efficiency: reducing deforestation is most cost efficient mitigation strategy for climate change	"solving the deforestation problem is a prerequisite for any effective response to climate change"	Operational cost of carbon accounting far outweighs other mitigation strategies
Win–win–win: REDD+ helps reduce emissions, improve forest conservation, and reduces poverty	"reduce emissions maximize cobenefits leading pathway towards sustainable development"	Does not improve forest conservation because lack of land and tree tenure nor poverty alleviation becausecarbon payments negligible and limited decision-making power
Market rationale: market is key to internalize environmental costs due to innovation in private sector	Public–private partnerships	Sustainable forestry management, not carbon, helpful to cacao business
Carbon accounting: forests are subject to management and control through technical advances	Monitoring, Verification, and Reporting (MRV)	Difficult and expensive to calculate
Technocratic rationale: societies can manage environmental cycles, so carbon becomes governmental/political domain	Results-based funding	Uncertainty in commodity Local deprioritization of carbon

The three storylines that are most firmly rooted in local/global and North/South power dynamics are not present in the official discourse. The only two storylines present in formal discourse are beyond markets and biodiversity. Civic environmentalism is most strongly present in the social and environmental safeguards such as "full and effective" participation and "consistent actions with national forest programs." However, these are also depoliticized. "Full and effective" participation is taken as a positive social safeguard, but it does not ensure decision-making power or governance by all stakeholders, particularly for communities. In practice, the REDD+ program does not provide adequate funding to ensure "full and effective" participation. As a result, the state maintains control of the REDD+ process. Additionally, environmental safeguards are depoliticized so co-benefits are only prioritized in practice when interests align between the state, private sector, and local communities.

Contrary to civic environmentalism discourse which loses its funding and prioritization in practice, ecological modernization becomes much stronger. The official discourse touts cost-efficiency, carbon accounting and market to reduce poverty through technocratic interventions. In practice, the operational cost through MRV to estimate carbon accounting for a successful carbon market outweighs all other costs, only to provide small community projects as benefits. It does not create a market incentive. Rather, it renders deforestation technical and limits funding for other sustainable forestry measures and representative participation. It channels available resources to technocratic experts rather than those conserving forests. Ecological modernization becomes the ultimate dominant discourse with the implementation of results-based funding.

3.7. State Expansion

Development projects form interventions in terms of geography and nature rather than powerlessness and social inequality, on both the national and international levels [39]. So, alternative solutions are not strictly managerial and technical, but social and political, a reworking of the actors to redistribute power and wealth [39]. Similarly, Scheba and Schebab conclude that conservation practice needs to be politicized in a way "that does not lose sight of structural inequalities alongside micro-political struggles" [19]. By depoliticizing critical discourse, it becomes easier in practice for the State to maintain and even acquire power. Establishing REDD+ in Ghana as a "fad" that implements technical, apolitical interventions and disjointed governance maintains a gap in which deforestation is never addressed. Ferguson's analysis of a development program and problematic in

Lesotho recognized this recurring gap as well, thereby arguing that the instrumental aim or effect of the program, then, was to expand disperse state power [37]. Similarly, Lund et al. note that the gap between discourse and practice "allow certain actors within development and conservation industry to tap into financial resources" [62].

In the case of REDD+ in Ghana, funding for the programs flow through state channels and largely stay at the high level where they are distributed to key powerful actors—traditional authorities, private sector, and certain civil society members. To access these funds, the State needs to adhere to international frameworks in a way that will ensure no loss of power over land, natural resources, or contradictory economic opportunities. Ghanaian national government uses REDD+ as "free money." At the same time that the State expands due to funding for REDD+, mainly through the REDD+ readiness program, it is able to maintain its revenue, power, and resources that fuel deforestation, such as lack of land and tree tenure for communities, high export timber quota, and contradictory forestry programs. The State does not have to sacrifice any interest in other revenue-creating industries in Ghana to implement the REDD+ program.

Fortress Conservation

As a result of State expansion, the key instrumental effect of concern for REDD+ Ghana is the cracking down on illegalities with methods of fortress conservation. To respond to illegal logging, Ghana policies have favored fortress conservation and enforcement of laws rather than more access to lumber or timber benefits, which would require reform in the sector and a loss in state power. As noted by Sefwi Wiawso officials, measures have been taken to disincentivize illegal logging by the State—including the Voluntary Partnership Agreement (VPA) that established a wood tracking system to make illegal timber transport more difficult and increase penalties for selling illegal logs to ten times the market price, which also would cancel your permit if left unpaid. These reforms have been put in place and have been used by the Forestry Commission (Sefwi Wiawso officials), but the stampage fees that are required for communities to benefit from timber are rarely enforced. At the two meetings between the Forestry Commission district officials, REDD+ national officials, and community members that I attended, the stampage fees were the greatest point of concern and frustration for community members. As mentioned earlier, REDD+ does not address the underlying, systemic, and political causes of illegal logging in Ghana, the State can use its newly acquired resources to further criminalize and prosecute illegal loggers and miners. The REDD+ program also allows the state to expand its governance and enforcement at the district levels, similar to the VPA program.

While enforcement of legal timber market is necessary, the tree tenure and legal, domestic market conditions are unfavorable for communities to have access or benefit from timber. As policies against illegal logging are adopted to deter locals from selling illegal timber and policies for timber companies to pay communities are not enforced, with the gap in domestic supply chain, communities are not likely to buy into Forestry Commission programs. Furthermore, without policies to address the underlying systemic problems with tree tenure and the market, deforestation will continue, and the State will adopt more and more aggressive stance against illegal logging.

4. Conclusions

In the case of REDD+ in Ghana, civic environmentalism discourse becomes weaker as it moves from narratives debates to official discourse and then practice. Social and environmental safeguards are rendered technical and lack funding, so in practice civic environmentalism discourse only serves to justify REDD+. At the same time, ecological modernism gains strength by defining results-based success and siphoning funding for commodification. This case shows the limits of critical discourse in international climate programs that largely derive from neoliberal environmentalism.

Official discourse ignored political and power dimensions and so rendered the intervention a purely technocratic process shaped by bureaucratic "governmentality", largely due to the state-centric nature of the REDD+ program. This "depoliticization" is reflected in official REDD+ discourse, which effectively enables the Ghanaian state to divert responsibility for deforestation, maintain decision-making power, and enact contradictory national policies in sustainable forestry. At the same time, financial resources are also directed towards the State for the infeasible actualization of the forest carbon commodity, leaving critical concerns for local communities unaddressed in practice.

As a result, the Ghanaian government reaps financial gains without addressing the primary causes of deforestation under the veil of social inclusion. Civic environmentalism, largely social and environmental safeguards in this case, then, legitimize expanding State power in practice. Without ownership and management rights of land and trees and market reform, local communities then are further criminalized by the State as it expands. This power is cyclical in nature, which expands with each new, depoliticized fad in forest governance. At the same time, because REDD+ is highly localized in specific geographical areas and aspects of forestry, the Ghanaian government can still pursue carbon emitting and deforestation activities for more financial gain.

Interviewees discussed several ways that this cycle of international frameworks implemented at the state level could improve. First, these programs cannot ignore the politics of natural resource governments and only tout technical solutions. Future programs should prioritize consistent legal reforms for community ownership, management, and use rights a prerequisite. Additionally, market-based solutions should not be considered as they are expensive, limit participation, and only value one aspect of sustainable forestry management. Funding, also, should not go exclusively to State governments, but rather directly to local communities and CSOs on the ground in a lump sum. Finally, instead of new international frameworks every few years that are seen as momentary silver bullets, international frameworks should ensure that sustainable forestry national programs and policies are consistent, fair, and enforced.

Ferguson contends that state expansion, as an instrumental effect, is not intentional nor centralized [37]. The state does not "rationalize and centralize" power relations, rather it "grabs and loops around existing power relations to cinch them together like a knot" ([37], p. 274). Bureaucratic state power, then, is a mode of power that relies on state institutions but exceeds them. REDD+ in Ghana operates in a similar way, as depoliticizing discourse and governmentality squash political challenges to the program and create disjointed policies at the national level which establishes REDD+ as an unsustainable conservation fad. The fad, then, allows for the expansion of the state to the detriment of local communities and the forest.

Funding: This research received no external funding.

Institutional Review Board Statement: The study was conducted according to the guidelines of the Declaration of Helsinki, and approved by the Institutional Review Board (or Ethics Committee) of The Graduate Institute of International and Development Studies (January 2019).

Informed Consent Statement: Informed consent was obtained from all subjects involved in the study.

Acknowledgments: I would like to thank my interviewees in Ghana, who referred me with good wishes to other informants and shared openly with me, as well as my advisor Shaila Galvin, who was readily available for support and feedback.

Conflicts of Interest: The author declares no conflict of interest.

References

1. Newell, P.; Paterson, M. *Climate Capitalism: Global Warming and the Transformation of the Global Economy*; Cambridge University Press: Cambridge, UK, 2010.
2. Pirard, R. Market-based instruments for biodiversity and ecosystem services: A lexicon. *Environ. Sci. Policy.* **2012**, *19*, 59–68. [CrossRef]
3. Zhang, B. Market-based solutions: An appropriate approach to resolve environmental problems. *Chin. J. Popul. Resour. Environ.* **2013**, *11*, 87–91. [CrossRef]

4. Wunder, S.; Brouwer, R.; Engel, S.; Ezzine-de-Blas, D.; Muradian, R.; Pascual, U.; Pinto, R. From principles to practice in paying for nature's services. *Nat. Sustain.* **2018**, *1*, 145–150. [CrossRef]
5. Arsel, M.; Büscher, B. Nature™ Inc.: Changes and Continuities in Neoliberal Conservation and Market-based Environmental Policy. *Dev. Chang.* **2012**, *43*, 53–78. [CrossRef]
6. McElwee, P. Payments for environmental services as neoliberal market-based forest conservation in Vietnam: Panacea or problem? *Geoforum* **2012**, *43*, 412–426. [CrossRef]
7. Adams, W. Sleeping with the enemy? Biodiversity conservation, corporations and the green economy. *J. Political Ecol.* **2017**, *24*, 243–257. [CrossRef]
8. Bakker, K. Neoliberalizing Nature? Market Environmentalism in Water Supply in England and Wales. *Ann. Assoc. Am. Geogr.* **2005**, *95*, 542–565. [CrossRef]
9. Bakker, K. The "Commons" Versus the "Commodity": Alter-globalization, Anti-privatization and the Human Right to Water in the Global South. *Antipode* **2007**, *39*, 430–455. [CrossRef]
10. OECD. *Green Investment Banks: Scaling up Private Investment in Low-carbon, Climate-resilient Infrastructure, Green Finance and Investment*; OECD Publishing: Paris, France, 2016. [CrossRef]
11. OECD. *Mobilising Bond Markets for a Low-Carbon Transition, Green Finance and Investment*; OECD Publishing: Paris, France, 2017. [CrossRef]
12. OECD. *Making Blended Finance Work for the Sustainable Development Goals*; OECD Publishing: Paris, France, 2018. [CrossRef]
13. OECD; The World Bank; UN Environment. *Financing Climate Futures: Rethinking Infrastructure*; OECD Publishing: Paris, France, 2018. [CrossRef]
14. UNFCCC. *Key Decisions Relevant for Reducing Emissions from Deforestation and Forest Degradation in Developing Countries (REDD+): Decision Booklet REDD+*; 2016; Available online: https://unfccc.int/files/land_use_and_climate_change/redd/application/pdf/compilation_redd_decision_booklet_v1.1.pdf (accessed on 1 October 2021).
15. McAfee, K. The Contradictory Logic of Global Ecosystem Services Markets. *Dev. Chang.* **2012**, *43*, 105–131. [CrossRef]
16. McDermott, C.; Levin, K.; Cashore, B. Building the Forest-Climate Bandwagon: REDD+ and the Logic of Problem Amelioration. *Glob. Environ. Politics* **2011**, *11*, 85–103. [CrossRef]
17. Streck, C. Who Owns REDD+? Carbon Markets, Carbon Rights and Entitlements to REDD+ Finance. *Forests* **2020**, *11*, 959. [CrossRef]
18. Feindt, P.H.; Oels, A. Does discourse matter? Discourse analysis in environmental policy making. *J. Environ. Policy Plan.* **2005**, *7*, 161–173. [CrossRef]
19. Scheba, A.; Scheba, S. REDD+ as 'inclusive' neoliberal conservation: The case of Lindi, Tanzania. *J. East. Afr. Stud.* **2017**, *11*, 526–548. [CrossRef]
20. Allan, J.I.; Dauvergne, P. The Global South in Environmental Negotiations: The politics of coalitions in redd +. *Third World Q.* **2013**, *34*, 1307–1322. [CrossRef]
21. Rowe, E.W. Locating international REDD+ power relations: Debating forests and trees in international climate negotiations. *Geoforum* **2015**, *66*, 64–74. [CrossRef]
22. Di Gregorio, M.; Gallemore, C.T.; Brockhaus, M.; Fatorelli, L.; Muharrom, E. How institutions and beliefs affect environmental discourse: Evidence from an eight-country survey on REDD+. *Glob. Environ. Chang.* **2017**, *45*, 133–150. [CrossRef]
23. Nielsen, T.D. The role of discourses in governing forests to combat climate change. *Int. Environ. Agreem. Polit Law Econ.* **2013**, *14*, 265–280. [CrossRef]
24. Stephan, B. Bringing discourse to the market: The commodification of avoided deforestation. *Environ. Politics* **2012**, *21*, 621–639. [CrossRef]
25. Christen, D.R.; Espinosa, M.G.; Reumann, A.; Puri, J. Results based payments for REDD+ under the green climate fund: Lessons learned on social, environmental and governance safeguards. *Forests* **2020**, *11*, 1350. [CrossRef]
26. Bayrak, M.M.; Marafa, L.M. Ten Years of REDD+: A Critical Review of the Impact of REDD+ on Forest-Dependent Communities. *Sustainability* **2016**, *8*, 620. [CrossRef]
27. Hajjar, R.; Engbring, G.; Kornhauser, K. The impacts of REDD+ on the social-ecological resilience of community forests. *Environ. Res. Lett.* **2021**, *16*, 024001. [CrossRef]
28. Satyal, P. Civil society participation in REDD+ and FLEGT processes: Case study analysis from Cameroon, Ghana, Liberia and the Republic of Congo. *For. Policy Econ.* **2018**, *97*, 83–96. [CrossRef]
29. Ameyaw, L.K.; Ettl, G.J.; Leissle, K.; Anim-Kwapong, G.J. Cocoa and Climate Change: Insights from Smallholder Cocoa Producers in Ghana Regarding Challenges in Implementing Climate Change Mitigation Strategies. *Forests* **2018**, *9*, 742. [CrossRef]
30. Isyaku, U.; Arhin, A.A.; Asiyanbi, A.P. Framing justice in REDD+ governance: Centring transparency, equity and legitimacy in readiness implementation in West Africa. *Environ. Conserv.* **2017**, *44*, 212–220. [CrossRef]
31. Saeed, A.-R.; McDermott, C.; Boyd, E. Examining equity in Ghana's national REDD+ process. *For. Policy Econ.* **2018**, *90*, 48–58. [CrossRef]
32. Asiyanbi, A.; Arhin, A.A.; Isyaku, U. REDD+ in West Africa: Politics of Design and Implementation in Ghana and Nigeria. *Forests* **2017**, *8*, 78. [CrossRef]
33. Jaeger, H.-M. "Global Civil Society" and the Political Depoliticization of Global Governance. *Int. Politics Sociol.* **2007**, *1*, 257–277. [CrossRef]

34. Blaikie, P.; Brookfield, H. *Land and Degradation and Society*; Methuen & Co Inc.: London, UK, 1987; p. 284.
35. Forsyth, T. *Critical Political Ecology: The Politics of Environmental Science*; Routledge: London, UK; New York, NY, USA, 2013.
36. Mouffe, C.; Laclau, E. *Hegemony and Socialist Strategy: Towards a Radical Democratic Politics*; Verso: New York, NY, USA, 1985.
37. Ferguson, J. *The Anti-Politics Machine: "Development," Depoliticization, and Bureaucratic Power in Lesotho*; Cambridge University Press: Cambridge, UK, 1990; in print.
38. Foucault, M. Truth and Power, an interview. *Crit. Anthropol.* **1979**, *4*, 131–137. [CrossRef]
39. Nasi, R.; Putz, F.E.; Pacheco, P.; Wunder, S.; Anta, S. Sustainable Forest Management and Carbon in Tropical Latin America: The Case for REDD+. *Forests* **2011**, *2*, 200–217. [CrossRef]
40. Ciplet, D.; Roberts, J.T. Climate change and the transition to neoliberal environmental governance. *Glob. Environ. Chang.* **2017**, *46*, 148–156. [CrossRef]
41. McCormick, S. The Brazilian Anti-Dam Movement. *Organ. Environ.* **2006**, *19*, 321–346. [CrossRef]
42. Dryzek, J.S. *The Politics of the Earth: Environmental Discourses*, 3rd ed.; Oxford University Press: Oxford, UK, 2013.
43. Bäckstrand, K.; Lövbrand, E. Planting Trees to Mitigate Climate Change: Contested Discourses of Ecological Modernization, Green Governmentality and Civic Environmentalism. *Glob. Environ. Politics* **2006**, *6*, 50–75. [CrossRef]
44. Mol, A.; Spaargaren, G.; Sonnenfeld, D. *Ecological Modernization Theory: Where Do We Stand?* Campus Verlag: Frankfurt, Germany, 2014; ISBN 9783593500614.
45. *Coalition for Rainforest Alliance in Papua New Guinea*, Conference May 2005. Available online: https://wrm.org.uy/articles-from-the-wrm-bulletin/section1/papua-new-guinea-rainforest-alliance-undermining-forest-conservation-efforts/ (accessed on 1 October 2021).
46. Ghana *REDD+ Strategy*, Republic of Ghana. 2016. Available online: https://www.oldwebsite.fcghana.org/userfiles/files/REDD%2B/Ghana%20REDD%2B%20Strategy.pdf (accessed on 1 October 2021).
47. Hajer, M.A.; Fischer, F.; Forester, J. Discourse Coalitions and the Institutionalization of Practice. In *The Argumentative Turn in Policy Analysis and Planning*; Duke University Press: Durham, CA, USA, 1993; pp. 43–76. [CrossRef]
48. Participatory Self-Assessment and Synthesis of Ghana's REDD+ Readiness Process (R-Package), July 2016. Accra, Ghana. Available online: https://www.forestcarbonpartnership.org/system/files/documents/Ghana%20RPackage%20Final%20for%20posting%20clean%20doc.pdf (accessed on 12 February 2019).
49. Agbosu, L. *Customary and Statutory Land Tenure, and Land Policy in Ghana*; Institute of Statistical, Social & Economic Research, University of Ghana: Legon, Ghana, 2007.
50. Concessions Act, Ghana Code. 1962. Available online: http://extwprlegs1.fao.org/docs/pdf/gha168936.pdf (accessed on 1 October 2021).
51. PAB Development Consultants. A Framework for Tree Tenure and Benefit Sharing Scheme in Ghana. 2016. Available online: https://www.oldwebsite.fcghana.org/userfiles/files/NREG/Draft%20Final%20Report_30_05_16%20.pdf (accessed on 19 February 2019).
52. Constitution, Ghana Code. 1992. Available online: https://www.constituteproject.org/constitution/Ghana_1996.pdf?lang=en (accessed on 12 June 2019).
53. Ghana Forest and Wildlife Policy, Ghana Code. 2012. Available online: http://extwprlegs1.fao.org/docs/pdf/gha144198.pdf (accessed on 1 October 2021).
54. UN-REDD Programme | How We Work. (n.d.). Available online: https://www.un-redd.org/how-we-work (accessed on 13 June 2019).
55. Tortajada, C. Nongovernmental Organizations and Influence on Global Public Policy. *Asia Pac. Policy Stud.* **2016**, *3*, 266–274. [CrossRef]
56. West, P. *Conservation Is Our Government Now: The Politics of Ecology in Papua New Guinea*; Escobar, A., Rocheleau, D., Eds.; Duke University Press: Durham, NC, USA, 2006.
57. Ghana FCPF ER Program Advanced Draft BSP. 2018. Available online: https://www.forestcarbonpartnership.org/system/files/documents/Ghana%20FCPF%20ER%20Program%20Advanced%20Draft%20BSP.pdf (accessed on 23 June 2019).
58. Seyller, C.; Desbureaux, S.; Ongolo, S.; Karsenty, A.; Simonet, G.; Faure, J.; Brimont, L. The 'virtual economy' of REDD+ projects: Does private certification of REDD+ projects ensure their environmental integrity? *Int. For. Rev.* **2016**, *18*, 231–246. [CrossRef]
59. Ghana ER Program GCFRP Benefit Sharing Plan. Available online: https://vdocument.in/advanced-draft-enefit-sharing-plan-ghana-ocoa-forest-redd-ghana-er-program-gcfrp.html (accessed on 1 October 2021).
60. Corbera, E.; Brown, K.; Adger, W.N. The Equity and Legitimacy of Markets for Ecosystem Services. *Dev. Chang.* **2007**, *38*, 587–613. [CrossRef]
61. Ghana Cocoa Forest REDD+ Programme. Available online: https://www.forestcarbonpartnership.org/system/files/documents/GCFRP_Carbon%20Fund_Final%20Draft_April%2022%202017-formatted.pdf (accessed on 14 May 2019).
62. Lund, J.F.; Sungusia, E.; Mabele, M.B.; Scheba, A. Promising Change, Delivering Continuity: REDD+ as Conservation Fad. *World Dev.* **2017**, *89*, 124–139. [CrossRef]

sustainability MDPI

Article

Governance of Urban Green Infrastructure in Informal Settlements of Windhoek, Namibia

Amayaa Wijesinghe [1,2,*] and Jessica P. R. Thorn [1,3]

1 African Climate and Development Initiative, University of Cape Town, Cape Town 7700, South Africa; jessica.thorn@york.ac.uk or jessicaprthorn@gmail.com
2 UN Environment Programme World Conservation Monitoring Centre (UNEP-WCMC), Cambridge CB3 0DL, UK
3 York Institute of Tropical Ecosystems, Department of Environment and Geography, University of York, Heslington YO10 5DD, UK
* Correspondence: amayaawijesinghe@gmail.com

Abstract: Facing increased rural-urban migration, population growth, climate change impacts, and cascading natural, security, and health hazards, many municipalities in sub-Saharan Africa are beginning to consider the benefits of urban green infrastructure for improving the resilience and wellbeing of residents living in informal settlements. However, present governance systems are often ill-equipped to deliver the scale of planning needed. Integration of urban green infrastructure into local government mandates, spatial planning and targeted action plans remains limited, further inhibited by scarce empirical research on the topic in Africa. Taking Windhoek, Namibia, and specifically Moses ||Garoëb, Samora Machel, and Tobias Hainyeko constituencies as a case study, we fitted key informant interview (n = 23), focus group (n = 20), and participant observation data into existing governance theory to investigate (a) benefits and trade-offs of present urban green infrastructure in Windhoek's informal settlements; (b) urban green infrastructure governance in terms of institutional frameworks, actors and coalitions, resources, and processes; and (c) the key desirable pathways for future urban green infrastructure governance in informal settlements. To this end, we used five green infrastructure initiatives to dissect governance intricacies and found diverse opportunities for innovative governance mechanisms. The urgent need for climate resilience in Namibia offers a policy and practice window to adopt context-specific approaches for multifunctional urban green infrastructure. However, for these initiatives to succeed, collaborative governance platforms and clearly delineated mandates are necessary, with explicit integration of urban green infrastructure into strategies for in-situ informal settlements upgrading and green job growth.

Keywords: green space; inclusive city; informality; local stewardship; participation; peri-urban settlements; policy instruments; self-governance; sub-Saharan Africa; right to the city

Citation: Wijesinghe, A.; Thorn, J.P.R. Governance of Urban Green Infrastructure in Informal Settlements of Windhoek, Namibia. *Sustainability* **2021**, *13*, 8937. https://doi.org/10.3390/su13168937

Academic Editor: Carlos Oliveira Cruz

Received: 31 May 2021
Accepted: 26 July 2021
Published: 10 August 2021

Publisher's Note: MDPI stays neutral with regard to jurisdictional claims in published maps and institutional affiliations.

Copyright: © 2021 by the authors. Licensee MDPI, Basel, Switzerland. This article is an open access article distributed under the terms and conditions of the Creative Commons Attribution (CC BY) license (https://creativecommons.org/licenses/by/4.0/).

1. Introduction

Sub-Saharan Africa (SSA) is among the world's fastest urbanising regions, and the global proportion of African urban dwellers is projected to rise from 11.3% in 2010 to 20.2% by 2050 [1]. Demographic changes driven by rural-urban migration and natural population growth interact with challenges such as widespread poverty and unplanned informal settlement expansion [2]. These challenges are exacerbated by governance systems ill-equipped to deliver the scale of planning needed to deal with burgeoning informal settlements [3], which feature high densities of temporary structures built with low-cost materials such as corrugated iron sheets. Occupants here usually lack formal employment, secure land rights, food security, good sanitation, and other basic services and infrastructure [4]. At a local government level, the lack of expertise, funds, and data on urbanisation continues to hinder efforts to improve living conditions and implement sustainable natural resource use [5].

Concurrently, cities are social-ecological systems, where development and rapid land use change cause unique disturbance regimes [6]. Urban natural areas provide humans and other species with important ecosystem services [7,8]. With the rise of movements to 'bring nature back into cities' [9], new possibilities for planning and managing urban natural areas are emerging. Globally, the increased interest in the nexus between global environmental change, cities, and natural ecosystems [10,11] has brought concepts such as multifunctional urban green infrastructure (UGI) to the fore [12]. However, there remains a research gap in understanding the applicability, utility, and practical implementation of UGI in an SSA context, as its multiple benefits and governance mechanisms are often framed from a predominantly 'Western' urban perspective [13,14].

Additionally, projections of climate change indicate precipitation decreases of 10–20%, more days at maximum temperature, longer dry spells, and more frequent extreme events, especially over Namibia, Botswana, northern Zimbabwe, and southern Zambia in southern Africa [15]. Increasing exposure to such climatic and non-climatic changes necessitates urgent interventions to improve the resilience of vulnerable people and ecosystems in peri-urban areas [16,17]. This presents a unique set of challenges [16] but also many opportunities for UGI integration into adaptation and mitigation plans [18,19]. Research on urban climate resilience in peri-urban SSA remains nascent but is critical to provide insights for context-specific pathways forward [5,20–22].

1.1. Governance of Urban Green Infrastructure in SSA's Informal Settlements

In this paper, we define UGI, following Thorn et al. [5] and Ahern et al. [23] (p. 255), as 'permeable, multifunctional and interconnected spaces that support environmental, recreational and wellbeing functions within a city, through provisioning, regulating, supporting and cultural ecosystem services.' Considered cost-effective, especially in the long term [24], and multifunctional in the co-benefits these spaces can provide [25,26], UGI also works effectively with engineered infrastructure as 'hybrid infrastructures' [27]. UGI is often referred to as 'green and blue spaces' or 'ecological infrastructure' [28] and falls under the broader umbrella of nature-based solutions [29]. Many ecosystem services derived from UGI are important for resilience of informal settlement residents, such as microclimate regulation through cooling, filtration of grey water, production of food (urban agriculture and home gardens), air quality improvements, energy supply (fuelwood), support for livelihoods, improved mobility, and better psycho-social wellbeing [18,19,28].

However, unlocking pathways for sustainable and inclusive development [30,31] of UGI relies on good governance. Here, we define governance as the 'processes, interactions, organisations, and decisions which enable stakeholders to control and coordinate their interconnected needs and desires, while interacting with the environment at different scales' ([32], as seen in [33] (p. 465)). Governance encompasses both state and non-state actors, formal and informal institutions, rules, mechanisms, processes, and scales [33]. For UGI, 'governance' encompasses multi-level interactions between local communities, the private sector, non-governmental actors, and local authorities working together in a polycentric system, and stands in contrast to 'government', where actors such as local authorities have the primary, often top-down responsibility for UGI [34].

In SSA, community-led or autonomous adaptations using UGI, implemented as a stress response to environmental changes, are already apparent (e.g., maintenance of home gardens to combat food insecurity [17]), with enormous potential for UGI to alleviate resource and financial constraints faced by municipal authorities in SSA [35,36]. Nevertheless, a pro-grey mindset and conceptualisation of environmental management as 'luxury' not 'necessity', especially for informal settlements [5,19], coupled with town planning approaches that are often colonial relics with little provision for equitable land and tenure policies [28], perpetuate institutional failures where UGI governance for low-income residents is deprioritised or overlooked. However, there is a growing body of evidence that concerted integration of UGI into settlement planning would simultaneously unlock pathways for inclusivity and social justice [14]. Most studies on UGI governance in SSA focus on South Africa, creating a distinct geographical bias [18]. We contribute to closing this research gap by examining dryland Windhoek, Namibia, as a case study, as it represents wider changes and governance challenges faced by cities across Africa and the Global South.

1.2. Windhoek, Namibia: Growth of Informality and Impact on Natural Urban Areas

Namibia has undergone an accelerated developmental transformation from a largely rural-based society to 47.9% of the population now living in urban centres [37]. The capital Windhoek has witnessed rapid urbanisation in the last 30 years, but town planning regulations, affordable housing provisioning for low-income groups, and flexible land tenure options have not kept abreast with these developments [38]. This has led to approximately 42.3% of households in Windhoek being temporary shack homes in informal settlements [37]. In-situ upgrading led by the City of Windhoek (CoW) and the Khomas Regional Council has not kept pace with this informal settlement growth.

At present, peri-urban expansion in Windhoek encroaches into natural areas such as riverbeds, hilly slopes, and other marginal lands [39]. Approximately 75% of informal settlement residents depend on riverbeds and peripheral green spaces as areas for open defecation, and the health impacts are evident in recent Hepatitis E outbreaks [5,40,41]. The pollution of water bodies is caused by unregulated dumping of solid waste and heavy metals, agricultural runoff, nutrient loading, and sedimentation [42]. In May 2019, President Hage Geingob declared the drought affecting Namibia, including Windhoek, as a natural disaster [43]. Meanwhile, flooding occurs in the rainy season, where fast currents cause drowning and loss of property in informal settlements. Predicted climate change impacts of drought, heat stress, and flooding may further interact with existing urban poverty, food and water insecurity, lack of sanitation, and disaster risk to perpetuate cycles of vulnerability and inequality [44].

Recognising this need for dryland-specific climate interventions, Windhoek was the first city in Namibia to formulate an Integrated Climate Change Strategy and Action Plan (ICCSAP) [21]. Developed to fill the gap in municipal policies addressing cumulative impacts of climate change on the city and its inhabitants, the ICCSAP is currently in draft form. Encouragingly, it contains explicit sections on 'human settlements' and 'biodiversity and ecosystem goods and services' [21,45]. Since inadequate planning and governance systems are a foremost barrier in implementing UGI in urban SSA [46], such draft plans as the ICCSAP, together with ongoing upgrading efforts, provide the policy and practice windows to integrate UGI more explicitly.

1.3. Conceptual Framework for Analysis of UGI Governance

Lawrence et al. [33] presents a useful framework to analyse governance of UGI in Windhoek's informal settlements. Developed to assess urban forestry governance, the framework draws from the policy arrangements approach [47]. This approach postulates how state and non-state actors work together to fulfil governance functions. The framework emphasises how change in one dimension affects change in another dimension ([34]; Figure 1). Applying this framework to Windhoek, we examine four adapted dimensions. Institutional frameworks relate to governmental policies, laws, and regulations that affect UGI, land ownership, and access rights. Actors and coalitions relate to active and supporting stakeholders, as well as formal partnerships established to deliver the UGI. Resources encompass funding, transfer and access of technical knowledge, and policy tools that support implementation, such as assigned staff or monetary incentives. Finally, processes are the ways in which actors are consulted through participatory mechanisms and include methods for monitoring, evaluation, and learning. Governance occurs along a continuum, ranging from local authorities (state) taking a leading role to collaborative governance with non-state actors and self-governance by non-state actors [33,34].

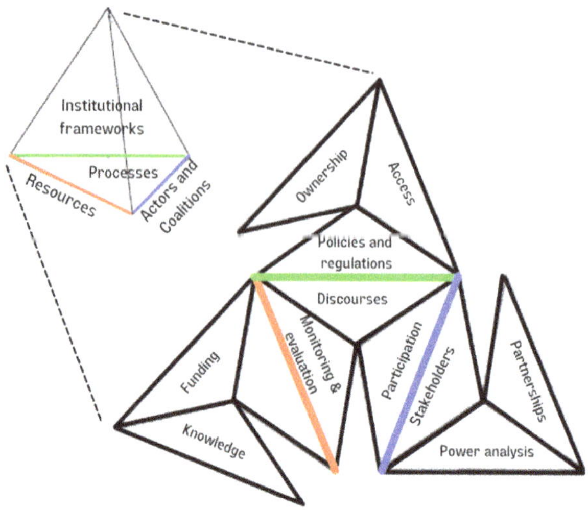

Figure 1. Dimensions of governance arrangements for Windhoek, adapted from Lawrence et al. [33]. If one dimension of the tetrahedron fails, then the integrity of the entire structure is compromised. The top left figure represents an aggregation of the categories shown in the bottom right figure. The colours represent the green, blue, and brown spaces and initiatives that make up urban green infrastructure (UGI), with brown spaces representing drylands.

The overarching aim of this paper is to determine present and potential governance mechanisms for UGI in Windhoek, while drawing out applicable insights for similar SSA informal settlements and dryland environments. We adapt the case study framework described (Figure 1) to investigate (a) benefits and trade-offs of present UGI in Windhoek's informal settlements; (b) UGI governance in terms of institutional frameworks, actors and coalitions, resources, and processes; and (c) the key desirable pathways for future UGI governance in informal settlements.

2. Materials and Methods

2.1. Study Site

Namibia is one of SSA's most arid countries. Approximately 70% of the land area, including its capital Windhoek, is classified as arid to semi-arid [48]. Table 1 provides a demographic and environmental profile of Windhoek. Our study focuses on the peri-urban areas near the northern and north-western boundaries of the city, adjacent to Katutura, the former black township during Namibia's apartheid era (Figure 2). Although segregation based on ethnicity ended with Namibia's independence in 1990, an economic gradient still exists from affluent neighbourhoods in the south to the north west of the city.

2.2. Data Collection

Our empirical research employed key informant interviews (n = 22), focus groups (n = 20), and rapid participant observation over two months of fieldwork in Windhoek, Namibia, between June and July 2019, which constitute the dry winter months in Central Namibia. Discussions were conducted in English or Oshiwambo with a translator.

Table 1. Basic profile of Windhoek (Sources: [21,37,49,50]).

Variable	Value
Region	Khomas
Area	5133 km^2
Elevation	1655 m
Population (2016)	415,780
Population percentage living in informal settlements (2016)	42.3%
Average rainfall	360 mm/a
Average evaporation	3300 mm/a
Threshold for drought year	<150 mm/a
Average maximum temperatures	28–32 °C
Average minimum temperatures	2–8 °C
Predicted climate change scenarios for Windhoek for 2040	2 °C warmer with twice as many hot days and 1/3 less rainfall 1.5–2 °C warmer with more rain coming later in the rainy season 1–1.5 °C warmer, with average rainfall remaining constant, but becoming more intense

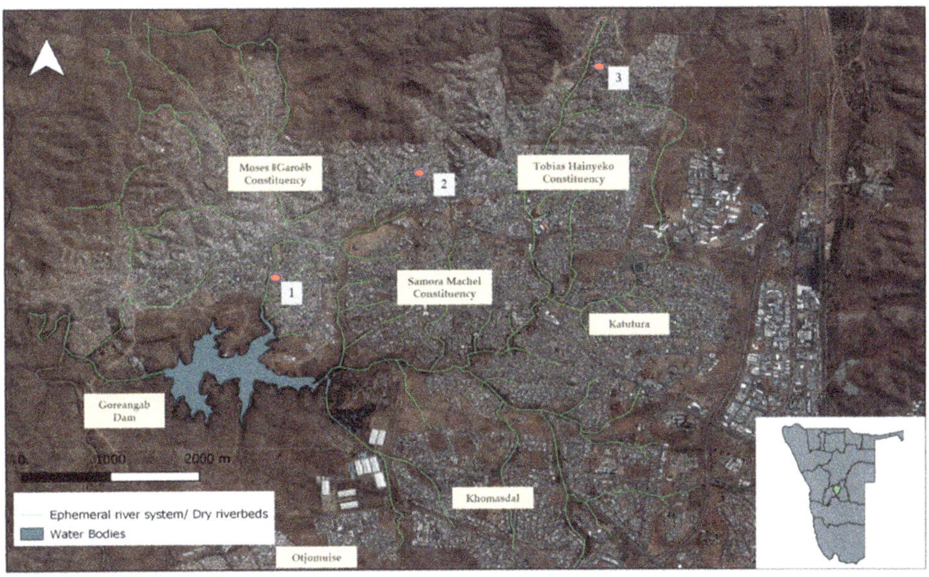

Figure 2. Map of study site in Windhoek, Namibia. The area of interest for this study lies in the informal settlements, which are located on the outskirts of the Moses ǁGaroëb, Samora Machel, and Tobias Hainyeko Constituencies. (**1**) Location of Greenwell Matongo C informal settlement, bordering Goreangab Dam, where one focus group was held. (**2**) Hakahana community centre near Havana informal settlement, where another focus group was held. (**3**) Okuryangava with surrounding informal settlements, one of the areas where transect walks were carried out. The image shows unplanned sprawl extending towards the north and northwest, as well as hilly terrain surrounding the city. (Base map provided by Development Workshop—Namibia, 2019).

2.2.1. Key Informant Interviews

To access key informants for semi-structured interviews, we first conducted actor network analysis and an online search of relevant reports, articles, and staff pages of organisations such as the CoW. Our sampling design employed snowball sampling, i.e., tapping into the network of known informants, and obtaining contact details and introductions through them [51]. Key informants were selected to be a representative cross section of actors with expertise or experiences related to UGI in informal settlements (Table 2). Interviews, which lasted approximately 45 mins, covered the following areas: diverse actors' understanding of natural ecosystems and their benefits in Windhoek, challenges of UGI implementation in informal settlements, identification of relevant actors and their roles in UGI governance, and optimal governance mechanisms for potential success (schedule in Appendix A).

Table 2. Sectors represented by key informants and focus group participants. The institution type assigned to participants is based on their role at the time of participation in the study, but many also drew insights from past multi-sectoral experiences.

Institution Type	Relevance to Study	n	No. of Females
Residents of informal settlements	Appointed community leaders of informal settlements, and residents who are active in the formal and informal labour force (e.g., food sellers, fodder collectors), as well as members of neighbourhood youth clubs.	20	14
Constituency	Elected head councillors of constituencies who have sections of informal settlements within their jurisdictions.	2	0
Local authority	Officials in City of Windhoek Divisions of Environmental Management and Health Services, Human Settlements, and Parks and Recreation of the City of Windhoek, with direct or indirect responsibility for governance and maintenance of UGI in formal and informal areas of the city.	6	1
NGO/Third Sector	NGOs and Civil Society Organisations (CSOs) who work directly with residents of informal settlements, broader coordinating bodies, and those active in rural areas, namely Shack Dwellers Federation Namibia—Namibia Housing Action Group (SDFN-NHAG), Development Workshop- Namibia, Namibian Chamber of Environment, and Desert Research Foundation of Namibia.	9	4
Private sector	Architects, planning consultants, and businesses that work closely with residents in informal settlements.	4	1
Academia	Lecturers and researchers at the Namibia University of Science and Technology (NUST), in the Department of Architecture and Spatial Planning.	2	0
		43	46.5%

2.2.2. Focus Groups

Two focus group discussions of 2–5 h were held in Greenwell Matongo C (n = 6 participants) and Hakahana (n = 14 participants) with residents of informal settlements. Introductions to the communities were facilitated by the Shack Dwellers Federation of Namibia and the Namibia Housing Action Group (SDFN-NHAG), an NGO partnership that works closely with residents to facilitate access to tenure and services through a savings group model [52].

2.2.3. Participant Observation

To validate and contextualise findings, qualitative data from focus groups and interviews were combined with rapid participant observation. With the formal permission of constituency councillors, and with a local translator and guide from the community, we conducted transect walks by visiting three informal settlements areas in Greenwell Matongo C and Hakahana in Moses ||Garoëb constituency and Okuryangava in Tobias Hainyeko constituency. Additionally, using an ethnographic approach, the first author spent two months living in Windhoek. While in Windhoek, she embedded herself in the day-to-day activities of residents of a middle-income neighbourhood of Khomasdal in Windhoek. She visited local markets and roadside selling points, attended SDFN-NHAG meetings in informal settlements, and had informal conversations with conservation and environmental practitioners, town planners, students, and residents across the city. Both authors continued engagement through 2019–2021 using in-person and remote means.

2.3. Analysis

Anonymised interview and focus group transcripts were deductively coded and clustered using NVivo (12.0.0.0) to generate key themes. Through an iterative process, we then assigned these to nodes corresponding to the dimensions of our adapted governance case study framework [53]. Using these themes and insights, and supported by the broader literature on UGI, we present a multi-dimensional snapshot of UGI governance in this complex setting.

3. Results

3.1. UGI in Windhoek's Informal Settlements

3.1.1. Benefits

UGI that holds value for informal settlement residents and other actors is predominantly riparian dryland-adapted vegetation (e.g., grasses, trees, shrubs) adjacent to the ephemeral Arebbusch and Gammams river networks that runs through Windhoek and drains into the Goreangab Dam. Other key forms of UGI are found in meeting areas, sports grounds, hilly slopes, urban farms, small garden plots, and green fences interspersed among dwellings (see also [5]).

Respondents derive several provisioning, regulating, supporting, and cultural ecosystem services from UGI (Figure 3). In terms of provisioning ecosystem services, fuelwood is usually collected from bushes and trees within the city, particularly around riverbeds, by residents who cannot afford gas or electricity in homes, or for informal businesses such as barbequed meat stands. For some residents, these form the basis of their cooking and livelihoods, while others resort to illegally tapping into the grid for energy needs. Grasses and *Acacia* pods are harvested from riverbeds and Camel Thorn trees by a few residents and sold to farmers as livestock feed, either directly on farms or by the roadside. A resident in a focus group mentioned that '*farmers contact these people directly, and then only a few people have this small business*' (June 2019). Urban agriculture, although not widespread, allows individuals to grow vegetables for household consumption and re-sale. Cultivation is usually on small plots near homes. Some residents grow plants for medicines, such as *Aloe vera* for burns and injurie, as well as *Moringa oleifera* (Drumstick Tree). These trees can be seen dotted around the settlements, especially in school yards or church gardens.

Regulating ecosystem services are provided by grasses, trees, and shrubs, by filtering and attenuating water; regulating microclimate; purifying air; and preventing soil erosion of riverbanks and hilly slopes. Intact riparian vegetation buffers against stormwater overflow and slows streamflow velocity. Shade provided by larger trees are prized by residents amidst intense land demands, and these trees are protected even when the surroundings are cleared. These are particularly important as communal meeting points and cooler spaces outside unventilated corrugated iron shacks, which lack electric cooling in summer months. Although they are considered an invasive species, *Prosopis* trees growing in common areas of informal settlements provide these services.

Supporting services include UGI providing habitats for snakes, rats, birds, and baboons found in dense riparian vegetation, especially around the Goreangab Dam. UGI filters grey water, particularly when well maintained. Residents are also aware that green vegetation captures atmospheric carbon, produces oxygen, and maintains air quality. Other supporting ecosystem services include soil regeneration and nutrient cycling.

UGI in Windhoek has much to provide in terms of cultural ecosystem services, such as recreational spaces, improved psychological wellbeing, and aesthetic beauty, but it has not been optimised or maintained for these services. As described by a CoW official from the Division of Environmental Management and Health Services in an interview: '*During the rainy season, the riverbeds can transform into beautiful water bodies*' (June 2019). Recreational spaces include school football fields and communal meeting areas, and the largest recreational space easily accessible to residents is the UN Plaza, a green park in Katutura. In addition, the Goreangab Dam offers aesthetic, recreational, and tourism value, as evidenced by operations such as Penduka Village and Lodge (a social enterprise benefitting women, including waterside accommodations and beadmaking). Water from the dam is used by Penduka to irrigate grass on the site, and although they run small-scale hydroponic greenhouses, the water is not used for food production, as it is contaminated.

Figure 3. Examples of ecosystem services in and around informal settlements in Windhoek, Namibia. (**a**) Riparian vegetation near the Goreangab Dam acts as habitat and corridors for landscape connectivity for biodiversity (supporting services). (**b**) Penduka is a social enterprise and small guest house located on the edge of the Goreangab Dam – with the potential as a multifunctional recreational area (cultural services). (**c**) A riverbed in Okuryangava, one of the informal settlements, which has some grasses growing on the slopes, accelerating breakdown of pollutants and providing bank stabilisation (regulating services). (**d**) Grass harvested from a riverbed in Khomasdal is stacked on the side of the road, with a phone number shown for interested buyers (provisioning services). (**e–i**) Urban food garden initiative spearheaded by Shack Dwellers Federation Namibia (SDFN) and Eloolo Permaculture / Farm Okukuna, deployed in 2020 in response to severe food insecurities brought on by the ongoing COVID-19 crisis. Over 600 plots, growing staples for Namibian cuisine such as spinach, have been established during the last year (images by first author during winter 2019, and SDFN, 2020–2021).

3.1.2. Trade-Offs Associated with UGI

Some green spaces such as riverbeds are linked to ecosystem disservices, which are functions of natural ecosystems that are perceived to negatively affect human wellbeing [54]. For instance, vegetation conceals criminal activity and increases risks of drowning in waterways during the rainy season, when people cross over makeshift bridges or build structures near watercourses in informal settlements. In general, riverbeds are considered malodorous and dirty, bringing grey water and industrial pollutants from other areas of the city to the informal settlements. Riverbeds are also widely used for open defecation. This causes outbreaks of water- and vector-borne diseases such as cholera and hepatitis E. Speaking about stagnant water in riverbeds, one resident mentioned that: '*When that water remains stuck, you can see the mosquitoes just lying there*' (July 2019). Residents living near the Goreangab Dam also complained of snakes in the riparian vegetation.

3.2. Complexities of UGI Governance in Windhoek's Informal Settlements

We now consider five UGI initiatives to dissect the complexities of governing UGI in and around informal settlements [27] (Table 3). The five case studies are Farm Okukuna, the Community-Led Total Sanitation (CLTS) Programme, existing green recreational spaces Goreangab Dam and UN Plaza, Fruitful Landscape in Katutura, and Windhoek Riverwalk. Farm Okukuna is a partnership between the NGO Eloolo Permaculture and the CoW, set up specifically to enhance food security in informal settlements of Windhoek. Farm Okukuna trainers run urban agriculture and community nutrition programmes, with a particular focus on women. A partnership between Development Workshop—Namibia, UN Children's Emergency Fund (UNICEF), and the Namibian Chamber of Environment, established the CLTS programme, in response to the sanitation crisis in informal settlements. One of the objectives of CLTS is to make riverbeds open defecation-free by installing toilets and cleaning riverbeds. The Goreangab Dam has picnic sites and the Penduka Village on its banks. UN Plaza is a large public park. These are the largest formal green spaces close to the informal settlements and are maintained by the Parks and Recreation Division of CoW. The site previously known as 'Fruitful Landscape' was an arrangement between the Namibia University of Science and Technology (NUST) and a private landholding in Katutura [55]. It was established as an academic training laboratory for NUST students to study integrated techniques for landscape restoration and agriculture. NUST restored 5 ha of land by diverting storm water flow and regenerated vegetation including dryland-adapted trees. The project demonstrated the successful use of contour ditches and bunds to trap rainwater and rehabilitate soil. However, the project has been discontinued. On the other end of the spectrum, the Windhoek Riverwalk is still in planning stages. If implemented, the Riverwalk will see the strategic rehabilitation of riverbeds through the city, creating opportunities for commerce, non-motorised mobility, and recreation.

Table 3. Description of five case studies that can be classified as UGI-related initiatives in and around Windhoek's informal settlements. Categories are interrelated and non-exhaustive.

Case	Farm Okukuna–Eloolo Permaculture Project	Community-Led Total Sanitation (CLTS)	Green Recreational Spaces (e.g., Goreangab Dam, UN Plaza)	Fruitful Landscape, Katutura	Windhoek Riverwalk
Description	NGO-run permaculture project, in partnership with CoW municipality	Urban sanitation and river rehabilitation programme within informal settlements	Multifunctional green spaces, close to residents in north-western Windhoek	Ecosystem and soil restoration initiative, designed as an academic living laboratory for students	Urban riverbed rehabilitation programme, connecting historically segregated neighbourhoods through green walking and cycling paths
Scale	Single site initiative	Neighbourhoods	Single site initiatives	Private landholding	Citywide
Vegetation	Permaculture project for urban agriculture, vegetables and fruit, and shade trees	Dryland riparian vegetation, shade trees, recreational areas	Dryland riparian vegetation, shade trees, recreational areas	Contour ditches and ponding bunds for rainwater infiltration, supporting dryland grasses and fruit trees	Dryland riparian vegetation, scrub, recreational areas, and economic nodes
Status	Ongoing	Ongoing	Ongoing	Discontinued	Planning

3.2.1. Institutional Frameworks

Policies, Planning, and Regulations

The CoW does not have policies, strategies and plans that explicitly recognise UGI, but UGI components are implicit. For example, the CoW Transformational Strategic Plan 2017–2022 makes provisions for a 'quality urban environment'. Meanwhile, the CoW's Development and Upgrading Strategy of 1999 includes planning for green spaces within in-situ informal settlement upgrading plans. Primary responsibility for public spaces and environmental management is in the hands of the CoW, according to the Local Government Act No. 23 of 1992 (amended in 2002). The Water Resource Management Act No. 11 of 2013 gives the CoW the mandate for wastewater management. However, the Act itself does not detail the specific regulations that the CoW should implement [56]. This gap leaves water management open to interpretation by the CoW, and citizens complain that the municipality is taciturn and ambiguous regarding regulations for rainwater harvesting and stormwater diversion. This presents irrigation barriers for urban agriculture. For instance, NUST's Fruitful Landscapes was discontinued in large part due to the difficulties of adhering to the ambiguous water diversion restrictions and inflexible zoning regulations of the CoW, among other reasons. However, it was a success in terms of ecosystem restoration, as described a researcher from NUST: '*it was interesting to see how nature healed itself. We achieved a lot of water infiltration, and the grasses grew so plentifully there. They formed a natural barrier and puffed up the soil underneath*' (July 2019).

Another regulation related to UGI is the Environmental Management Act No. 7 of 2007. This Act is important for biodiversity conservation and allows the CoW to address habitat fragmentation and loss of green space in cities [57]. The Act stipulates that Environmental Impact Assessments (EIAs) must be carried out prior to settlement upgrading initiatives to assess the potential social, economic, , and ecological impacts of the proposed developments. However, externally appointed consultants usually carry this out, and public participation in the EIA process is lacking. A housing NGO representative explained, *'if you cannot really identify with the community what are the environmental assets they can use and utilise, environmental impact assessments in upgrading are a useless exercise'* (June 2019).

A separate regulatory mechanism that influences UGI and low-income residents' livelihoods is a permit from the CoW's Division of Health and Environment that allows for harvesting wood. Many residents live on unserviced land, relying on bought or collected fuelwood for cooking. The permit prevents unlawful harvesting of fuelwood and *Acacia* pods. However, the monitoring and enforcement of these regulations are not apparent, and although this footprint is small compared to clearing for developments and housing, it has driven green space reduction in north-western Windhoek in the last 20 years [5]. Consequently, residents now travel much further to the city's periphery to collect wood, even venturing into protected areas such as the Daan Viljoen Reserve. As described by an NGO representative: *'In terms of change, we have seen informal urbanisation spread, and we have seen the indigenous vegetation being cleared to allow that. This has a big local impact on people's quality of life'* (June 2019). Comparatively, grass collection for sale as animal fodder, particularly from riverbeds after the rainy season, does not require a permit, as the CoW prefers grasses to be shorter in riverbeds for security purposes. Therefore, it allows grass sellers to cut these grasses prior to the CoW's dredging activities.

Ownership

UGI in informal settlements is generally found in land with contested land tenure rights, and according to results of a study by Thorn et al. [5], 98% of informal residents in Windhoek do not have formal tenure. This 'undeclared' or 'unzoned' nature of informal settlements is particularly problematic for CoW divisions such as Parks and Recreation, Disaster Risk Management, and Roads and Stormwater, which manage various components of public open spaces in formal areas of the city. According to the Local Government Act of 1992 and the Windhoek Town Planning Scheme, these divisions do not have the mandate for acting in 'undesignated areas', leaving a gap in management, irrigation, maintenance, and monitoring of green spaces in informal settlements. Often, one-off attempts at planting street trees are undermined by instances of vandalism because non-state actors do not have the human resource and financial capacity for long term maintenance. Furthermore, there is no evidence of environmental NGOs active in informal settlements in Windhoek. This factor has prompted NGOs such as SDFN-NHAG and Development Workshop—Namibia to take the environmental mandate upon themselves, expanding their current work in land and housing.

Ownership has been shown in other studies to have a strong connection to sense of place as well [13]. In Windhoek, the lack of secure land tenure and affordable housing options and unstable income streams lead to a diminished sense of ownership and belonging. This demotivates informal residents from investing scarce financial resources to maintain or restore UGI (see also [58]).

Access

In neighbouring South Africa, a 'green apartheid' has occurred, where more affluent, historically white-designated areas have significantly more green spaces than historically predominant Black African, Coloured, or Indian areas, with little indication of this trend being reversed [28]. Windhoek underwent the same physical apartheid segregation [52] and shows a similar pattern of greening across the city. It is apparent that there is need for green spaces to act as freely accessible social enablers [28], further explored in sections on the Riverwalk. However, where UGI is publicly available and accessible in Windhoek (Table 4), ecosystem disservices are common (Section 3.1.2) and negatively impact recreational and other cultural benefits [59].

Table 4. Comparison of institutional frameworks that underpin initiatives that can be classified as UGI-related in and around Windhoek's informal settlements. Categories are interrelated and non-exhaustive.

Case	Farm Okukuna—Eloolo Permaculture Project	Community-Led Total Sanitation (CLTS)	Green Recreational Spaces (e.g., Goreangab Dam, UN Plaza)	Fruitful Landscape, Katutura	Windhoek Riverwalk
Policies	CoW Transformational Strategic Plan 2017–2022	CoW Development and Upgrading Strategy, 1999	CoW Transformational Strategic Plan 2017–2022	CoW Transformational Strategic Plan 2017–2022	CoW Sustainable Urban Transport Master Plan
Planning and regulations	CoW drainage and stormwater regulations	Riverbeds in formal areas are under the mandates of Parks and Recreation, Environmental Management, Human Settlements, and Roads and Stormwater Divisions of CoW	These spaces in formal areas are under the mandate of Parks and Recreation Division of CoW	CoW drainage and stormwater regulations, and zoning regulations	CoW zoning regulations
Ownership	Mixed: NGO lease of land from CoW	Mixed: Riverbeds are under CoW, with resident management and NGO expertise in this instance	Public: CoW	Private: agreement between university and private landholding	Public: CoW
Access and use rights	Private: Restricted access through organised trainings or tours	Public access	Public access	Private: Restricted access	Public access

3.2.2. Actors and Coalitions

Stakeholders and Partnerships

The unplanned, transient, and heterogeneous nature of peri-urban areas [17] means that a variety of actors assume responsibility for maintaining different types of UGI (Table 5). For instance, neighbourhood groups of twenty houses often self-govern as 'blocks' to carry out functions like mending fences, reinforcing slopes, and maintaining meeting areas marked by larger trees. Many private households informally manage UGI, such as flower gardens and fences surrounding homes. Some schools and clinics maintain playgrounds and gardens, and few elected constituency councillors have urban food gardens in their office premises. A community-appointed water point committee oversees the maintenance and management of the public taps and associated tariffs, endorsed by the CoW.

In 2020, a non-state collaborative partnership was formed to address growing food insecurity during COVID-19 lockdowns. The SDFN joined with Farm Okukuna and philanthropic donors to train residents in urban gardening techniques. The partnership established over 600 garden plots near informal residents' homes, starting March 2020. A key enabler was the free provisioning of water for informal settlements by the CoW to mitigate impacts of unhygienic practices during the pandemic, whereas previously there were water tariffs. Residents use permaculture and water-sparing techniques taught by Farm Okukuna trainers to grow produce such as spinach, tomatoes, carrots, and lemons for household consumption and resale (Figure 3e–i).

Table 5. Comparison of actors and coalitions within initiatives that can be classified as UGI-related in and around Windhoek's informal settlements. Categories are interrelated and non-exhaustive.

Case	Farm Okukuna—Eloolo Permaculture Project	Community-Led Total Sanitation (CLTS)	Green Recreational Spaces (e.g., Goreangab Dam, UN Plaza)	Fruitful Landscape, Katutura	Windhoek Riverwalk
Primary stakeholders	Eloolo Permaculture Initiative NGO, CoW	Residents of informal settlements, CoW, NGOs	Residents, CoW, especially Parks and Recreation Division	Namibia University of Science and Technology (NUST), landowners	CoW, NUST, Environmental NGOs, business owners (for economic nodes), private architecture firm, citizens
Partnerships	Between NGO and CoW, where municipality provides land and water	Numerous, e.g., youth organisations, volunteer network, and NGOs	Future partnerships could include CoW and Riverwalk, for instance	Informal agreements to sell produce to intermediaries	Public private partnership twinned with Riverwalk in City of Austin, Texas, USA
Power analysis (those with high influence)	Led by Eloolo Permaculture Initiative, oversight by CoW	High influence of elected councillors and community leaders for resident buy-in	Management could be through voluntary citizen committees, requires CoW leadership	Privately led, but discouraged by CoW due to conflicts on water regulations	Conceptualised by private entity, now transferred to CoW for public ownership

Initiatives around Windhoek's main dams also offer examples of partnerships for UGI management, albeit with differing levels of success. Avis Dam, in the city's southern extent, is maintained by a voluntary non-profit NGO named Greenspace [60]. The NGO provides readily accessible recreational opportunities to residents in southern Windhoek. Goreangab Dam in the northwest is an area of scenic beauty. Yet the lack of strategic governance hampers the equitable sharing of the dam's recreational benefits to low-income residents. A public private coalition for the management of this dam, the Goreangab Action Committee, used to exist but is now defunct due to coordination difficulties. There remains potential to resurrect such partnerships, with citizens and industry working together with the CoW Parks and Recreation Division. However, current plans to develop a Goreangab Waterfront as a leisure centre and property development risks gentrification [61].

Coalitions also exist to implement city-wide UGI projects, such as the proposed Riverwalk Initiative. The initiative aims to connect southeast to northwest Windhoek by rehabilitating 200 ha of riverbeds to frame a 20 km green recreational biking and walking corridor, with strategic economic nodes in between. Riverwalk brings together the CoW, NUST, Barnard Mutua Architects, NGOs, and citizens to form a public private partnership. The Goreangab Dam is proposed as the western-most point of the initiative, but the planned initiative does not include informal settlements. However, implementation of the Riverwalk is hampered by the competing priorities of land necessary for motorised transport, ambiguity of responsibility for management of riverbeds, funding gaps, and fears of compromised security.

Power Analysis

By their very nature of operation outside formal administrative systems, power structures prove difficult to decipher in informal settlements [17,62]. Stakeholders have differing degrees of power and agency to influence decision making and associated access to green spaces. Residents have agency over their immediate surroundings in terms of daily maintenance (e.g., planting, irrigation, soil management). Beyond the household level, religious leaders, schools, members of Constituency Development Committees (CDCs), and nominated community leaders have a significant influence on residents' behaviours and perceptions. The food garden initiative (in section on Stakeholders and Partnerships above) is an example of collective action by non-state actors overcoming the lack of individual agency within informal settlements. Additionally, resource management programmes such as CLTS have been championed by elected constituency councillors and the CoW, which is critical for community buy-in and longevity.

Despite these examples, the illegality of residents' occupation makes them vulnerable to eviction, which is a major barrier to investment in UGI. We found that residents generally expect municipal guidance and action for public services such as solid waste management, sanitation, and UGI implementation. However, the CoW's divisions that should be responsible for these services do not have clear mandates or strategies for actions in informal settlements. This frustrates residents and leads to conflicts over unmet needs and lack of momentum.

3.2.3. Resources

Funding

Individuals frequently report a lack of funding to restore and maintain UGI, such as to obtain irrigation water, specialised equipment, seeds, pesticides, and cold storage facilities for perishable agricultural produce, in addition to lack of land. At the local authority level, the CoW is financially autonomous from the central government [56] and uses rates and income tax to maintain green spaces and dredge riverbeds in formal parts of the city. The CoW Human Settlements Division estimates that NAD 3–4 billion (c. USD 200–300 million) would be required for complete servicing and upgrading of the informal settlement areas, but this estimate only includes provisioning of land tenure and basic services and excludes ecosystem restoration and maintenance. Officials emphasised that informal settlement residents do not pay rates and taxes. Because of the political and economic influence associated with such payments, and despite acknowledged equity concerns, UGI restoration is generally prioritised for wealthier areas.

Nevertheless, other governance approaches exist that prevent the financial burden from falling solely on residents or the CoW, such as the mixed financing model Eloolo Permaculture (Table 6). The NGO runs Farm Okukuna as a social enterprise with donor and CoW support. Meanwhile, SDFN-NHAG savings groups have monetary mechanisms that could be adapted for UGI investment [52]. Examples exist of smaller NGOs, such as Family of Hope Services nursery school in Moses ||Garoëb, using donations to install and maintain drip irrigation for an urban food garden, feeding children daily through meal programmes. Riverwalk hopes to obtain private investment and donations, while using rental income and public funds. In the long term, there is significant potential for UGI to support informal livelihoods, as evidenced by one resident taking part in the SDFN food garden initiative: *'Since I started growing spinach outside my shack, I have had neighbours begging me to sell them some of my produce. I have sold more than 300 NAD worth of spinach so far, and I used some of that money to buy more seeds and tools for my garden'* (May 2021).

Table 6. Comparison of resources and types of knowledge within initiatives that can be classified as UGI-related in and around Windhoek's informal settlements. Categories are interrelated and non-exhaustive.

Case	Farm Okukuna—Eloolo Permaculture Project	Community-Led Total Sanitation (CLTS)	Green Recreational Spaces (e.g., Goreangab Dam, UN Plaza)	Fruitful Landscape, Katutura	Windhoek Riverwalk
Funding	CoW, external fundraising, income from produce and some training programmes	Current financial support from European Union funding	Presently CoW; Could include philanthropy, CSR, nominal access fees	University	CoW, private donations, businesses hiring premises
Knowledge	Knowledge of permaculture practices specific to dryland areas	Knowledge of riparian ecosystem restoration, floodlines, and disaster risk reduction techniques are required, but not applied	Landscape architecture and ecology, and sustainable management models are required	Expertise on land restoration, contours and drainage, urban agriculture, including dryland techniques	Expertise on landscape architecture and ecology, innovative business model for economic nodes

Knowledge and Information

Scaling UGI requires knowledge of the local community social structures, as well as ecological processes underpinning the landscape. Scaling UGI also requires scientific and non-scientific expertise and experiences of diverse actors [63]. The national government and CoW Disaster Risk Management and Environmental Management divisions have assessed climate risks and vulnerabilities in informal settlements, although this data is not publicly available yet. These efforts have also not been linked to potential UGI initiatives. Recently, programmes such as Future Resilience for African Cities and Landscapes (FRACTAL) have made advances in mainstreaming climate change, including ecosystem-based adaptation, into local government planning [21]. Meanwhile Urban Ecolution and Peri Urban Resilient Ecosystems were the first research programmes to focus on UGI and climate risk in informal settlements [5,22]. Despite these initiatives, engineered solutions, such as bunds and culverts for flooding, and electronic cooling solutions for higher temperatures take prominence over nature-based solutions. This is further inhibited by a lack of data on UGI efficacy in this context. Encouragingly, the initiatives mentioned above demonstrate that expertise in Windhoek on soil restoration, climate resilience, and urban agriculture exists.

The modes of disseminating knowledge and raising awareness in informal settlements represent another barrier. CDCs, with members nominated by residents and councillors, are meant to be conduits of information between the residents, councillors, and the CoW. However, residents say that influence and power dynamics surrounding these CDCs hinder true representation, instead making them gatekeepers of information, forming information bottlenecks. Ad hoc community meetings take place in the absence of more formal platforms.

3.2.4. Processes

Discourses

The discourse on UGI, urban ecosystem services, and nature-based solutions in Windhoek's informal settlements is nascent at best and, perhaps justifiably, overshadowed by the crippling issues of land tenure and service provision in peri-urban areas (Table 7). The discourse is also strongly centred on the leadership being provided by the CoW, with NGOs, the private sector, and residents playing a supporting role. The current discourse on water security and water as a 'right' or 'commodity' [64] has significant implications for UGI. One of the city's most pressing issues relates to projected changes in climate and their impact on already strained water resources [20,65]. Windhoek depends on a combination of water reclamation, dams, surface reservoirs, and boreholes to supply water to its inhabitants. By 2042, it is projected that the annual rainfall will decline [20,21,65]. The ongoing considerations for water security interact with the discourse on food insecurity, housing, land, and resilience. The cultural services discourse is also prominent, where residents emphasise the need for recreational spaces such as football fields. More recently, as we have shown, the discourse landscape on UGI governance has started to shift from local governmental regulation to collaborative, community-driven governance models for urban agriculture and open defecation-free riverbed initiatives.

Table 7. Comparison of processes driving or inhibiting initiatives that can be classified as UGI-related in and around Windhoek's informal settlements. Categories are interrelated and non-exhaustive.

Case	Farm Okukuna–Eloolo Permaculture Project	Community-Led Total Sanitation (CLTS)	Green Recreational Spaces (e.g., Goreangab Dam, UN Plaza)	Fruitful Landscape, Katutura	Windhoek Riverwalk
Discourses	In danger of discontinuation due to developmental pressures for alternative land uses; cultural attitudes that agriculture is difficult in the city	Concerns whether riverbeds can be rehabilitated from the level of solid waste disposal, sanitation, encroachment taking place; open defecation-free status achieved in some areas	Concerns due to lack of political willingness to promote green spaces	Perception that Windhoek conditions are unsuitable for agriculture; feasibility of restoration	A green vision for a connected, prosperous, and healthy city
Participation	NGO active in planning, management, and advocacy programmes, working with housing NGOs and resident groups	Local community groups, with momentum from councillors and NGOs; door-to-door volunteers; supported by CoW	Participatory method of management would mean communities' needs are considered in implementation of these green and blue spaces	Conflicts and low acceptance of ideas by peripheral actors, lack of support from CoW, vandalism and theft are barriers to participation	Citizen groups, businesses, and town planners have been consulted; lack of engagement of peri-urban residents
Monitoring and evaluation	Annual reporting; CoW supervision	Public meetings in community areas; quarterly reporting	Annual reporting	Site assessments by university	Actors' meetings

Participation and Monitoring and Evaluation

Legitimised forms of community participation in decision-making about informal settlement green spaces is scant. However, the sustained efforts of Eloolo through Farm Okukuna offer an example of how technical advice and consistent advocacy over a longer period have encouraged uptake of urban garden plots among informal settlement residents, particularly women, by training organisations such as SDFN to engage their savings' group members in deploying gardens. Historically, many perceived the CoW as adopting a top-down approach with limited willingness to engage with informal settlements, but this precedent has also started changing since 2017. The intent to improve engagement is represented in that Human Settlements Division of the CoW now has a section for 'public engagement'. However, the formulation of the ICCSAP has only minimally engaged informal settlement actors so far, primarily through constituency councillors, and there remains a pressing need to facilitate building of accountability, transparency, and access to information. Monitoring and evaluation is another key challenge, especially for long-term mitigation of UGI encroachment, open defecation, and solid waste disposal.

4. Discussion

Our objective in this paper was to analyse current governance mechanisms for UGI in Windhoek. Using a case study analysis framework, we isolated five examples of UGI implementation. In doing this, we address the paucity of empirical research on UGI and ecosystem services governance outside South Africa in SSA [13,18], and in informal settlement or dryland contexts [13]. Moreover, with its focus on equity and inclusivity of residents in peri-urban areas, our study contributes to ensuring that the mainstreaming of UGI does not perpetuate historical inequalities in access to nature and green spaces in cities [66].

This research is timely, given the urgent need for governments to adopt innovative, local solutions to combat the pervasive impacts of climate change and urban expansion [67]. The challenges and opportunities presented by complex governance configurations need to be part of this discussion [68], where UGI offers a suite of benefits to mitigate climate change [69], alleviate flood risk [70], improve public health [71], be economically affordable [14], and be delivered at a scale accounting for administrative and ecological boundaries [72–74]. For municipality planners and managers, UGI measures often prove more cost-effective than grey infrastructural measures, with options for more citizen-centred, collaborative governance configurations that are uniquely suited to SSA [14,27]. Our study provides insights that fill critical gaps in this scholarship.

In the following section, using this reflection of the past and present, we distil key desirable pathways for future UGI governance in peri-urban areas.

4.1. *The Need for Collaborative Governance Platforms for UGI*

Information asymmetry, lack of community consultation, and absence of collaborative governance are major barriers for settlement planning, including integration of UGI and ecosystem services. Community participation and stewardship are essential for UGI schemes in Windhoek's informal settlements to succeed, not only to account for benefits that residents perceive (e.g., trees acting as barriers against wind and dust, snakes eliminating rodents) but also to reflect the true dynamics of how the informal economy and survival strategies relate to the natural environment. Cognition of ecosystem services is an enabler of participation, and therefore, awareness and capacity development programmes, as well as environmental education, should form the bedrock of UGI [27].

A window of opportunity exists to enhance UGI consideration through public private partnerships. For example, SDFN-NHAG is in the process of updating its strategies for upcoming work in Windhoek, together with the CoW. SDFN-NHAG has existing participatory mechanisms in place across Namibia to directly engage residents in upgrading and securing tenure through the Community Land Information Programme [52], and other co-production processes involving green space design [75]. Processes such as these could complement EIAs for in-situ upgrading and consultation for climate resilience. More active engagement of environmental NGOs in Windhoek's peri-urban areas would also help raise awareness of biodiversity conservation in urban centres, which is currently a gap in Namibia [76], while linking to global discourse and actions on the UN Decade for Ecosystem Restoration 2021–30 [77].

To this end, an opportunity lies in establishing a focal body for UGI coordination - whether community-driven, municipality-driven, or using a hybrid strategy – to help shift perspectives to viewing informality as an opportunity to deploy innovative UGI approaches that are not possible in formal areas of the city [70]. Reviving multi-stakeholder partnerships could help promote inclusivity and accessibility in the planning, design and management of UGI, while improving local stewardship and valuing of green spaces –

as seen in the case of informal settlements in Nairobi [27]. Political leadership from the CoW will continue to underpin UGI implementation [19], but this must be coupled with an involved, informed community and supported by NGOs, the private sector, universities such as NUST, and others who work through purpose-built collaborative governance platforms. Local committees for UGI, like existing self-organised water point committees, will be important for maintenance. For Windhoek, these will also operationalise principles espoused in the Sustainable Development Goals, especially Goal 17 on partnerships [78], as shown by Cumming et al. [79] for South Africa's National Development Plan.

4.2. The Need to Integrate Informal Settlements and UGI into Municipal Climate Change Strategies

Windhoek is now poised to implement a multi-scale climate adaptation plan through the ICCSAP [21]. The ICCSAP provides an ideal opportunity for UGI to be mainstreamed into policy at the local authority level. The disaster risk reduction potential of UGI identified in this study is supported by a growing body of evidence worldwide, such as slope stabilisation, stormwater management [70,80], microclimate regulation [81], urban agriculture [82], and swales [83]. These not only increase the resilience of informal settlement residents to climate change but also feed into wider benefits by improving quality of life, while making the entire city more accessible and inclusive [71].

Furthermore, novel funding could be sought from sources such as the Green Climate Fund, the Global Environmental Facility, other governments, or private capital. Nationally, niche schemes such as the First Lady's 'One Nation Fund'—a microfinancing scheme for low-income entrepreneurs—can be leveraged for urban agriculture schemes. Various sustainability and climate change related projects have already been funded in rural Namibia, but peri-urban areas have often been overlooked. Another way to provide financial momentum is through committees formed by informal residents that could charge a mutually agreed upon contribution in cash or kind. For instance, constituency councillors emphasized the potential of nominal fee-based access rights for future recreational spaces to generate funds to maintain these spaces. Success of future UGI and climate adaptation programmes would hinge on local support, including CoW recognition and legitimisation, as well as partnerships with new actors such as the private sector through corporate social responsibility. Most importantly, clarifying mandates, roles, and modes for collaborative UGI governance in peri-urban, informal areas, and formalising these in future policies, strategies, plans, and programmes remains paramount.

4.3. The Need to Include UGI in Integrated Development Plans such as the Human Settlements Upgrading Policy

The Human Settlements Division of the CoW was established in 2017 with the aim of strategically coordinating the in-situ upgrading of informal settlements. It is currently in the process of updating its Human Settlements Upgrading Strategy to create a policy. This presents an opportunity to include explicit provisions for UGI within settlement upgrading policy. Some priority areas hold promise, namely UGI-based recreational opportunities, riparian restoration, urban agriculture, and dryland-specific techniques. As trees and vegetation surrounding the peri-urban areas dwindle, the need for pre-emptive action and planning is evident. Schemes have been considered but not implemented at scale. Initiatives such as planting indigenous street trees for mitigating urban heating (replacing invasive species), greywater reclamation for irrigation, sunken planting pits, and using shade structures for seedlings to reduce moisture loss through evapotranspiration should be explored [84,85].

The riverbeds which turn into ephemeral rivers in the rainy season, form an intersection between health, climate adaptation, and ecological outcomes. They must urgently be cleaned, restored, and maintained as multifunctional ecological assets. Encouragingly, the CLTS programme that has succeeded in making some parts of the Moses ||Garoëb and Samora Machel constituencies open defecation free [86] is government-supported and aligns with the Harambee Prosperity Plan II (2021-2025). Taking CLTS as an example, there are opportunities to convert riverbeds to multifunctional UGI, while improving walkability and creating recreational areas [87].

Furthermore, financially viable alternatives for cooking with firewood are needed to curb the overharvesting of vegetation, and examples could include solar heaters or subsidised gas provisions, leveraging existing platforms such as the 'Think Namibia' climate-smart campaign operating in rural areas [88].

Food insecurity and malnutrition is widespread in Windhoek. Urban agriculture is only adopted by a small percentage of the population for fear of theft of produce, lack of land, or lack of skill or interest in urban agricultural activities [58]. Even in our study, the servicing, upgrading, and provisioning of secure tenure was the highest priority for all informants from informal settlements. *'Once land is serviced, then the constituency can say ok now you can grow tomatoes or spinach'* (June 2019), said a constituency councillor. The feasibility of urban agriculture must be carefully assessed [89], not least due to a dismissive attitude seen towards urban agriculture [90]. However, a precedent has been set by initiatives such as the SDFN food gardens and the Eloolo Permaculture Initiative, showing that training and advocacy are critical. A recent study by Shikangalah and Mapani [65] found that precipitation in Windhoek follows an approximate pattern of showing years with high rainfall, followed by drought years, indicating that rainwater harvesting in peak rainfall years is viable to help irrigate urban agriculture, in tandem with greywater reclamation. Therefore, we recommend the CoW revise and clearly communicate stormwater and grey water regulations for households to further enable urban agriculture schemes.

Going beyond policy mainstreaming, innovative approaches such as starting small, 'safe-to-fail' pilot schemes within a learning-by-doing approach [23], collaboratively developed with informal residents through 'urban learning labs' used elsewhere in Namibia, Southern Africa [21,91], and SSA [92], will be important in this dynamic setting. Digital messaging platforms can also help overcome information bottlenecks and achieve wider innovation.

4.4. The Need to Consider UGI in the Informal Economy and for Green Jobs

Previous studies in Windhoek have mainly focused on the contribution of ecosystem services in the informal food economy [58,93], but our findings show that trade in firewood, grasses, *Acacia* pods, and reeds harvested from the surroundings forms a part of many residents' livelihoods. Green job creation is a significant opportunity that can be explored for Windhoek, as involving residents in UGI implementation, maintenance, and management could provide consistent sources of low-skill employment through initiatives such as public works programmes [94].

More recently, there is a growing movement to greening the recovery from COVID-19 [95,96]. South African cities, for instance, have committed to green urban recovery post COVID-19, with explicit provisions for nature-based solutions including UGI in these strategies [97]. Namibian national and local governments, working collaboratively with communities, have a unique opportunity to model UGI initiatives in a similar way.

5. Conclusions

This paper analysed the ecosystem services, benefits, trade-offs, and governance structures of UGI in and around Windhoek's informal settlements. Our results reveal that Windhoek is facing an escalating crisis of unplanned urban sprawl, climate change, and lack of basic infrastructure in peri-urban areas, and residents are disproportionately exposed to social and environmental risks. Opportunities exist to leverage UGI for climate resilience and to enhance socio-economic wellbeing and quality of life. UGI initiatives are often most effective when deployed in a complementary manner with grey or engineered solutions, but much greater recognition of UGI's multiple benefits needs to be mainstreamed in decision-making [27]. Collaborative governance platforms and clearly delineated mandates are necessary, with explicit integration of UGI into strategies for climate adaptation, informal settlement upgrading, and green job growth.

This research contributes much needed empirical evidence from Africa to the growing global body of research on nature-based solutions for cities [98]. Future research could focus on the impact of climate change on the natural environment of Windhoek and how this would interact with the sustainability, governance, and feasibility of UGI initiatives. Further research is needed to determine which governance configurations will allow the most inclusive and participatory approach for UGI in peri-urban Windhoek, with a particular focus on gender [62]. Research institutions have a role to play as knowledge brokers in this context [99]. As rural-urban migration accelerates and climate impacts intensify, it is ever more critical that local authorities and other actors with the responsibility to meet decentralised developmental commitments integrate agendas of equitable development and environmental justice through UGI in policy, planning, and urban design.

Author Contributions: A.W. and J.P.R.T.: Conceptualization, methodology, validation, investigation, formal analysis, funding acquisition, resources, writing and editing, visualization, and project administration. A.W.: software. J.P.R.T.: supervision. All authors have read and agreed to the published version of the manuscript.

Funding: We acknowledge funding from the School of Geography and the Environment, University of Oxford, and Linacre College, Oxford, and from UK Research and Innovation's Global Challenges Research Fund (UKRI GCRF) through the Development Corridors Partnership project (project number: ES/P011500/1). This work was also supported through the African Women in Climate Change Science Fellowship of the African Institute of Mathematical Sciences and the Next Einstein Forum and the Climate Research for Development (CR4D) Postdoctoral Fellowship CR4D-19-21 implemented by the African Academy of Sciences (AAS) in partnership with the United Kingdom's Department for International Development (DfID), the Weather and Climate Information Services for Africa (WISER) Programme, and the African Climate Policy Centre (ACPC) of the United Nations Economic Commission for Africa (UNECA). Statements made and views expressed in this work are solely the responsibility of the authors.

Institutional Review Board Statement: The study was approved by the Central University Research Ethics Committee (CUREC) of The University of Oxford, UK (SOGE-1A19-83, May 2019).

Informed Consent Statement: Informed consent was obtained from all subjects involved in the study.

Acknowledgments: This research is a part of a larger programme of work known as the Urban Ecolution project, implemented jointly by the York Institute of Tropical Ecosystems at the Department of Environment and Geography at the University of York and the African Climate and Development Initiative at the University of Cape Town. The authors would like to thank the facilitators of community engagement, the Shack Dwellers' Federation of Namibia and Namibia Housing Action Group (SDFN-NHAG) and acknowledge the insight and technical assistance from the Development Workshop—Namibia, City of Windhoek municipality, Namibia University of Science and Technology (NUST), and research assistants. We also thank all participants who gave their time and experience, especially the residents and authorities in informal settlements. This work also benefited from the kind support of the Namibian Chamber of Environment, who hosted the lead author during fieldwork, and from Chloe Montes Strevens at the University of Oxford who provided early guidance and supervision.

Conflicts of Interest: The authors declare no conflict of interest. The funders had no role in the design of the study; in the collection, analyses, or interpretation of data; in the writing of the manuscript; or in the decision to publish the results.

Appendix A Themes and Open-Ended Questions from Semi-Structured Interviews and Focus Groups (Survey Tool)

Give a brief description of your role/job in Windhoek?

What are the changes in climate that you have observed in Windhoek over the last twenty years? If there are any, what are the main impacts of these changes?

Do you think that natural ecosystems can play a part in helping people in the informal settlements? If yes, how do natural ecosystems play a part in helping people? Please give examples.

Are you aware of a concept called ecosystem services?

Here's my definition of ecosystem services: [give definition]. What do you think are the most important ecosystem services received by residents living in Windhoek?

Are there any that are particularly vital to the wellbeing of the informal settlement residents?

What will be the most pressing issues (non-climatic) within the informal settlements in the next 30 years?

Do you think any of these issues can be feasibly addressed by using natural ecosystems? What would the alternatives be? Please describe any examples that come to mind.

Are you aware of any green space and natural ecosystem management plans currently being carried out in Windhoek, particularly focused on informal settlements?

Can you describe them briefly, and tell me who oversees their implementation?

What are some of the important opportunities for development within the informal settlements in the next 30 years?

What are the strengths within (a) the community (b) the authorities (c) broader stakeholders to incorporate ecosystem-based strategies?

Community ownership of the initiatives is vital to the success of multifunctional green spaces. Who do you think will be most likely to accept and work towards conserving ecosystems in order to gain adaptation benefits?

Are there any barriers that you can think of in the short-term (2030) which will challenge implementation of urban green spaces and urban green infrastructure? Any in the long term (2063)?

What is your view about the impact that climate change may have on Windhoek in the short-term (e.g., 2030) and long-term (e.g., 2063)?

What would be the impact of climate change on the informal settlements in the peri-urban areas of Windhoek, and how does this differ from other formal areas?

References

1. United Nations Human Settlements Programme (UN-Habitat). *The State of African Cities 2014: Re-Imagining Sustainable Urban Transitions*; United Nations Human Settlements Programme (UN-Habitat): Nairobi, Kenya, 2014.
2. Kessides, C. The Urban Transition in Sub-Saharan Africa: Challenges and Opportunities. *Environ. Plan. C Gov. Policy* **2007**, *25*, 466–485. [CrossRef]
3. Satterthwaite, D.; Archer, D.; Colenbrander, S.; Dodman, D.; Hardoy, J.; Patel, S. *Responding to Climate Change in Cities and in Their Informal Settlements and Economies*; International Institute for Environment and Development: Edmonton, AB, Canada, 2018.
4. Satterthwaite, D. How Can Urban Centers Adapt to Climate Change with Ineffective or Unrepresentative Local Governments? *Wiley Interdiscip. Rev. Clim. Chang.* **2011**, *2*, 767–776. [CrossRef]
5. Thorn, J.; Biancardi Aleu, R.; Wijesinghe, A.; Mdongwe, M.C.; Marchant, R.; Shackleton, S. Mainstreaming Nature-Based Solutions for Climate Resilient Infrastructure in Peri Urban Sub-Saharan Africa. *Landsc. Urban Plan.*. in review.
6. Pickett, S.T.A.; Cadenasso, M.L.; Grove, J.M.; Boone, C.G.; Groffman, P.M.; Irwin, E.; Kaushal, S.S.; Marshall, V.; McGrath, B.P.; Nilon, C.H.; et al. Urban Ecological Systems: Scientific Foundations and a Decade of Progress. *J. Environ. Manag.* **2011**, *92*, 331–362. [CrossRef] [PubMed]
7. Elmqvist, T.; Fragkias, M.; Goodness, J.; Güneralp, B.; Marcotullio, P.J.; McDonald, R.I.; Parnell, S.; Schewenius, M.; Sendstad, M.; Seto, K.C.; et al. *Urbanization, Biodiversity and Ecosystem Services: Challenges and Opportunities: A Global Assessmentnull*; Springer Nature: Berlin/Heidelberg, Germany, 2013; ISBN 9783319161044.
8. Gómez-Baggethun, E.; Barton, D.N. Classifying and Valuing Ecosystem Services for Urban Planning. *Ecol. Econ.* **2013**, *86*, 235–245. [CrossRef]

9. Mata, L.; Ramalho, C.E.; Kennedy, J.; Parris, K.M.; Valentine, L.; Miller, M.; Bekessy, S.; Hurley, S.; Cumpston, Z. Bringing Nature Back into Cities. *People Nat.* **2020**, *2*, 350–368. [CrossRef]
10. Frantzeskaki, N. Seven Lessons for Planning Nature-Based Solutions in Cities. *Environ. Sci. Policy* **2019**, *93*, 101–111. [CrossRef]
11. Geneletti, D.; Zardo, L. Ecosystem-Based Adaptation in Cities: An Analysis of European Urban Climate Adaptation Plans. *Land Use Policy* **2016**, *50*, 38–47. [CrossRef]
12. Connop, S.; Vandergert, P.; Eisenberg, B.; Collier, M.J.; Nash, C.; Clough, J.; Newport, D. Renaturing Cities Using a Regionally-Focused Biodiversity-Led Multifunctional Benefits Approach to Urban Green Infrastructure. *Environ. Sci. Policy* **2016**, *62*, 99–111. [CrossRef]
13. Lindley, S.; Pauleit, S.; Yeshitela, K.; Cilliers, S.; Shackleton, C. Rethinking Urban Green Infrastructure and Ecosystem Services from the Perspective of Sub-Saharan African Cities. *Landsc. Urban Plan.* **2018**, *180*, 328–338. [CrossRef]
14. Titz, A.; Chiotha, S.S. Pathways for Sustainable and Inclusive Cities in Southern and Eastern Africa through Urban Green Infrastructure? *Sustainability* **2019**, *11*, 2729. [CrossRef]
15. Hoegh-Guldberg, O.; Jacob, D.; Taylor, M.; Bindi, M.; Brown, S.; Camilloni, I.; Diedhiou, A.; Djalante, R.; Ebi, K.L.; Engelbrecht, F.; et al. Impacts of 1.5 °C of Global Warming on Natural and Human Systems. In *Global Warming of 1.5 °C*; IPCC: Geneva, Switzerland, 2018; p. 138.
16. Satterthwaite, D.; Archer, D.; Colenbrander, S.; Dodman, D.; Hardoy, J.; Mitlin, D.; Patel, S. Building Resilience to Climate Change in Informal Settlements. *One Earth* **2020**, *2*, 143–156. [CrossRef]
17. Thorn, J.; Thornton, T.F.; Helfgott, A. Autonomous Adaptation to Global Environmental Change in Peri-Urban Settlements: Evidence of a Growing Culture of Innovation and Revitalisation in Mathare Valley Slums, Nairobi. *Glob. Environ. Chang.* **2015**, *31*, 121–131. [CrossRef]
18. du Toit, M.J.; Cilliers, S.S.; Dallimer, M.; Goddard, M.; Guenat, S.; Cornelius, S.F. Urban Green Infrastructure and Ecosystem Services in Sub-Saharan Africa. *Landsc. Urban Plan.* **2018**, *180*, 249–261. [CrossRef]
19. Roy, M.; Shemdoe, R.; Hulme, D.; Mwageni, N.; Gough, A. Climate Change and Declining Levels of Green Structures: Life in Informal Settlements of Dar Es Salaam, Tanzania. *Landsc. Urban Plan.* **2018**, *180*, 282–293. [CrossRef]
20. Scott, D.; Iipinge, K.N.; Mfune, J.K.E.; Muchadenyika, D.; Makuti, O.V.; Ziervogel, G. The Story of Water in Windhoek: A Narrative Approach to Interpreting a Transdisciplinary Process. *Water* **2018**, *10*, 1366. [CrossRef]
21. Jack, C.D.; Jones, R.; Burgin, L.; Daron, J. Climate Risk Narratives: An Iterative Reflective Process for Co-Producing and Integrating Climate Knowledge. *Clim. Risk Manag.* **2020**, *29*, 100239. [CrossRef]
22. Thorn, J.P.R.; Kavonic, J.; Hejnowicz, A.; Marchant, A.; Cinderby, S.; Mueller, A.; Ajala, O. *Dryland Nature Based Solutions for Informal Settlement Upgrading Schemes in Africa*; ICLEI Africa: Cape Town, South Africa, 2021; Available online: https://africa.iclei.org/iclei_publications/policy-brief-dryland-nature-based-solutions-for-informal-settlement-upgrading-schemes-in-africa/ (accessed on 31 May 2021).
23. Ahern, J.; Cilliers, S.; Niemelä, J. The Concept of Ecosystem Services in Adaptive Urban Planning and Design: A Framework for Supporting Innovation. *Landsc. Urban Plan.* **2014**, *125*, 254–259. [CrossRef]
24. Jaffe, M. Environmental Reviews & Case Studies: Reflections on Green Infrastructure Economics. *Environ. Pract.* **2010**, *12*, 357–365. [CrossRef]
25. Kim, D.; Song, S.-K. The Multifunctional Benefits of Green Infrastructure in Community Development: An Analytical Review Based on 447 Cases. *Sustainability* **2019**, *11*, 3917. [CrossRef]
26. Gaffin, S.R.; Rosenzweig, C.; Kong, A.Y.Y. Adapting to Climate Change through Urban Green Infrastructure. *Nat. Clim. Chang.* **2012**, *2*, 704. [CrossRef]
27. Mulligan, J.; Bukachi, V.; Clause, J.C.; Jewell, R.; Kirimi, F.; Odbert, C. Hybrid Infrastructures, Hybrid Governance: New Evidence from Nairobi (Kenya) on Green-Blue-Grey Infrastructure in Informal Settlements. *Anthropocene* **2020**, *29*, 100227. [CrossRef]
28. Venter, Z.S.; Shackleton, C.M.; Van Staden, F.; Selomane, O.; Masterson, V.A. Green Apartheid: Urban Green Infrastructure Remains Unequally Distributed across Income and Race Geographies in South Africa. *Landsc. Urban Plan.* **2020**, *203*, 103889. [CrossRef]
29. Badiu, D.L.; Nita, A.; Iojă, C.I.; Niţă, M.R. Disentangling the Connections: A Network Analysis of Approaches to Urban Green Infrastructure. *Urban For. Urban Green.* **2019**, *41*, 211–220. [CrossRef]
30. Kithiia, J. Climate Change Risk Responses in East African Cities: Need, Barriers and Opportunities. *Curr. Opin. Environ. Sustain.* **2011**, *3*, 176–180. [CrossRef]
31. McDonald, R.I.; Mansur, A.V.; Ascensão, F.; Colbert, M.; Crossman, K.; Elmqvist, T.; Gonzalez, A.; Güneralp, B.; Haase, D.; Hamann, M.; et al. Research Gaps in Knowledge of the Impact of Urban Growth on Biodiversity. *Nat. Sustain.* **2019**, *3*, 16–24. [CrossRef]
32. Tacconi, L. Developing Environmental Governance Research: The Example of Forest Cover Change Studies. *Environ. Conserv.* **2011**, *38*, 234–246. [CrossRef]
33. Lawrence, A.; de Vreese, R.; Johnston, M.; Bosch, C.C.K.; van den Sanesi, G. Urban Forest Governance: Towards a Framework for Comparing Approaches. *Urban For. Urban Green.* **2013**, *12*, 464–473. [CrossRef]
34. Buizer, M.; Elands, B.; Mattijssen, T.J.M.; van der Jagt, A.; Ambrose, B.; Gerőházi, É.; Santos, A.; Møller, M.S. *The Governance of Urban Green Spaces in Selected EU-Cities*; Green Surge: Copenhagen, Denmark, 2015.

35. van der Jagt, A.P.N.; Elands, B.H.M.; Ambrose-Oji, B.; Gerőházi, É.; Møller, M.S.; Buizer, M. Participatory Governance of Urban Green Spaces: Trends and Practices in the EU. *Nord. J. Archit. Res.* **2017**, *28*, 3.
36. Adegun, O.B. Green Infrastructure in Informal Unplanned Settlements: The Case of Kya Sands, Johannesburg. *Int. J. Urban Sustain. Dev.* **2019**, *11*, 68–80. [CrossRef]
37. Namibia Statistics Agency. *Namibia Inter-Censal Demographic Survey 2016 Report*; Namibia Statistics Agency: Windhoek, Namibia, 2016.
38. Remmert, D.; Ndhlovu, P. *Housing in Namibia: Rights, Challenges and Opportunities*; Institute for Public Policy Research: London, UK, 2018.
39. Weber, B. Addressing Informal Settlement Growth in Namibia. *Namib. J. Environ.* **2017**, *1*, B-26.
40. Karuaihe, S.T.; Wandschneider, P.R. Limited Access to Services for the Urban Poor in Windhoek, Namibia. *Dev. S. Afr.* **2018**, *35*, 466–479. [CrossRef]
41. Bustamante, N.D.; Matyenyika, S.R.; Miller, L.A.; Goers, M.; Katjiuanjo, P.; Ndiitodino, K.; Ndevaetela, E.-E.; Kaura, U.; Nyarko, K.M.; Kahuika-Crentsil, L.; et al. Notes from the Field: Nationwide Hepatitis E Outbreak Concentrated in Informal Settlements—Namibia, 2017–2020. *MMWR Morb. Mortal. Wkly. Rep.* **2020**, *69*, 355–357. [CrossRef]
42. Labbe, N.; McBride, N.; Ray, E. *Flooding and Erosion Control in the Informal Settlements of Windhoek, Namibia*; Worcester Polytechnic Institute: Worcester, MA, USA, 2006.
43. Shikangalah, R.N. The 2019 Drought in Namibia: An Overview. *J. Namib. Stud. Hist. Polit. Cult.* **2020**, *27*, 35–58.
44. Williams, D.S.; Máñez Costa, M.; Sutherland, C.; Celliers, L.; Scheffran, J. Vulnerability of Informal Settlements in the Context of Rapid Urbanization and Climate Change. *Environ. Urban.* **2019**, *31*, 157–176. [CrossRef]
45. Iipinge, K. *Windhoek Third Learning Lab Report*; FRACTAL: Windhoek, Namibia, 2018.
46. Shackleton, C.M.; Blair, A.; De Lacy, P.; Kaoma, H.; Mugwagwa, N.; Dalu, M.T.; Walton, W. How Important Is Green Infrastructure in Small and Medium-Sized Towns? Lessons from South Africa. *Landsc. Urban Plan.* **2018**, *180*, 273–281. [CrossRef]
47. Arts, B.; Leroy, P.; van Tatenhove, J. Political Modernisation and Policy Arrangements: A Framework for Understanding Environmental Policy Change. *Public Organ. Rev.* **2006**, *6*, 93–106. [CrossRef]
48. Ministry of Environment and Tourism. *National Climate Change Strategy and Action Plan 2013–2020*; Ministry of Environment and Tourism: Windhoek, Namibia, 2012.
49. IECN. *Let's Act to Adapt: Dealing with Climate Change*; Ministry of Environment and Tourism, Namibia: Windhoek, Namibia, 2011.
50. Garrard, S.; Heyns, P.; Pfaffenthaler, M.; Schneider, G. *Environmental Awareness for Sustainable Development: A Resource Book for Namibia*; Hanns Seidel Foundation: Windhoek, Namibia, 2017; ISBN 978-99945-79-89-1.
51. Sadler, G.R.; Lee, H.-C.; Lim, R.S.-H.; Fullerton, J. Research Article: Recruitment of Hard-to-Reach Population Subgroups via Adaptations of the Snowball Sampling Strategy. *Nurs. Health Sci.* **2010**, *12*, 369–374. [CrossRef] [PubMed]
52. Chitekwe-Biti, B. Co-Producing Windhoek: The Contribution of the Shack Dwellers Federation of Namibia. *Environ. Urban.* **2018**, *30*, 387–406. [CrossRef]
53. Mihas, P. Qualitative Data Analysis. Available online: https://oxfordre.com/education/view/10.1093/acrefore/9780190264093.001.0001/acrefore-9780190264093-e-1195 (accessed on 31 May 2021).
54. Shackleton, C.M.; Ruwanza, S.; Sinasson Sanni, G.K.; Bennett, S.; De Lacy, P.; Modipa, R.; Mtati, N.; Sachikonye, M.; Thondhlana, G. Unpacking Pandora's Box: Understanding and Categorising Ecosystem Disservices for Environmental Management and Human Wellbeing. *Ecosystems* **2016**, *19*, 587–600. [CrossRef]
55. Zimmermann, I. *Arrangements to Convert Degraded Rangeland into Fruitful Landscape [Namibia]*; World Overview of Conservation Approaches and Technologies (WOCAT): Windhoek, Namibia, 2016.
56. Dubbeling, M. *Policy Review for Urban and Peri-Urban Agriculture Development in Namibia*; RUAF Foundation: Windhoek, Namibia, 2016.
57. Rastandeh, A.; Jarchow, M. Urbanization and Biodiversity Loss in the Post-COVID-19 Era: Complex Challenges and Possible Solutions. *Cities Health* **2020**, 1–4. [CrossRef]
58. Crush, J.; Nickanor, N.; Kazembe, L. Informal Food Deserts and Household Food Insecurity in Windhoek, Namibia. *Sustainability* **2019**, *11*, 37. [CrossRef]
59. Giombini, V.; Thorn, J.P.R. Urban green spaces in a post-apartheid city: Challenges and opportunities for nature-based solutions. In *Exploring the Multiple Values of Nature—Connecting Ecosystems and People across Landscapes*; in press.
60. Odendaal, N. Claiming greenspace: From oppositional practice to co-production in windhoek, namibia. In *Urban Space*; Hernández-García, J., Cárdenas-O'Byrne, S., García-Jerez, A., Beza, B.B., Eds.; Experiences and Reflections from the Global South; Sello Editorial Javeriano: Cali, Colombia, 2018; pp. 91–116. ISBN 978-958-54-5338-8.
61. Cole, H.V.S.; Garcia Lamarca, M.; Connolly, J.J.T.; Anguelovski, I. Are Green Cities Healthy and Equitable? Unpacking the Relationship between Health, Green Space and Gentrification. *J. Epidemiol. Community Health* **2017**, *71*, 1118–1121. [CrossRef] [PubMed]
62. Kovacic, Z.; Musango, J.K.; Ambole, L.A.; Buyana, K.; Smit, S.; Anditi, C.; Mwau, B.; Ogot, M.; Lwasa, S.; Brent, A.C.; et al. Interrogating Differences: A Comparative Analysis of Africa's Informal Settlements. *World Dev.* **2019**, *122*, 614–627. [CrossRef]
63. Faehnle, M.; Bäcklund, P.; Tyrväinen, L.; Niemelä, J.; Yli-Pelkonen, V. How Can Residents' Experiences Inform Planning of Urban Green Infrastructure? Case Finland. *Landsc. Urban Plan.* **2014**, *130*, 171–183. [CrossRef]

64. Renaud, C.; Scott, D.; Muchadenyika, D.; Iipinge, K.; Macavele, H.; Maure, G.; Mwalukanga, B.; Pinto, I.; Siame, G. *Preliminary Mapping of Water and Climate Change Governance in Lusaka, Windhoek and Maputo*; FRACTAL: Windhoek, Namibia, 2018; p. 7.
65. Shikangalah, R.N.; Mapani, B. Precipitation Variations and Shifts over Time: Implication on Windhoek City Water Supply. *Phys. Chem. Earth* **2019**, *112*, 103–112. [CrossRef]
66. Tozer, L.; Hörschelmann, K.; Anguelovski, I.; Bulkeley, H.; Lazova, Y. Whose City? Whose Nature? Towards Inclusive Nature-Based Solution Governance. *Cities* **2020**, *107*, 102892. [CrossRef]
67. Midgley, S.J.E.; Esler, K.J.; Holden, P.B.; Rebelo, A.J.; Stuart-Hill, S.I.; Cullis, J.D.S.; Methner, N. Typologies of Collaborative Governance for Scaling Nature-Based Solutions in Two Strategic South African River Systems. *Ambio* **2021**, *50*, 1587–1609. [CrossRef]
68. Mell, I.C. Green Infrastructure: Reflections on Past, Present and Future Praxis. *Landsc. Res.* **2017**, *42*, 135–145. [CrossRef]
69. Hobbie, S.E.; Grimm, N.B. Nature-Based Approaches to Managing Climate Change Impacts in Cities. *Philos. Trans. R. Soc. B Biol. Sci.* **2020**, *375*, 20190124. [CrossRef]
70. Mguni, P.; Herslund, L.; Jensen, M.B. Sustainable Urban Drainage Systems: Examining the Potential for Green Infrastructure-Based Stormwater Management for Sub-Saharan Cities. *Nat. Hazards J. Int. Soc. Prev. Mitig. Nat. Hazards* **2016**, *82*, 241–257. [CrossRef]
71. Adegun, O.B. Green Infrastructure in Relation to Informal Urban Settlements. *J. Arch. Urban.* **2017**, *41*, 22–33. [CrossRef]
72. Forman, R.T.T. Town Ecology: For the Land of Towns and Villages. *Landsc. Ecol.* **2019**, *34*, 2209–2211. [CrossRef]
73. Kambites, C.; Owen, S. Renewed Prospects for Green Infrastructure Planning in the UK. *Plan. Pract. Res.* **2006**, *21*, 483–496. [CrossRef]
74. Thomas, K.; Littlewood, S. From Green Belts to Green Infrastructure? The Evolution of a New Concept in the Emerging Soft Governance of Spatial Strategies. *Plan. Pract. Res.* **2010**, *25*, 203–222. [CrossRef]
75. Delgado, G.; Muller, A.; Mabakeng, R.; Namupala, M. Co-Producing Land for Housing through Informal Settlement Upgrading: Lessons from a Namibian Municipality. *Environ. Urban.* **2020**, *32*, 175–194. [CrossRef]
76. Güneralp, B.; Lwasa, S.; Masundire, H.; Parnell, S.; Seto, K.C. Urbanization in Africa: Challenges and Opportunities for Conservation. *Environ. Res. Lett.* **2017**, *13*, 015002. [CrossRef]
77. Young, T.; Schwartz, M. The Decade on Ecosystem Restoration Is an Impetus to Get It Right. *Conserv. Sci. Pract.* **2019**, *1*. [CrossRef]
78. Haywood, L.K.; Funke, N.; Audouin, M.; Musvoto, C.; Nahman, A. The Sustainable Development Goals in South Africa: Investigating the Need for Multi-Stakeholder Partnerships. *Dev. S. Afr.* **2019**, *36*, 555–569. [CrossRef]
79. Cumming, T.L.; Shackleton, R.T.; Förster, J.; Dini, J.; Khan, A.; Gumula, M.; Kubiszewski, I. Achieving the National Development Agenda and the Sustainable Development Goals (SDGs) through Investment in Ecological Infrastructure: A Case Study of South Africa. *Ecosyst. Serv.* **2017**, *27*, 253–260. [CrossRef]
80. Dhakal, K.P.; Chevalier, L.R. Managing Urban Stormwater for Urban Sustainability: Barriers and Policy Solutions for Green Infrastructure Application. *J. Environ. Manag.* **2017**, *203*, 171–181. [CrossRef] [PubMed]
81. Salmond, J.A.; Tadaki, M.; Vardoulakis, S.; Arbuthnott, K.; Coutts, A.; Demuzere, M.; Dirks, K.N.; Heaviside, C.; Lim, S.; Macintyre, H.; et al. Health and Climate Related Ecosystem Services Provided by Street Trees in the Urban Environment. *Environ. Health* **2016**, *15*, S36. [CrossRef]
82. Lwasa, S.; Mugagga, F.; Wahab, B.; Simon, D.; Connors, J.; Griffith, C. Urban and Peri-Urban Agriculture and Forestry: Transcending Poverty Alleviation to Climate Change Mitigation and Adaptation. *Urban Clim.* **2014**, *7*, 92–106. [CrossRef]
83. Zhou, W.; Qian, Y.; Li, X.; Li, W.; Han, L. Relationships between Land Cover and the Surface Urban Heat Island: Seasonal Variability and Effects of Spatial and Thematic Resolution of Land Cover Data on Predicting Land Surface Temperatures. *Landsc. Ecol.* **2014**, *29*, 153–167. [CrossRef]
84. Jama, B.A.; Mohamed, A.M.; Mulatya, J.; Njui, A.N. Comparing the "Big Five": A Framework for the Sustainable Management of Indigenous Fruit Trees in the Drylands of East and Central Africa. *Ecol. Indic.* **2008**, *8*, 170–179. [CrossRef]
85. Wheeler, S.M.; Abunnasr, Y.; Dialesandro, J.; Assaf, E.; Agopian, S.; Gamberini, V.C. Mitigating Urban Heating in Dryland Cities: A Literature Review. *J. Plan. Lit.* **2019**, *34*, 434–446. [CrossRef]
86. Namibian Chamber of Environment Development Workshop—Namibia. *Update: January–March 2021 Programme for Appropriate Low-Cost Urban Sanitation DW-Namibia*; Namibian Chamber of Environment: Windhoek, Namibia, 2021.
87. Zuniga-Teran, A.A. Green infrastructure in walkable neighborhoods: A climate change adaptation strategy for cities in drylands. In *Climate Change Sensitive Cities: Building Capacities for Urban Resilience, Sustainability, and Equity*; Delgado Ramos, G.C., Ed.; PINCC, UNAM: Mexico City, Mexico, 2017.
88. Chioreso, E.; Begbie-Clench, B. *Fact Sheet on: Land Degradation—Implications for Food Security in Namibia*; Think Namibia; Desert Research Foundation of Namibia: Windhoek, Namibia, 2015.
89. Crush, J.; Hovorka, A.; Tevera, D. Food Security in Southern African Cities: The Place of Urban Agriculture. *Prog. Dev. Stud.* **2011**, *11*, 285–305. [CrossRef]
90. Thornton, A. Beyond the Metropolis: Small Town Case Studies of Urban and Peri-Urban Agriculture in South Africa. *Urban Forum* **2008**, *19*, 243–262. [CrossRef]
91. Bobbins, K.; Culwick, C. Green Growth Transitions through a Green Infrastructure Approach at the Local Government Level: Case Study for the Gauteng City-Region. *J. Public Adm.* **2015**, *50*, 32–49.

92. Adegun, O.B. Green Infrastructure Can Improve the Lives of Slum Dwellers in African Cities. *Front. Sustain. Cities* **2021**, *3*. [CrossRef]
93. Pendleton, W.; Crush, J.; Nickanor, N. Migrant Windhoek: Rural–Urban Migration and Food Security in Namibia. *Urban Forum* **2014**, *25*, 191–205. [CrossRef]
94. Moyo, T. Local Government and Green Jobs Creation: Exploring Opportunities in Selected Metropolitan Municipalities in South Africa. *J. Public Adm.* **2015**, *50*, 70–89. [CrossRef]
95. Gulati, M.; Becqué, R.; Godfrey, N.; Akhmouch, A.; Cartwright, A.; Eis, J.; Huq, S.; Jacobs, M.; King, R.; Rode, P. *The Economic Case for Greening the Global Recovery through Cities: 7 Priorities for National Governments*; Coalition for Urban Transitions: London, UK, 2020.
96. UNICEF. *Reimagining Our Future: Building Back Better from COVID-19*; United Nations Children's Emergency Fund (UNICEF): New York, NY, USA, 2020.
97. IISD South Africa and WRI Collaborate for Green, Equitable Urban Recovery—Sustainable Recovery. 2020. Available online: https://www.iisd.org/sustainable-recovery/news/south-africa-and-wri-collaborate-for-green-equitable-urban-recovery/ (accessed on 29 May 2021).
98. Escobedo, F.J.; Giannico, V.; Jim, C.Y.; Sanesi, G.; Lafortezza, R. Urban Forests, Ecosystem Services, Green Infrastructure and Nature-Based Solutions: Nexus or Evolving Metaphors? *Urban For. Urban Green.* **2019**, *37*, 3–12. [CrossRef]
99. Ambole, A.; Musango, J.K.; Buyana, K.; Ogot, M.; Anditi, C.; Mwau, B.; Kovacic, Z.; Smit, S.; Lwasa, S.; Nsangi, G.; et al. Mediating Household Energy Transitions through Co-Design in Urban Kenya, Uganda and South Africa. *Energy Res. Soc. Sci.* **2019**, *55*, 208–217. [CrossRef]

sustainability

MDPI

Article

Climate Change, Security, and the Resource Nexus: Case Study of Northern Nigeria and Lake Chad

Samuel Stephen Wakdok [1] and Raimund Bleischwitz [1,2,*]

[1] Bartlett School of Environment, Energy and Resources, University College London, London WC1H 0NN, UK; samuel.wakdok.19@ucl.ac.uk

[2] Institute for Integrated Management of Material Fluxes and of Resources, United Nations University UNU FLORES, 01067 Dresden, Germany

* Correspondence: r.bleischwitz@ucl.ac.uk

Abstract: This paper analyses the impacts of climate change and its implications for human security for the regions of Northern Nigeria and Lake Chad. The introduction identifies a gap between evidence on global environmental change and interactions on the ground; it positions the scope for a deeper understanding of the climate–security–resource nexus in Northern Nigeria and Lake Chad and consequences for the implementation of SDGs (Sustainable Development Goals). The section on methods describes the nexus concept and justifies adopting it. As a result of analysing the region, the paper sheds light on the conflict pathways triggered by failures in land grazing policy, which is further evidenced by a short comparison with Northern Kenya. A potentially novel contribution is discussed in terms of scaling up collaboration and green markets for the future of Lake Chad, along with an integrated agricultural nexus policy, both of which are ambitious in the spirit of mission-oriented policies and delivering on the SDGs.

Keywords: climate change; human security; resource nexus; Northern Nigeria; Lake Chad

Citation: Wakdok, S.S.; Bleischwitz, R. Climate Change, Security, and the Resource Nexus: Case Study of Northern Nigeria and Lake Chad. *Sustainability* **2021**, *13*, 10734. https://doi.org/10.3390/su131910734

Academic Editors: Adam P. Hejnowicz and Jessica P. R. Thorn

Received: 8 June 2021
Accepted: 17 September 2021
Published: 27 September 2021

Publisher's Note: MDPI stays neutral with regard to jurisdictional claims in published maps and institutional affiliations.

Copyright: © 2021 by the authors. Licensee MDPI, Basel, Switzerland. This article is an open access article distributed under the terms and conditions of the Creative Commons Attribution (CC BY) license (https://creativecommons.org/licenses/by/4.0/).

1. Introduction

Climate change is a reality and disproportionately impacts vulnerable societies in the Global South. The continuous rise in temperatures is a driver for changing weather patterns, shifting vegetation zones and sea level rise. People's livelihoods in the Global South are highly dependent on the biosphere. The recent Global Environmental Outlook (GEO 6), published in 2019 by the United Nations Environment Programme (UNEP), the latest Intergovernmental Science-Policy Platform on Biodiversity and Ecosystem Services (IPBES) Assessment Report, dated 2019, the ongoing work from the Intergovernmental Panel on Climate Change (e.g., IPCC AR 6 2021) and the World Bank's report series *Turn down the Heat* (2012–2014) clearly articulate evidence-based concerns over rising insecurity over water and food; they address risks for regions that might become inhabitable and trigger mass migration, with further repercussions on regional and global security. Altogether, these impacts will raise challenges in meeting the UN's Sustainable Development Goals (SDGs)—and, in particular, an alignment of goals related to environmental policy and an inclusive sustainable growth.

Our paper addresses a gap in the literature between evidence on global environmental change and interactions on the ground. Based on the notion of climate change as a stress multiplier [1], our paper starts from assumptions about multiple existing environmental stress factors affecting livelihoods and governance performance today. The paper adopts a nexus approach, which has been defined as a "set of context-specific critical interlinkages between two or more natural resources used in delivery chains towards systems of provision for water energy, food, land, and materials" [2,3]. The nexus approach is an emerging concept at the interface of academia and international organisations that is considered useful to assess risks and explore opportunities enabling the delivery of a range of SDGs.

Although numerous regional nexus case studies have been written, there is no such study yet for Northern Nigeria and Lake Chad addressing our scope.

The objective of our article is a deeper understanding of the climate–security–resources nexus in Northern Nigeria and Lake Chad and the consequences for the implementation of SDGs. In line with Biggs et al. [4], our paper underlines the relevance of land use in the nexus concept as well as for livelihoods. It is also in line with the IPBES' new work programme seeking to address the interlinkages among biodiversity, water food, and health in the context of climate change. This regional case study seeks to obtain evidence on impacts of climate change and security challenges in an interaction with environmental policy failures and future challenges. Taking into account the long-term character and the multi-level dimensions of related SDG policies, the paper will also develop three policy-relevant socio-environmental scenarios. Again, this is a novel contribution given a gap in the literature in futures thinking based on a geographic context with outlooks on the SDGs [5]. A recent paper by Thorn et al. [6] identifies such a gap after a literature review on participatory planning to envision the futures of mountain social-ecological systems around the themes of governance, economy, land use changes and biodiversity. In a similar research perspective, Miedzinski et al. [7] propose policy roadmaps and mission-oriented policies to help deliver the SDGs. Our article thus seeks to add novelty by (i) providing a multi-disciplinary case study on Northern Nigeria and Lake Chad and by (ii) developing exploratory scenarios about an inclusive and green growth in the region.

Addressing these gaps via our case is relevant because Nigeria is a large emerging economy confronting a range of security issues recently, and Lake Chad, which was one of Africa's largest freshwater bodies, has shrunk by some 90%. Over 10 million people across the region need emergency assistance. The United Nations has termed the Lake Chad crisis as "one of the worst in the world". We argue below that climate change was not initially high on the government agenda in Northern Nigeria, although it had already been disrupting lives and places around the world. Obviously, there were early proactive warnings about climate change becoming a major threat to global security due to the escalating competition for increasingly scarce resources [8] with impacts on human security [9,10], but only few scholars in Nigeria proactively postulated that climate change would pose a challenge to the natural resource base and policies in the country.

Testing our approach, the paper analyses the case with a specific focus on land use and grazing policies in Nigeria; we explore a security and conflict pathway from agricultural practices, land use policy and livelihoods, combined with long-standing competition over resources between farmers and migrating herdsmen. Doing so allows us to test the hypothesis of climate change being a stress multiplier rather than a direct driver of conflicts. Using the analytical nexus framework, the paper also compares our findings with a brief case study in Northwest Kenya written by Daher et al. [1]; conclusions are drawn with care. In line with our aim to use such analysis to help in delivering the SDGs, this article develops a vision and scale-up scenarios for green markets in the region by presenting three scenario narratives that aim to explore potential future transformations of Lake Chad. The article hopes to contribute to envisioning opportunities and risks that can be assessed using our nexus approach towards more inclusive and sustainable growth beyond 2030.

The next section elaborates on the nexus concept and how it can be used as a method for the scope of this paper. Subsequent Sections 3 and 4 analyse the case and develop scenarios. Afterwards, we summarise results and discuss findings related to our scope and in the wider context of inclusive green policies in Africa and the SDGs. Our conclusions develop messages from our work and recommendations for policy and research.

2. Methods: The Nexus Concept as a Novel Approach

The nexus concept has been emerging since the Bonn 2011 Conference, "The Water Energy and Food Security Nexus—Solutions for the Green Economy", with significant involvement from the UN and other international organisations. It has been defined as a "set of context-specific critical interlinkages between two or more natural resources used in delivery chains towards systems of provision" [3]. From a policy perspective, it looks at delivery chains of resources, such as water and energy, in a polycentric manner, i.e., as independent providers based on ecosystem services with interlinkages across delivery stages, but without a presumed hierarchy among those dimensions. Thus, water, energy and food are seen as interrelated and of equal priority for the SDGs, considering the specific conditions of their provision and the strategic interests of relevant actors.

Our paper draws on this nexus concept for the following reasons: it helps in understanding the interface between global and local drivers for environmental risks; it looks at the interface of using multiple natural resources and their delivery chains in a regional context; it contributes to a holistic understanding of the SDGs and strengthens those goals that cross-cut inclusive and sustainable growth (SDG 8), life on land (SDG15) and strong institutions (SDG 16).

While the nexus offers a promising conceptual approach, the development and use of specific rigorous methods to systematically evaluate interlinkages or support policy development has been limited. There is no specific nexus method yet, but rather a mix of different methodologies that are applied according to the scope of research, such as Input-Output Analysis based on water or other "footprints" and the use of Sankey diagrams, and a range of qualitative and semi-quantitative tools (e.g., Water Evaluation Planning, WEAP, Long-range Energy Alternatives Planning, LEAP). This paper uses the following methods in relation to the nexus approach: (i) a case study approach based on a scoped literature survey and interdisciplinary insights transcending environmental research and policy analysis, (ii) a cautiously crafted comparative approach with a paper analysing our scope for Kenya [1], where we seek to counter risks of oversimplification across both cases by key references underlining specific circumstances and (iii) a scenario approach addressing gaps in the literature as stated above that follows selected nexus papers on this topic and recent foresight literature [1,2]. While being a research-based concept, it is important to underline that the nexus does inform implementation strategies, for instance via a Water-Energy-Food Knowledge-Action Network hosted by Future Earth and the WEF security resource platform. As a bottom line, it is thought that the nexus concept helps to overcome a silo mentality leading to the SDGs being pursued in isolation, where interlinkages are potentially underassessed or even overlooked.

3. The Case of Northern Nigeria

Nigeria, with a population of about two hundred million people and thirteen million cattle, is faced with the impacts of both climate change and unsustainable population growth, as this population is expected to double by 2050. Nigeria has a land mass of about 923,800 km^2 and a total surface area of about 91 million hectares [11]. However, Nigeria's large and rapidly growing population is putting dire pressure on the environment and its limited resources, which are also threatened by climate change.

Writers like Okoli et al. [12] argue that desert encroachment triggered the disappearance of grazing and fertile lands, thus indicating a causation between climate change and rising human insecurity in Nigeria. But Benjaminsen et al. [13] opine that political failures such as allowing for ungoverned spaces, rent seeking and intrusion are the main factors underlining a lack of authority and legitimacy as drivers for conflicts and weakening security. Ironically, the latter did not consider the role of climate change as a possible reason for scarcity of resources and the incursion on farmlands by migrating herdsmen. We propose that resource scarcity, worsened by climate change and institutionally determined access to resources, have aggravated conflicts and insecurity in Northern Nigeria.

To illustrate this perspective, we briefly look back on the last twenty years. As the twentieth century came to an end, while some urban areas in Nigeria faced rising crime and social insecurity due to rapid unplanned urbanization [14], rural areas were generally safer in terms of social and physical security, with most of the dwellers earning their living from subsistence agriculture. There was generally a low level of crime, conflicts and physical insecurity among the rural dwellers. For quite a long time, the northern part of the country featured the Sahel savanna for farming and pastoralism. The central part, with its lowland rain forest and savanna, served as the food basket of the country, attracting nomadic herdsmen who moved southwards during dry seasons in search of pasture and water, and then northwards during the wet season. The southern part, with its heavy rain and mangrove forests, had plantation farmers, fishermen and some Fulani pastoralists (also called herdsmen); the latter have been moving across the northern and southern parts of the country between seasons. During those previous years, security was not beyond government control, while the citizens could engage in their socio-economic activities freely.

However, the changing climate now makes shifting weather patterns and water insecurity more unpredictable, extreme and stressful; those factors started to exceed the intimate understanding of natural rhythms associated with climate and weather across different temporal scales within the different groups competing for access to land. Rising competition by Fulani nomads for a depleting grazing land caused by climate change and overgrazing pitched the migrating herdsmen against indigenous farming communities [15]. According to Amobi and Onyishi [16], Nigeria, with its location and unique ecology, is now highly susceptible to the fluctuating effects of climate change, which further extends the insecurity as nomadic herdsmen from the north move downwards towards the central and southern states in search of grazing land for their cattle.

In terms of community cohesion and livelihood sustainability, communities have become increasingly insecure in a tightening competition over scarce resources, as the Sudan savanna of the northern and middle parts of the country transits to pure Sahel, and the influence of the Sahara increases southwards. As the ecology of the Guinea savanna gives way to Sudan savanna grassland, the nomadic herdsmen of the lower Sahel and Sudan savanna ecosystems migrate to the Guinea savanna and forest belt of the South [17]. This can be seen as a regime shift in the regional vegetation and land use cover, and part of a larger transition, or potentially an escalation, towards insecurity and conflicts.

Following Slettebak [18], our identification of a regime shift in land use in northern Nigeria is at a risky intersection between human security and conflicts. The migrating herders sometimes displace communities and farmers in search of perpetual grazing lands while their livestock ravage crops and farmlands, resulting in conflicts or rising insecurity. This is a departure from the past, when the herders stayed intermittently and both parties coexisted peacefully, without any threat of existentialism. Lacking enough grazing land to return northwards to during the wet season, the herders' sojourn southward is no longer temporary. States with more green vegetation such as Plateau, Benue, Taraba, Adamawa and Kaduna have seen various forms of confrontations and violent conflicts between migrating herders and communities over access to scarce resource of land and water exacerbated by climate change.

While Sayne [19] believes that no one knows the full security implications of climate change, it seems also fair to say that the government and people hitherto did not consider the full implication of climate change on security in Nigeria. A relevant observation in our case is the increasing ruthlessness of some attacks. Amidst different attacks on farming communities, an estimated 500 villagers were reportedly killed by suspected armed herdsmen in the farming community of Dogo Nahawa in 2010, and about 100 people were also killed in Barkin Ladi in June 2018, again by suspected armed herdsmen, all in Plateau State [20,21]. There have also been incidences of cattle rustling and killings of herdsmen in farming communities, thereby perpetuating the security breach.

We agree with Conroy [22] and Sayne [19] that climate change patterns leading to low rainfalls and approaching deserts, particularly in the last three decades, are responsible for the upsurge in the southward migration of the nomads searching for pasture and water. However, "owing largely to its plurality and ethnic divide, environment-induced migration creates volatile contact and competition between groups of highly conflicting natural resource-dependent livelihood systems" [23]. Because the herdsmen are mainly Fulani and predominantly Muslim, while several farming communities in the central part are largely

Christian, the conflicts sometimes assume a religious dimension [23]; this seems in line with De Juan and Heinze [24], who conclude on ethnic polarisation being relevant, but not a primary driver for conflicts. Ultimately, climate-induced and resource-driven migration substantially endangers security and creates socio-economic disorders. In our case, there is evidence of such movements escalating fierce competitions and skirmishes over access to natural resources between farming communities and the migrating herdsmen. We wouldn't deny that climate migration also leads to beneficial adaptive outcomes through, for example, allowing people to enter seasonally into the cash economy and send back remittances to their rural homesteads, which promotes development. Our case study, however, reveals little evidence for such benefits of migration, but instead points towards a different direction of increasing conflicts. The next section looks at underlying land use policy patterns that could potentially either drive conflicts or facilitate risk mitigation.

3.1. Land Use Policy Matters: An Appraisal of Grazing Policies in Nigeria

Land and water are key natural resources worldwide which are also used for the cultivation of crops and grazing of livestock. It is evident that the wide range of lands which preserve the earth's biodiversity also offer a means for sustenance all around the world. Though both resources and their usage are essential to human existence and ought to be utilized sustainably, human activities such as climate change contribute significantly to the availability or scarcity of land and water. Taiye [25] documents that there are in Nigeria, "about 210 persons and 180 grazing animals per kilometre square of land and 15,000 persons and 12,500 grazing animals per kilometre square of water, leading to high demand for food, water and fodder and stress on the environment". These statistics give a visual interpretation of how transhumance grazing, deforestation and irrigation reinforced by climate change are damaging the environment and biodiversity in Nigeria.

Climate change and other environmental factors also threaten the sustainability of land and water in Nigeria, as elsewhere, with serious implications for food, livelihoods, security, peace and sustainable development. This paper aligns with Okoli and Atelhem [12] on the eco-violence theory, whose key assumption is that insecurity and conflicts are being induced by competition over scarce natural resources. Climate variability is also rapidly worsening resource scarcity, hence creating a conflict pathway driven by scarcity, agricultural practices and migration. Next, we will analyse migration in the context of grazing policies in Nigeria.

3.2. Previous Grazing Policies

The southward migration of Fulani herdsmen and their encroachment on farmlands in search of freshwater and grassland for cattle, and the depletion of grazing areas because of unsustainable population growth, have exacerbated the conflicts between the herdsmen and farmers in Northern Nigeria. To address the violence over resources, the Nigerian government initiated or conceptualized different grazing policies: the grazing reserve, grazing route, anti-open grazing reserve, rural grazing area (RUGA) and the National Livestock transformation plan policies.

- Grazing Reserves and Routes Policy:

In 1964, Nigeria's first law on Grazing Reserve was introduced to settle the Fulani nomads on lands with pastures for their livestock, but the policy was poorly implemented. Envisaging the impending resource crisis and a low productivity, the then military government in 1988 decreed the National Agricultural Policy of 1988. At least 10% of the republic's approximately 10 million acres was to be reserved for grazing. A policy failure led to the acquisition of a dismal 3% of the proposed areas [26]. Again in recent years, there was an effort to create grazing routes and reserves in some selected states of the federation through the National Grazing Route and Reserve Commission Bill of 2011 [11]. It was rejected by the federal legislators who argued that it was unconstitutional for the federal government to set up grazing routes and reserves across states in a federal system [26]. This policy was also opposed in the central and southern states of Nigeria.

- RUGA Policy:

The Rural Grazing Area (RUGA) policy was designed to organize and settle pastoralists on lands providing basic amenities, infrastructure and markets. It is a prototype of 40 units of huts for 10 farmsteads on at least 20 hectares of land, mostly in the central and southern states [27]. However, fears from the people of these states led to uproars and fierce resistance. In questioning the policy and calling for its better handling, Nigeria's only Nobel laureate Wole Soyinka added his voice to the unpopularity of the policy [28]. Amid the uproar from some quarters, the government announced the suspension of the RUGA policy, claiming inconsistency with the approved national livestock transformation plan [29].

- National Livestock Transformation Plan (NLTP):

Consistent with the nation's public announcement trends, Nigeria's Vice President announced that the NLTP would modernise livestock production using a mix of nomadic breeding and ranching that would serve a modernised dairy and meat processing industry. Unfortunately, controversy still trailed the NLTP due to its similarity with the unpopular RUGA policy, because many people believed that the NLTP was an alternate method of implementing the RUGA policy.

- Anti-Open Grazing Policy:

Ekiti and Benue states were the first states to introduce the anti-open grazing law to tackle the menace of open grazing and the accompanying conflicts. Against the odds, the 2016 Ekiti state anti-grazing policy barred the grazing of cattle and other animals on any land in the state which has not been authorized for ranching by the governor [30]. The Benue state government, effective November 2017, also banned open grazing and requested the setting up of ranches [23]. While states with high casualty like Plateau and Adamawa do not have an anti-grazing law, Miyetti Allah Cattle Breeders Association of Nigeria (MACBAN), an ethno-cultural umbrella body of the Fulani herdsmen, decried the law as contravening the nation's constitution on freedom of movement and the right to settle anywhere in Nigeria [31].

We conclude here on the long-standing experience with migration and grazing policy in Nigeria; what has changed is the more permanent migration induced by resource scarcity over land and water, as well as the increasing violence and security ramifications. The recent anti-open grazing policies in some states may postpone conflicts there, if implemented properly; however, they are likely to lead to a regional shift with increased conflicts in other states, and will need proper enforcement. The next section, therefore, looks at a case with similar environmental challenges in a comparable geographical setting but slightly different institutional mechanisms and development challenges.

3.3. A Brief Comparative Analysis of Nigeria and Kenya

Following a similar nexus approach, Daher et al. [1] assess the climate-security-resource nexus in line with other work on the nexus interlinkages and ramifications across utilization of resources [3]. Northwestern Kenya comprises seven counties and is considered arid and semi-arid land, where the predominant livelihood is pastoralism. This same region in Kenya is affected by climate change and, consequently, a rising level of climate induced insecurity [1]. Much like in Nigeria, climate change is manifested in the semi-arid northwestern and northeastern states, but the security implications linked to it due to desertification and low rain falls extend mostly to the north central states of Benue, Plateau, Nasarawa, parts of Kaduna and some northeastern states like Adamawa and Taraba. The first quarter of 2018 witnessed attacks in Benue and Plateau states by alleged armed herdsmen as they sought to expand downwards in search of land and water for grazing [32]. We do not claim to search for specific reasons but observe that, unlike in Kenya's case, the climate security nexus in Nigeria is transported beyond the original location where the climate impact occurs. Just to note, one could argue that in northwestern Kenya, climate impacts are also quite pronounced in different regions from where the impacts are directly seen; for example, around Lake Turkana where 300,000 people live on the border of the Omo valley in Ethiopia, Uganda and South Sudan and have had food, water and energy security impacted by climate change as it becomes more difficult to access grazing land and water. If [33] was right, that nearly 64% of the Nigerian landmass is threatened by desertification which is likely to affect almost 65 million people directly or indirectly, we can then postulate that this figure considered the people both in the climate endangered zone and those in other climate-induced insecurity zones.

Although Nigeria operates a Land Use Act promulgated in 1978 which vested power of lands on the subnational governments [26], most lands in the communities are either privately owned or ancestrally inherited. Most rural lands are not communally owned in Nigeria, unlike the lands in northwestern Kenya, which are jointly owned by the community under group ranches [34]. So far, there are fewer negative reactions to the privatization of lands in Nigeria than in Kenya, based on the article by [1]. However, there are also various tenure regimes in northwestern Kenya with varying degrees of tenure security. This is particularly the case with increasing privatization around Lake Turkana, where there is the largest investment in windfarms in Africa. The same is true in other areas in the north, where since the 1930s there has been a lot of land acquisition by British settlers. The problem in Nigeria, however, is the displacement of farmers or communities from their ancestral lands and villages by armed herdsmen, who sometimes invade the communities or encroach on farmlands to graze their cattle. In reaction to the Federal Government's attempt to create grazing reserves around the country [35], there were calls for the private acquisition of land for ranching by interested herders.

There is an increasing community-based conservation in Kenya's northwestern region [1], which in our opinion has dual implications. As the region battles with climate change, conservation appears to become an appropriate response—although it may limit land available for grazing in the short run. This puts land use under additional stress. There is quite extensive degradation and deforestation in northwest Kenya—leading to soil nutrient mining, soil erosion, the loss of livestock weight and thus the quality and quantity of meat and milk, as well as loss of biodiversity [36,37]. Comparing this deterioration with Nigeria, the migrating herders and the farmers are both deforesting the region, which has dual implications with potentially severe impacts over the next years [38]. The accompanying loss of ecosystem services exposes the region to the approaching desertification and intensifies the conflict, as both farmers and pastoralists compete for waning land and water resources. The long run implication is worsening human and environmental security as land and water become scarcer.

While the region in Kenya is undergoing massive development leading to transformation [1], the same cannot be said of the region in Nigeria. Plummeting crude oil revenue, affecting government expenditure and wider insecurity, have led to a lull in development activities in these areas with consequences for livelihoods, further undermining security as manifested in the sharp rise in kidnappings for ransom. Nigeria faces a more perilous security situation than Kenya, as Table 1 shows. Nigeria is currently the third most terrorized country in the world and the first in Sub-Saharan Africa, while Kenya holds the 23rd position globally and the 10th in Sub-Saharan Africa. Terrorism is defined in the Global Terrorism Index (GTI), published by the US Institute for Economics and Peace (IEP), as "the threatened or actual use of illegal force and violence by a non-state actor to attain a political, economic, religious, or social goal through fear, coercion or intimidation" [39].

Table 1. Comparison of Nigeria and Kenya on SSA Global Terrorism Index (GTI) 2002–2019 (GTI 2020 (p. 50), US IEP).

No.	Country	Overall Rank	Regional Rank	Change 2002–2019	Change 2018–2019	Overall Score
1	Nigeria	3	1	4.805	−0.286	8.314
2	Kenya	23	10	1.011	−0.100	5.644

According to the Global Terrorism Index, Nigeria had the worst terrorist attacks in Africa, with deaths from the alleged armed herdsmen accounting for 26% of terror-related deaths in 2019 [39], an indication of increasing insecurity and resource conflicts between the herdsmen and farming communities. We refer to [40] for a statement on climate change and terrorism reinforcing each other through a response loop; however, we also point to the need for more in-depth research on causal loops to also address religious and political issues in a conflict pathway analysis. We conclude that the resource nexus security analytical framework used by [1] in Kenya is useful to apply to our case study of Nigeria to assess impacts of climate change and challenges for governance arising through changing patterns of land use and migration; however, we note that the scope and context differ, and both the violence and the vulnerability seem larger in northern Nigeria due to those specific conditions.

4. The Climate-Security-Resources-Livelihoods Nexus in the Lake Chad Region

Given the challenges in the region and expected severe impacts of climate change, we are posing the question of whether it is possible to restore Lake Chad to the stable basin that it was several decades ago and, if so, would that be sufficient to rectify some of these challenges and issues that the subsequent section identifies.

4.1. The Transformation of Lake Chad

Lake Chad is an exceptional freshwater lake located on the edge of the arid African Sahara. For 1000 years, until around 60 years ago, Lake Chad was the fourth largest lake and one of the largest bodies of freshwater in Africa, providing sustainable livelihoods for more than 35 million people through farming, fisheries and livestock (Figure 1) [41]. Ref. [42] reports that Lake Chad, home to 120 species of fishes, 372 birds, 44 algae and reedbeds was a key contributor to global biodiversity. Situated in the northern central part of Africa, Lake Chad surrounds four countries—Chad, Cameroon, Niger and Nigeria, with the areas around the lake in these countries referred to as the Lake Chad region. However, the Lake Chad Basin spreads over eight countries: Algeria, Cameroon, Central African Republic, Chad, Libya, Niger, Nigeria and Sudan, covering nearly 8% of Africa. Geographically, the Chari River, which flows through Chad from the Central African Republic, provides Lake Chad with 90% of its water, and the Logone River, coming from Cameroon, meets the Chari River somewhere in N'Djamena before connecting with Lake Chad. The level of the lake's water is associated with the variation in the flow from the tributaries which depend on rainfall [43].

However, because of climate change and droughts starting from the 1970s, the inhabitants of the Lake Chad region started following the receding shores as the waters retreated, thereby causing a mass migration and convergence around the lake and its numerous islands, inevitably putting pressure on the lake through irrigation and over utilization of the lake's resources. While [42] and Odada et al. in 2005 [44] agree that Lake Chad shrank to 2000 km^2 from its 1960s peak of 25,000 km^2 and is on the verge of vanishing, [45] contend that the lake has since expanded to roughly 14,000 km^2. However, despite improved rainfall since 2002, Lake Chad has not recovered, evidently because of Cameroon's Maga dam which diverted 70% of the Logone River from flowing into the lake to rice farms [45]. Whatever opposing views exist, the consensus is that climate change led to the splitting of the endangered lake into two uneven water bodies, the northern and southern pools, spotting a drought-prone, shrunken landscape within the arid and semi-arid Sahel strip [46]. Addressing this data uncertainty is beyond the scope of this paper.

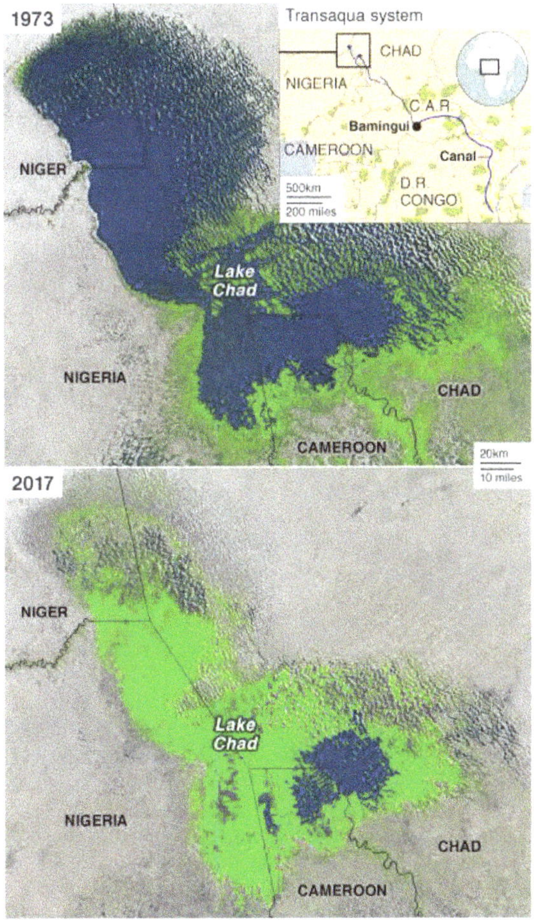

Figure 1. Lake Chad, 1973 and 2017; Source: UN.org.

4.2. Upcoming Security Challenges in the Region

We now analyse the transboundary security character of our case. The Boko Haram insurgency makes it difficult for climate change adaptability due to dislodgement of the people, limitations of movement and denied access to resources. This worsened the situation in the Lake Chad region as the insurgency, which started on the Nigerian side in 2009, spilled over to Chad, Niger and Cameroon, triggering a multifaceted humanitarian crisis [47]. While there is limited evidence to show causation between the climate change-induced shrinking of the lake and the insurgency, there is however a strong correlation, because of the loss of livelihood opportunities and the need for the inhabitants of the Lake Chad region to travel further in search of water and fodder due to, in part, climate change and the diminishing water levels of the lake. Hence, many men and youths of the region were easily recruited into the violent insurgency and terrorism that has killed over 20,000 people, displaced almost three million, exposed more than five million to hunger, put eleven million people in need of humanitarian aid and destroyed infrastructure in the region [48]. A full investigation of those recruitment patterns, however, is beyond the scope of this paper; it may well be that negative impacts on livelihoods and ways of life causes disenchantment and disenfranchisement from the status quo, and more variables would need to be assessed.

Nigeria retained the position of the third most terrorized country in the world for the fifth consecutive year in 2019, as depicted in Table 2. In 2018, there was a 33% rise in deaths, while those linked to the extremist herdsmen groups rose by 261% [39]. However, in 2019 there was a 39% decrease in total terrorism deaths and a 27% fall in terror related incidents. The decline was ascribed to a substantial fall in violence linked to Fulani extremists, which are linked to the armed herdsmen [39] (see also Section 3 above).

Table 2. Global Terrorism Index 2014–2019 and the rank of Nigeria. (GTI Reports 2015–2020, US IEP).

S/No	Year	Rank	Score	Change
1	2019	3	8.314	-
2	2018	3	8.597	-
3	2017	3	9.009	-
4	2016	3	9.314	-
5	2015	3	9.123	−1
6	2014	4	8.58	

Terrorism has affected the Nigerian economy negatively (Table 2), costing it 2.4% GDP in 2019, down from 2.7% in 2018. In 2018 alone, armed Fulani groups, which are part of the resource conflict, accounted for 1158 killings. This greatly reduced in 2019 to 325 fatalities, indicating a fall of 72% [39]. While climate variability affects land, the current security situation makes it difficult for those living in volatile areas like the Lake Chad area to explore dry season farming, using the stream for irrigation. In the worst case when the streams dry up, they can no longer access the lake itself, sealing off any means of livelihood.

We emphasise a pathway to insecurity with climate change as an underlying indirect driver while security ramifications are triggered by land use policy practices and by the emergence of organized crime and terrorism with "intrastate and regional interstate conflict pattern" threatening the stability of resource-dependent states such a Nigeria. This is in line with, e.g., [49] and the proposed drivers of a new vulnerability, as well as findings provided by [10] about institutional factors in analysing armed conflicts related to climate change. In accordance with our aims to be policy-relevant and contribute to foresight, we therefore see a need to conceptualise climate justice from the Green Climate Fund, with conditions capturing livelihood and land use policy practices as well as armed conflict patterns; it is not about compensating fossil fuel dependent states but strengthening basic institutions on the ground and building capacity [5,10]. We argue below that such funds or new financial mechanisms could be used to potentially recharge Lake Chad, halt desertification and deforestation, and reclaim desert lands to provide enough land for grazing and farming to tackle resource scarcity.

4.3. Three Scenarios to Explore Possible Futures for Lake Chad

In line with a stated lack of scenario-based studies in [5] and recent thinking on transformative sustainability research [50], we propose three diverse, plausible desired visions for Lake Chad. The aim of these scenarios is to explore the risks and opportunities of how capital, technology and collaboration could turn the region into an integrated hub as a future destination for sustainable finance and capacity building. Methodologically, it follows a mission-oriented road mapping approach to deliver the SDGs [7]. Singapore's successful "swamp to skyscrapers" transition makes this vision achievable. The reality this vision seeks to create is a safe place where people, planet and possibilities are assured of survival, security and sustainability with one of the scenario models.

Three scenarios are built for this vision. The scenarios are developed by a combination of literature review and expert judgement; a next step could be a participatory co-creation process with stakeholders. While the first scenario is limited in scope, the third scenario is more out of the box and ambitious and extends the scale of the second scenario through a more multi-regional approach. The sources of data include: [42,49,51–54].

Scenario 1—Better water engineering driven by Nigeria

This is the lowest cost scenario proposed for Nigeria. The Federal Government of Nigeria channels water from Rivers Niger and Benue, that cause perennial flooding in many states of the country, to the Lake Chad via the Hadejia, Jema'are, Komadugu and Yobe Rivers, which currently contribute about 3% of the flow into the lake. Other tributaries will actively contribute if properly channelled to feed into the major artery of the lake. The target year to achieve this is 2027 at a hypothetical cost of $5 billion, mainly being applied to the transfer of water, although the increased availability of water will not adequately tackle the main cause of shrinking of the lake if the over-exploitation is not first addressed. However, in the absence of other options, this offers a quick stop gap to the shrinking of the lake, especially at its northern pool, which is vanishing and provides some livelihood to people at the mouth of the lake.

Scenario 2—Regional collaboration

This is a possible scenario built upon the ambitious vision of an interbasin water transfer from the Congo Basin's Ubangi River, over 2400 km from the Democratic Republic of Congo. It requires a regional collaboration among the four countries of the Lake Chad region, Cameroun, Chad, Niger and Nigeria, who must pool resources and raise international capital. "The Ubangi River basin straddles the equator from 8° North to about 12° South of latitude, its geographic position straddling the equator, and its large size makes this basin relatively unaffected by the seasonal variations in its flow" [42]. With a length over 1000 km, the River Congo's largest tributary is the most renewable and sustainable source of recharging Lake Chad if diverted to the Chari River, which provides 95% of Lake Chad's water at the southern pool. This scenario halts the shrinking of Lake Chad and boosts its water to a minimum of 90% of its pre-historical level. Eventually, this will return livelihoods in fishing, farming and livestock across West and Central Africa and enhance sustainability. This intake from the Ubangi River amounts to less than 5% of the water in the river and only 0.25% of the water from the Congo River that empties into the Atlantic Ocean. For the sake of our scenario, it is estimated to take about 5 years to refill Lake Chad, so this scenario can be achieved by the year 2030 at an estimated cost of $23 billion according to the Lake Chad Basin Commission.

Scenario 3—Scaling up collaboration and green markets

This is the most preferred, though most expensive, scenario. It is built on the bolder, ambitious vision of scaling up scenario 2 beyond recharging the lake to a total quality vision of transfiguring Lake Chad. The pathway for this scenario is, in addition to scenario 2, create an international river with associated transport system that can aid the movement of people, farm produce and other goods to lift trade. Using the Suez Canal model will open the region around the land-locked countries of Chad and Niger, with those of Nigeria and Cameroon thereby creating a conurbation to form an international maritime city. This scenario is built to incorporate the other countries of the Lake Chad Basin Commission, i.e., Libya, Sudan, Algeria and the Central African Republic.

This scenario requires constructing almost 3000 km of navigable channel, which will serve as a realistic alternative to the Lagos–Mombasa Highway meant to connect East and West Africa, but has suffered many topographical challenges. There will be a flow control dam at Palambo between Congo, Bangui and the Democratic Republic of Congo, and construction of a 160 km long underpass to link the Chad and Congo basins. It needs river dredging which can discharge close to 1000 m^3 of water per second and will reconnect the northern and southern pools of Lake Chad.

An extra 500 km overflow of water would irrigate the Lake Chad border arid lands and repel encroachment of the Sahara Desert. A sustained tree planting drive of about 100 million trees a year for forestation and afforestation would be needed, along with other measures to enhance carbon sinks in the region and reduce emissions as well as to enhance ecosystem services. Wind and solar farms will be built, capable of generating about 10,000 megawatts of power to be distributed pro rata among the countries. The target year for full completion is 2040 and, according to the UNDP, is projected at a cost of $50 billion. However, the capital for this scenario can only be sourced by an alliance of International Consortia from the Lake Chad Commission countries, multi-lateral lending agencies and international financial institutions. The stakeholders could use the Design-Build-Finance-Maintain-Operate model for a return on their investments. The advantage of raising international capital is the additional buy-in from international stakeholders, eventually even with provision of international security from a trans-national military with naval, air force and marine bases, to drive and safeguard this huge investment and help address the security challenges in the region through additional policy learning.

5. Results

Our analysis leads to the following results:

5.1. A Nexus Approach Is Useful for a Deeper Understanding of Climate Change–Land Use–Security

Our case study confirms the usefulness of the nexus approach, assessing critical interlinkages and looking beyond particular sectors. Especially, the extension of the nexus to land use and related policies helps to understand the drivers of insecurity. The critical threshold is our identification of a regime shift in land use. Nomadic herdsmen of the lower Sahel and Sudan savanna ecosystems migrate from northern Nigeria to the Guinea savanna and forest belt of the South [17]. Together with displacements, this can be seen as a dangerous intersection with human security. Such a regime shift in the regional vegetation and land use cover can become part of a larger escalation towards insecurity and conflicts. Conflicts are aggravated through a religious dimension [23] because the herdsmen are mainly Fulani and predominantly Muslim, while several farming communities in the central part are largely Christian. At the same time, there is an increasing ruthlessness to some attacks and an enhanced involvement with the terrorism of Boko Haram. The recent anti-open grazing policies in some states may postpone conflicts there, if implemented properly; however, they will need proper enforcement and are likely to lead to more regional shifts with increased conflicts in other states. In relation to the broader debate on climate and security, as put forward by Uexkuell and Bulhaug [5], it is therefore the intersection of climate change with land-based ecosystem services, agricultural and migration patterns, and slow and inadequate governance responses, that put people at risk and drive conflicts in the region. Our paper adds nuances on land use policies, long-standing migration patterns and socio-cultural norms that act as triggers for conflicts and potential ramifications in the future.

5.2. A Comparison Is Difficult, but Can Underline Findings on Land Use Policy and Socio-Cultural Norms

Using the same resource nexus security analytical framework as used by [1], we see similar impacts of climate change and challenges for governance in Kenya arising through changing patterns of land use and migration. Assessing ecosystem services properly and providing access to land for livelihoods is key in both regions. However, we note that the scope and context differ. Both the violence and the vulnerability seem larger in northern Nigeria due to the displacements as an outcome of land use policy practices and by the emergence of organized crime and terrorism. Our brief comparison thus underlines that it is the displacement issue as a consequence of land use policy and socio-cultural norms and not the ownership models that make the difference. As a consequence, Nigeria and the Lake Chad region appear at larger risk, despite new global players arising in Kenya in the form of oil companies who may trigger corruption and other resource governance challenges. Therefore, we emphasise a pathway to insecurity starting with impacts from climate change but triggered by local and regional drivers of a new vulnerability stemming from access to grazing land, food and water insecurities and displacements, all combined with institutional failures and socio-cultural factors. This is in line with [50] as well as findings provided by [10]. Those regional contexts and policies seem to have a stronger impact compared to global drivers and global players.

5.3. Despite Challenges, Scenarios Can Open Opportunities

Despite pointing at challenges and potential escalation of conflicts in the future, this paper is combining a nexus approach with mission-oriented policies to identify more sustainable pathways. While issues of water and land use interact quite severely and require expertise in engineering and technology, our results underline the need to address policies and socio-cultural factors, too. Our third and most ambitious scenario transcends water engineering and combines it with access to food and energy and international collaborations. It overcomes the risks of siloed approaches inherent to single sectoral approaches and opens opportunities for a range of SDGs. Proposed in an integrated manner, a strength is the long-term vision for the year 2040 with an International Maritime Zone as a catalyst to a Continental Green Agro-allied Economic Zone. In line with our other results, political buy-in from the Lake Chad Commission countries, other regional stakeholders and international organisations will be key to establishing a platform and mobilising investments for such a policy roadmap to be sustained.

6. Discussion

6.1. A Systemic Policy Option on Grazing: A More Integrated Agricultural Nexus Policy

Grazing policies are key to resolving the challenges in Nigeria and around Lake Chad. Our identification of such policies as a critical variable on how resource interlinkages are unfolding should lead to thoughts about response options. A straightforward yet difficult policy option available to Nigeria could be an outright ban of open land grazing, which is not popular with the nomadic herdsmen and could lead to increased tensions or conflicts. The more moderate option is the development and implementation of a ranching policy which will flow from the former, but the trade-off is changing the lifestyle of the nomads and the huge monetary cost of providing pastures in the ranches. The third option, on which we elaborate more below, is to review and delineate modern grazing routes in a manner to guide open grazing and prevent encroachment into farms and destruction of the ecosystem while having the ability to provide some sort of open ranching.

This paper suggests using the nexus approach to formulate an Integrated Agricultural Nexus Policy to sustainably tackle the growing scarcity of land and water for crop farming and animal grazing, and the conflicts that arise. This is a follow-on based on recent findings on climate and security [5,10] and will allow for an innovative land use change that integrates forestry, crop farming and livestock grazing. This integrated system will benefit immensely from the interbasin water transfer which comes with the recharging of Lake Chad, especially as detailed in the third scenario developed in this paper. The advantages of this policy include the eco-friendly nature of systems integration, addressing the challenges and needs of multiple stakeholders and the environment. As tested in Brazil, this paper agrees that the four types of system integration can be successfully adopted methodically or comprehensively in Nigeria based on location specifics and availability of funds. "The crop-livestock system which integrates production of grains, grasses and animals; the livestock- forestry system which integrates production of grasses, animals and trees; the crop-forestry systems which integrates production of grains and trees, and finally the crop-livestock-forestry system which integrates production of grains, grasses, animals and trees" [55]. Such policy, if properly developed and adopted, will reclaim land and check desertification. An Integrated Agricultural System will facilitate more sustainable agriculture, which ultimately allows for recycling of nutrients and natural resource efficiency, and also supports incorporating local participation in all the policy making stages as a critical success factor [56]. At a more advanced level, agrovoltaics could also contribute to electricity supply [57]. The proposed green zone could also include sustainable cities, industrial parks, free trade zones and eco-tourism locations. Known for droughts and famine, the region will henceforth have resilient levels of water and food security, more energy security, and the people could be granted special citizens' status of "Lake Chad Sustainable Multi-Cities" across the member countries.

The merits of this policy roadmap include responses to climate change, a more optimal and fair utilization of land, forest and water resources, improved food production, increased yield from livestock, agricultural symbiosis, reduced hunger, prevention of conflicts over resources, and potential energy supply, all of which will counter the pathways from resource scarcity to conflicts and will contribute to implementing a range of SDGs. It is also in line with a recently launched programme at the Consortium of International Agricultural Research Centers (CGIAR) on the nexus across Water-Energy-Food-Forest-Biodiversity systems that will be conducted in other regions and may allow for lessons learned; we also note comparability to a study on climate adaptation and governance in an Alpine region conducted by Cattivelli [58]. Our results also feed into IPCC work on co-benefits and ongoing IPBES work on benefits people can obtain from ecosystem services at IPBES.

6.2. Challenges in Scenario Development and Impact Assessments

The ambitious vision of transforming Lake Chad comes with some challenges that must be addressed. The first challenge is funding, because countries of the Lake Chad region are low-income countries which do not have the capacity to mobilize such huge domestic resources in terms of funds, technology and expertise. International capital and technical assistance are needed to achieve this vision. While there have been some pledges by international donors, the outbreak of Covid-19 poses a danger, as countries and the global economy face an unprecedented health pandemic and economic downturn. Our case, however, does raise issues of environmental justice, as large parts of Africa suffer disproportionately from the impacts of climate change while having the lowest footprint per capita of 1 t per year and an insignificant share of about 3.8% to the world-wide greenhouse gas emissions [11].

Our proposal to reassess the Green Climate Fund and add conditions capturing livelihood and land use policy practices as well as armed conflict patterns would seek to overcome the finance challenge. Another challenge is the insecurity earlier described as the ongoing insurgency ravaging the region, which poses a great threat to lives and investment needed during the construction around the four littoral states. Adding capacity on policy learning about security and merits of collaboration will therefore be pivotal.

There is also the challenge of colonial history and language differences. Nigeria is the only Anglophone country of the four Lake Chad region countries. The Francophone countries have more affinity to France, which may lead to interventions, misunderstandings and red tape. In many instances, corruption and policy inconsistency/reversals are challenges that such capacity would need to encounter. It would also need to address mitigating environmental and social risks the project may pose during the construction of dams and dredging of rivers, which can lead to flooding, displacement and disruption of lives and livelihoods if managed based on limited knowledge and in a non-collaborative manner. To address these challenges and succeed, the Lake Chad Basin Commission could be reactivated and properly funded by member countries to carry out its mandate.

Our findings stress the relevance of conducting both environmental and socio-economic impact assessments for the proposed changes; however, our emphasis on financing and institutional factors rather than technical optimisation is in line with recent analysis on transition management and modelling [2,6,51,59]. Doing so for countries such as Nigeria and our case study region will need to integrate land use policy in pathways to sustainability and policy roadmaps. It could feed into recent programmes on climate-compatible growth and funding lifelines of resilient infrastructures carried out by international donors and the World Bank.

7. Conclusions and Recommendations

Studying the intersection of climate change, resource scarcity and security is pivotal to understanding multiple existing environmental stress factors and how they affect livelihoods and governance. Based on a nexus approach, our paper confirms failures in land grazing policy in northern Nigeria and the Lake Chad region as a main driver for displacements, which in turn lead to regime shifts in agriculture practices, deterioration of ecosystem services and violent conflicts. In comparison, climate change is more of an underlying driver and stress multiplier. Our case study demonstrates pathways of how environmental impacts, land use policy and socio-cultural factors induce migration, rise tensions and accelerate security challenges in the region and beyond. This is a more inclusive nexus compared to water-energy-food, which clearly shows the need to perform policy analysis with a security angle, studying implications for resources and livelihoods in northern Nigeria.

The implication for research and policy, however, does not need to be pessimistic. In contrast, our approach to developing nexus scenarios for the future in the Lake Chad region and thoughts on an Integrated Agricultural System Policy can be considered novel and can also be utilized to deliver a range of SDGs, especially the SDGs related to food, water, land use and inclusive and sustainable growth. Going beyond the timeframe of the SDGs, transforming the vanishing Lake Chad by 2030–2040 would significantly benefit about 50 million residents of the Lake Chad region. It will revive Lake Chad, will provide livelihoods and security, transform the economy of the lake Chad basin, restore biodiversity and could contribute to reversing main climate impacts. Other policy implications of this paper are the implementation of the suggested Integrated Agricultural System Policy in Nigeria, which will rely on the recharging of Lake Chad and the immediate implementation of the African Continental Free Trade Agreement by the four countries of the Lake Chad region, and possibly the eight members of the basin. If properly formulated and implemented, the Integrated Agricultural Nexus policy can lead to future developments that will see Nigeria becoming a large exporter of crops, livestock and dairies given its size, location and people.

We however realise the limitations of our research and advise that further data-driven and stakeholder-oriented research could be conducted and propose collaborations between academics and policy makers, among member countries and with the international community, to assess risks and address the challenges around collective capacity building associated with our SDG vision for Lake Chad and northern Nigeria. The strands of transition management, participatory modelling and mission-oriented policies should be able to inspire more research at the intersection of climate change, security and the resource nexus in the future.

Author Contributions: Conceptualization S.S.W. and R.B.; methodology R.B. and S.S.W.; analysis, investigation and data S.S.W., writing original draft preparation S.S.W., writing review and editing R.B. and S.S.W., supervision R.B. All authors have read and agreed to the published version of the manuscript.

Funding: This research received no external funding.

Institutional Review Board Statement: Not applicable.

Informed Consent Statement: Not applicable.

Data Availability Statement: Data sharing not applicable.

Acknowledgments: We wish to thank Kate Litwinczuk for support with editing and we express our gratitude to the UCL MSc programme on Sustainable Resources for bringing us together. We also wish to thank the guest editor, Adam Hejnowicz, for his patience with our manuscript and his useful advice.

Conflicts of Interest: The authors declare no conflict of interest.

References

1. Daher, B.; Lee, S.; Mohtar, R.H.; Asaka, J.O.; Van Deveer, S.D. Security, climate change and the resource nexus. In *Routledge Handbook of the Resource Nexus*, 1st ed.; Bleischwitz, R., Hoff, H., Spataru, C., Van Der Voet, E., Van Deveer, S., Eds.; Routledge: London, UK, 2017.
2. Bleischwitz, R.; Spataru, C.; VanDeveer, S.D.; Obersteiner, M.; van der Voet, E.; Johnson, C.; Andrews-Speed, P.; Boersma, T.; Hoff, H.; Van Vuuren, D.P. Resource nexus perspectives towards the United Nations sustainable development goals. *Nat. Sustain.* **2018**, *1*, 737–743. [CrossRef]
3. Andrews-Speed, P.; Bleischwitz, R.; Boersma, T.; Johnson, C.; Kemp, G.; VanDeveer, S.D. *Want, Waste or War? The Global Resource Nexus and the Struggle for Land, Energy, Food, Water and Minerals*; Routledge: London, UK, 2011.
4. Biggs, E.M.; Bruce, E.; Boruff, B.; Duncan, J.M.; Horsley, J.; Pauli, N.; Imanari, Y. Sustainable development and the water–energy–food nexus: A perspective on livelihoods. *Environ. Sci. Policy* **2015**, *54*, 389–397. [CrossRef]
5. Von Uexkuell, N.; Buhaug, H. Security implications of climate change: A decade of scientific progress. *J. Peace Res.* **2021**, *58*, 3–17. [CrossRef]
6. Thorn, J.P.R.; Klein, J.A.; Steger, C.; Hopping, K.A.; Capitani, C.; Tucker, C.M.; Marchant, R. A systematic review of participatory scenario planning to envision mountain social-ecological systems futures. *Ecol. Soc.* **2020**, *25*, 6. [CrossRef]
7. Miedzinski, M.; Mazzucato, M.; Ekins, P. A Framework for Mission-Oriented Innovation Policy Roadmapping for the SDGs: The Case of Plastic-Free Oceans. Available online: https://www.ucl.ac.uk/bartlett/public-purpose/wp2019--03 (accessed on 17 May 2021).
8. Stirrup, J. British Defence in a Changing World. *RUSI J.* **2007**, *152*, 20–25. [CrossRef]
9. Adger, W.N.; Pulhin, J.M.; Barnett, J.; Dabelko, G.D.; Hovelsrud, G.K.; Levy, M.; Oswald Spring, Ú.; Vogel, C.H. Human security. In *Climate Change 2014: Impacts, Adaptation, and Vulnerability. Part A: Global and Sectoral Aspects*; Field, C.B., Barros, V.R., Dokken, D.J., Mach, K.J., Mastrandrea, M.D., Bilir, T.E., Chatterjee, M., Ebi, K.L., Estrada, Y.O., Genova, R.C., et al., Eds.; Cambridge University Press: Cambridge, UK; New York, NY, USA, 2014; pp. 755–791.
10. Mach, K.J.; Kraan, C.M.; Adger, W.N.; Buhaug, H.; Burke, M.; Fearon, J.D.; von Uexkull, N. Climate as a risk factor for armed conflict. *Nature* **2019**, *571*, 193–197. [CrossRef] [PubMed]
11. Amusan, L.; Abegunde, O.; Akinyemi, T.E. Climate change, pastoral migration, resource governance and security: The Grazing Bill solution to farmer-herder conflict in Nigeria. *Environ. Econ.* **2017**, *8*, 35–45. [CrossRef]
12. Okoli, A.C.; Atelhe, G.A. Nomads against Natives: A Political Ecology of Herder/Farmer Conflicts in Nasarawa State, Nigeria, Okoli, Al Chukwuma and Atelhe George Atelhe. *Am. Int. J. Contemp. Res.* **2014**, *4*, 79.
13. Benjaminsen, T.A.; Alinon, K.; Buhaug, H.; Buseth, J.T. Does climate change drive land-use conflicts in the Sahel? *J. Peace Res.* **2012**, *49*, 97–111. [CrossRef]
14. Udeuhele, G.I. Urbanization and Insecurity in Nigeria: The Issues, Challenges and Prospect for National Development. *Middle-East J. Sci. Res.* **2018**, *26*, 68–77. [CrossRef]
15. Odoh, S.I.; Chilaka, F.C. Climate change and conflict in Nigeria: A theoretical and empirical examination of the worsening incidence of conflict between Fulani herdsmen and farmers in Northern Nigeria. *Arab. J. Bus. Manag. Rev.* **2012**, *34*, 1–15. [CrossRef]
16. Amobi, D.; Onyishi, T. Governance and climate change in Nigeria: A public policy perspective. *J. Policy Dev. Stud.* **2015**, *289*, 1–12. [CrossRef]
17. Fasona, M.J.; Omojola, A.S. Climate change, human security and communal clashes in Nigeria. In Proceedings of the International Workshop on Human Security and Climate Change, Oslo, Norway, 21–23 June 2005; pp. 21–23, Organisers: Centre for the Study of Civil War, International Peace Research Institute, Oslo (PRIO) & Centre for International Environmental and Climate Research at the University of Oslo (CICERO) for the Global Environmental Change and Human Security Program (GECHS).
18. Slettebak, R.T. Don't blame the weather! Climate-related natural disasters and civil conflict. *J. Peace Res.* **2012**, *49*, 163–176. [CrossRef]
19. Sayne, A. Climate Change Adaptation and Conflict in Nigeria. Available online: https://www.usip.org/sites/default/files/Climate_Change_Nigeria.pdf (accessed on 18 December 2019).
20. Nossiter, A. Toll from Religious and Ethnic Violence in Nigeria Rises to 500. *New York Times*, 8 March 2010. Available online: https://cfec.typepad.com/files/article---central-nigeria-violence---3-08-10---ny-times.pdf (accessed on 30 December 2019).
21. Bearak, M. Retrieved from The Washington Post. Available online: https://www.washingtonpost.com/news/world/wp/2018/12/10/feature/the-ordinary-people-keeping-the-peace-in-nigerias-deadly-land-feuds/ (accessed on 30 December 2019).
22. Conroy, S. Land conflicts and lethal violence in Nigeria: Patterns, mapping and evolution (2006–2014). *IFRA-Niger. Work. Pap. Ser.* **2014**, *38*, 1–38.
23. Kwaja, C.M.; Ademola-Adelehin, B.I. *The Implications of the Open Grazing Prohibition and Ranches Establishment Law on Farmer-Herder Relations in the Middle Belt of Nigeria*, 1st ed.; Search Common Ground: Washington, DC, USA, 2017; p. 8.
24. De Juan, A.; Hänze, N. Climate and cohesion: The effects of droughts on intra-ethnic and inter-ethnic trust. *J. Peace Res.* **2021**, *58*, 151–167. [CrossRef]
25. Taiye, F. Controlling the advancement of Savanna into southwestern Nigeria. *Zonas Áridas* **2007**, *11*, 251–259.
26. Muhammed, I.; Ismaila, B.A.; Bibi, M.U. An assessment of farmer-pastoralist conflict in Nigeria using GIS. *Int. J. Eng. Sci. Invent.* **2015**, *4*, 23–33.

27. Ibirogba, F. National Livestock Plan is Ruga in Disguise, Nigerians Warn. Available online: https://guardian.ng/features/agro-care/national-livestock-plan-is-ruga-in-disguise-nigerians-warn/ (accessed on 17 December 2019).
28. Soyinka, W. Ruga: Cattle Cannot Take Priority over Human Beings. Available online: https://www.channelstv.com/2019/07/02/ruga-cattle-cannot-take-priority-over-human-beings-says-soyinka/ (accessed on 18 December 2019).
29. Channels Television. Federal Government Suspends Planned RUGA Programme. Available online: https://www.channelstv.com/2019/07/03/breaking-news-federal-government-suspends-planned-ruga-programme/ (accessed on 18 December 2019).
30. Ogo-Oluwa, O. Anti-Grazing Policy and Conflict Resolution between Fulani Herdsmen and Farmers in Ekiti State. *Asian Res. J. Arts Soc. Sci.* **2017**, *4*, 1–13. [CrossRef]
31. Abuh, A. Miyetti-Allah Urges FG to Establish Livestock Ministry, Insists Open Anti-Grazing Law Unconstitutional. Available online: https://guardian.ng/news/miyetti-allah-urges-fg-to-establish-livestock-ministry-insists-open-anti-grazing-law-unconstitutional/ (accessed on 17 December 2019).
32. Ukoji, V.N.; Ayodokun, A.V.; Eze, V.C. Nigeria Watch: Eight Report on Violence in Nigeria. Ph.D. Thesis, IFRA-Nigeria, University of Ibadan, Ibadan, Nigeria, 2019.
33. Olagunju, T. Drought, desertification and the Nigerian environment: A review. *J. Ecol. Nat. Environ.* **2015**, *7*, 196–209. [CrossRef]
34. Wayumba, R. Developing Land Information Management Systems for County Governments in Kenya. *IOSR J. Eng.* **2017**, *7*, 42–49. [CrossRef]
35. Okeke, O.E. Conflicts between Fulani herders and farmers in central and southern Nigeria: Discourse on proposed establishment of grazing routes and reserves. *AFRREV IJAH An. Int. J. Arts Humanit.* **2014**, *3*, 66–84.
36. Maitima, J.M.; Mugatha, S.M.; Reid, R.S.; Gachimbi, L.N.; Majule, A.; Lyaruu, H.; Mugisha, S. The linkages between land use change, land degradation and biodiversity across East Africa. *Afr. J. Environ. Sci. Technol.* **2009**, *3*, 300–325.
37. Boles, O.J.C.; Shoemaker, A.; Courtney Mustaphi, C.J.; Petek, N.; Ekblom, A.; Lane, P.J. Historical ecologies of pastoralist overgrazing in Kenya: Long-term perspectives on cause and effect. *Hum. Ecol.* **2019**, *47*, 419–434. [CrossRef]
38. Ogunwale, A.O. Deforestation and Greening the Nigerian Environment. Available online: http://eprints.covenantuniversity.edu.ng/5327/ (accessed on 13 May 2021).
39. Institute for Economics & Peace. Global Terrorism Index 2019: Measuring the Impact of Terrorism, Sydney, November 2019. Available online: http://visionofhumanity.org/reports (accessed on 3 January 2020).
40. Asaka, J.O. Climate Change-Terrorism Nexus? A Preliminary Review/Analysis of the Literature. *Perspect. Terror.* **2021**, *15*, 81–92.
41. Vivekananda, J.; Wall, M.; Sylvestre, F.; Nagarajan, C.; Brown, O. *Shoring up Stability: Addressing Climate and Fragility Risks in the Lake Chad Region*; Adelphi: Berlin, Germany, 2019.
42. Ifabiyi, I.P. Recharging the Lake Chad: The hydropolitics of national security and regional integration in Africa. *Afr. Res. Rev.* **2013**, *7*, 196–216. [CrossRef]
43. Usigbe, L. From Africa Renewal: December 2019–March 2020. Available online: https://www.un.org/africarenewal/magazine/december-2019-march-2020/drying-lake-chad-basin-gives-rise-crisis (accessed on 8 April 2020).
44. Kabir, B.G.; Kagu, A.; Yerima, F.A. Agriculture, Cultural Interplay and Environmental Degradation on the Shores of the Lake Chad. In *Issues in the Geography of Borno State*; Adamu Joji Publishers: Kano City, Nigeria, 2009; pp. 43–49.
45. Genthon, P.; Hector, B.; Luxereau, A.; Descloitres, M.; Abdou, H.; Hinderer, J.; Bakalowicz, M. Groundwater recharge by Sahelian rivers—Consequences for agricultural development: Example from the lower Komadugu Yobe River (Eastern Niger, Lake Chad Basin). *Environ. Earth Sci.* **2015**, *74*, 1291–1302. [CrossRef]
46. Pearce, F. A Dam of Shame. Available online: http://www.waterafrica.co.za/index.php/contributors/67-fred-pearce/109-a-dam-shame (accessed on 11 April 2020).
47. Romero Hicks Murakami, L.F. Forced Migration, Conflict, and Climate Change in the Lake Chad Region: Perspectives on International Dynamics and the Anthropological Response to Enhancing Migrant Resilience and Community Cohesion. Master Thesis, Cornell University, Ithaca, NY, USA, 2020.
48. United Nations Office for the Coordination of Humanitarian Affairs (UNOCHA). Lake Chad Basin Crisis Overview. Available online: https://reliefweb.int/report/nigeria/lake-chad-basin-crisis-overview-23-january-2019 (accessed on 12 April 2020).
49. Bleischwitz, R.; Johnson, C.M.; Dozler, M.G. Re-Assessing resource dependency and criticality. Linking future food and water stress with global resource supply vulnerabilities for foresight analysis. *Eur. J. Futures Res.* **2014**, *2*, 34. [CrossRef]
50. Caniglia, G.; Luederitz, C.; von Wirth, T.; Fazey, I.; Martin-López, B.; Hondrila, K.; Lang, D.J. A pluralistic and integrated approach to action-oriented knowledge for sustainability. *Nat. Sustain.* **2021**, *4*, 93–100. [CrossRef]
51. Okpara, U.; Stringer, L.; Dougill, A. Lake drying and livelihood dynamics in Lake Chad: Unravelling the mechanisms, contexts and responses. *Ambio* **2016**, *45*, 781–795. [CrossRef]
52. Hall, E.L. Conflict for resources: Water in the Lake Chad Basin. *Army Command. Gen. Staff. Coll. Leavenworth Ks Sch. Adv. Mil. Stud.* **2009**, *15*, 308–325.
53. Vichi, M. Transferring Water from the Congo to Lake Chad: The Transaqua project. *Exec. Intell. Rev.* **2011**, *38*, 31–36.
54. Sayan, R.C.; Nagabhatla, N.; Ekwuribe, M. Soft Power, Discourse Coalitions, and the Proposed Interbasin Water Transfer between Lake Chad and the Congo River. *Water Altern.* **2020**, *13*, 752–778.
55. Gil, J.; Siebold, M.; Berger, T. Adoption and Development of Integrated Crop-Livestock-Forestry Systems in Mato Grosso, Brazil. *Agric. Ecosyst. Environ.* **2015**, *199*, 395. [CrossRef]

56. Moraine, M.; Duru, M.; Nicholas, P.; Leterme, P.; Therond, O. Farming System Design for Innovative Crop-livestock Integration in Europe. *Animal* **2014**, *8*, 1204–1217. [CrossRef] [PubMed]
57. Barron-Gafford, G.A.; Pavao-Zuckerman, M.A.; Minor, R.L.; Sutter, L.F.; Barnett-Moreno, I.; Blackett, D.T.; Macknick, J.E. Agrivoltaics provide mutual benefits across the food–energy–water nexus in drylands. *Nat. Sustain.* **2019**, *2*, 848–855. [CrossRef]
58. Cattivelli, V. Climate Adaptation Strategies and Associated Governance Structures in Mountain Areas. The Case of the Alpine Regions. *Sustainability* **2021**, *13*, 2810. [CrossRef]
59. Li, F.; Strachan, N. Take me to your leader: Using socio-technical energy transitions (STET) modelling to explore the role of actors in decarbonisation pathways. *Energy Res. Soc. Sci.* **2019**, *51*, 67–81. [CrossRef]

sustainability

MDPI

Article

Defining Pathways towards African Ecological Futures

Peter Scheren [1,*], Peter Tyrrell [2,3], Peadar Brehony [4], James R. Allan [5], Jessica P. R. Thorn [6,7], Tendai Chinho [1], Yemi Katerere [8], Vanessa Ushie [9] and Jeffrey S. Worden [1]

1 Regional Office for Africa, Worldwide Fund for Nature, P.O Box 62440-00200, Nairobi, Kenya; tchinho@wwf.org.zw (T.C.); jworden@wwfint.org (J.S.W.)
2 University of Oxford's Wildlife Conservation Research Unit (WildCRU), Oxford OX13 5QL, UK; peterdavidtyrrell@gmail.com
3 Department of Geography & Environmental Studies, University of Nairobi, PR98+VGV, Nairobi, Kenya
4 Department of Geography, University of Cambridge, Cambridge CB2 3EN, UK; peadar_b@hotmail.com
5 Institute for Biodiversity and Ecosystem Dynamics (IBED), University of Amsterdam, 1090 GE Amsterdam, The Netherlands; j.r.allan@uva.nl
6 Department of Environment and Geography, University of York, York YO10 5NG, UK; jessica.thorn@york.ac.uk
7 African Climate and Development Initiative, University of Cape Town, Cape Town 7700, South Africa
8 Right and Resources Institute, Washington, DC 20007, USA; talk2yemi@yahoo.com
9 African Natural Resources Centre, African Development Bank, Abidjan 01, Côte d'Ivoire; V.USHIE@afdb.org
* Correspondence: pscheren@wwfint.org

Abstract: Africa has experienced unprecedented growth across a range of development indices for decades. However, this growth is often at the expense of Africa's biodiversity and ecosystems, jeopardizing the livelihoods of millions of people depending on the goods and services provided by nature, with broader consequences for achieving the United Nations Sustainable Development Goals. Encouragingly, Africa can still take a more sustainable path. Here, we synthesize the key learnings from the African Ecological Futures project. We report results from a participatory scenario planning process around four collectively-owned scenarios and narratives for the evolution of Africa's ecological resource base over the next 50 years. These scenarios provided a lens to review pressures on the natural environment, through the drivers, pressures, state, impacts, and responses (DPSIR) framework. Based on the outcomes from each of these steps, we discuss opportunities to reorient Africa's development trajectories towards a sustainable path. These opportunities fall under the broad categories of "effective natural resource governance", "strategic planning capabilities", "investment safeguards and frameworks", and "new partnership models". Underpinning all these opportunities are "data, management information, and decision support frameworks". This work can help inform collaborative action by a broad set of actors with an interest in ensuring a sustainable ecological future for Africa.

Keywords: sustainable development; social–ecological systems; biodiversity; participatory scenario planning; governance; strategic planning; investment; decision support frameworks; green infrastructure; Africa

Citation: Scheren, P.; Tyrrell, P.; Brehony, P.; Allan, J.R.; Thorn, J.P.R.; Chinho, T.; Katerere, Y.; Ushie, V.; Worden, J.S. Defining Pathways towards African Ecological Futures. *Sustainability* **2021**, *13*, 8894. https://doi.org/10.3390/su13168894

Academic Editors: Carlos Oliveira Cruz and Hossein Azadi

Received: 15 June 2021
Accepted: 30 July 2021
Published: 9 August 2021

Publisher's Note: MDPI stays neutral with regard to jurisdictional claims in published maps and institutional affiliations.

Copyright: © 2021 by the authors. Licensee MDPI, Basel, Switzerland. This article is an open access article distributed under the terms and conditions of the Creative Commons Attribution (CC BY) license (https://creativecommons.org/licenses/by/4.0/).

1. Introduction

Africa's natural capital is immense-from the forests and minerals of the Congo, the diamonds of western and southern Africa, the water towers in Guinea, to the wildlife-packed savannahs and coral reefs of East Africa [1]. Perhaps the most unique aspect of Africa's natural capital is its biodiversity. The continent contains the world's most diverse and abundant megafaunal populations, which have been largely exterminated elsewhere in the world [2,3]. At the same time, African people rely heavily on the services that natural ecosystems provide such as clean water, firewood, protein from fisheries and wildlife, building materials, and revenue from wildlife-based tourism. Many economies continue to be dominated by agricultural production, either for export as in the case of cocoa production from Ghana, coffee and tea from Kenya, citrus fruits from Morocco, or wine from South Africa or for national consumption in subsistence economies. Tourism and other natural resource dependent sectors such as forestry and energy production add to this dependency on the continent's strong natural resource base. Hence, nature and natural resources remain the foundation for Africa's current and future development. It is therefore concerning that the continent's current development trajectory is undermining its ecosystems, as evidenced by declines in wildlife populations and habitats, and degrading freshwater systems, land, and other critical parts of ecosystems [1,4,5].

Despite strong economic advances, in particular over the past decade [6,7], Africa can still improve on many development indicators. For example, less than 40% of the continent's population has access to electricity, and internet usage is at 28.2% in Africa as compared to Europe, who are the highest users at 82.5% [8]. Many landlocked countries lack railways and remain reliant on inefficient trucking freight transport, while only a third of the rural population has access to roads (e.g., especially in West Africa) [6]. Moreover, only 11% of African urban dwellers have access to sewer connections and 59% to piped water, down from 67% in 2003 [9]. Furthermore, the number of people living in poverty in Africa is still increasing [10], a sign of inequality in wealth distribution.

Africa's rapid economic development, population growth, and associated needs for access to natural resources, in particular where a large section of poverty-driven population remains highly dependent on such resources for their livelihoods and survival, comes at a cost to the continent's rich natural capital. Deforestation rates are increasing, particularly in nations with dry forests [11], wildlife populations are declining and becoming more isolated [4], and marine and freshwater fish stocks are plummeting [12–14]. Mega infrastructure projects are penetrating previously remote areas, catalyzing the unsustainable exploitation of natural resources, to feed growing local needs and the ever-increasing demands of the global economic system [15]. The consequent decrease in ecosystem resilience further aggravates the impacts of climate change, which is predicted to drive approximately 68 million to 132 million into poverty by 2030—mostly in sub-Saharan Africa [16]. If this largely unstrategic and ecologically blind development continues, it will threaten the future development and prosperity of the continent, with particularly severe impacts on many of the most marginalised, vulnerable people who depend on natural resources for their livelihoods [17].

Africa's human population is projected to double by 2050 to 2.5 billion people, and then double or triple again by 2100 [18,19]. Africa is also one of the fastest urbanising regions and its burgeoning middle classes will grow from 355 million in 2010 to 1.1 billion in 2060 [20]. This rapid population expansion, urbanisation and the growing middle class indicates a need to vastly increase investments in infrastructure (e.g., for health, education, transport, energy, housing), and increase demands for food production, energy, water and other resources [21]. Finding a way to economically develop while maintaining the integrity of the environment which people depend on is therefore one of the great challenges faced by the continent.

However, Africa still has an opportunity to determine its own development trajectory, and chart new pathways to sustainability. Most African nations have in place national development plans and have committed to global targets such as the United Nations Sustainable Development Goals (SDGs) [22], the Convention on Biological Diversity's Strategic Plan [23], and the UNFCCC Paris Climate Agreement [24] (Figure 1). Additionally, all countries come together under the African Union, a multilateral governance platform promoting continental unity and cooperation. One of the key pillars of the African Union's ambitious Agenda 2063 is "environmentally sustainable and climate resilient economies and communities" [25]. Although not ubiquitous, the ambition is clear. To achieve it, Africa must look forward and plan accordingly.

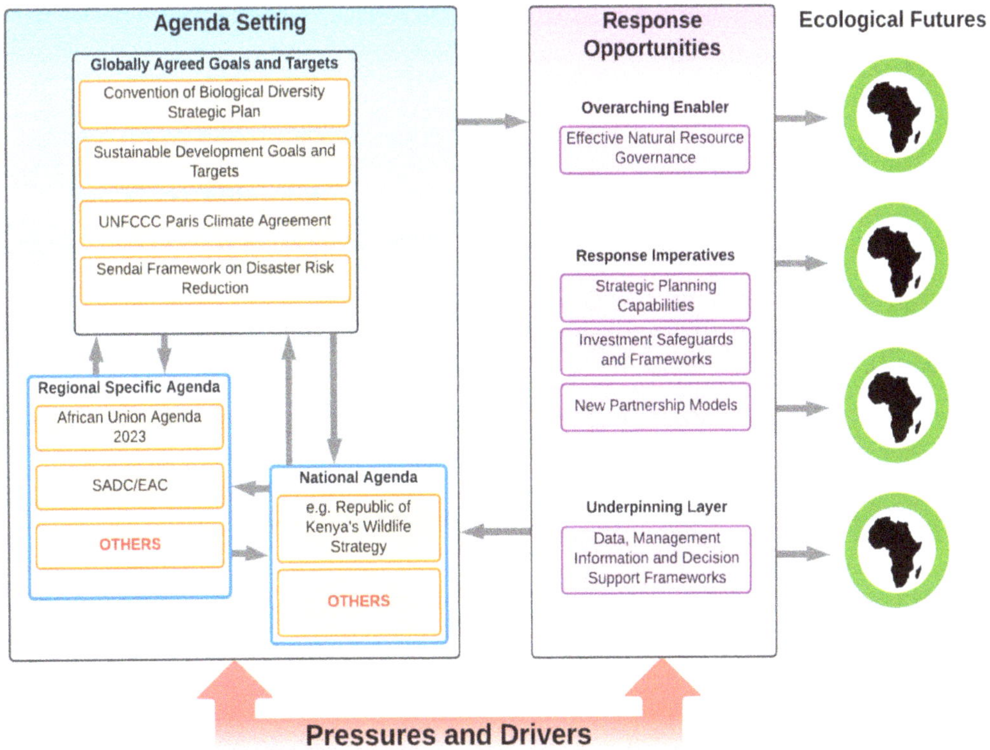

Figure 1. Agenda setting can drive response opportunities and define Africa's Ecological Futures.

"Futures thinking" involves taking a future-oriented approach to strategic planning [26], backed by an understanding of the developmental and social–ecological history. The term "futures" covers a range of techniques for understanding and anticipating, rather than trying to predict the future. It is usually pluralised since many different futures are possible. The aim of futures thinking is to create pathways towards ambitious long-term goals, and it is regarded as an effective technique to inform decision making that balances long-term planning with present action. One approach of futures thinking is scenario planning-which present plausible descriptions about how the future may develop, based on a coherent, internally consistent set of assumptions (or logic) about key relationships and their driving forces [27,28]. When developed in a participatory manner, the process includes a diverse array of relevant stakeholder views, which can improve the feasibility, validity, uptake, and concreteness of scenarios [29–31].

Increasing the capacity for futures thinking and generating scenarios that galvanise transformative change are recognised as important endeavours for achieving sustainability worldwide. However, in Africa the availability of necessary expertise to apply futures thinking is low, and scenario analyses in the environmental field have been underutilised [32]. Moreover, developing a coherent set of scenarios for the continent is challenging, considering the rich cultural diversity, vast geographic heterogeneity which no doubt, hosts many different, and sometimes conflicting visions of the future.

The African Ecological Futures (AEF) planning project, a joint initiative between the African Development Bank (AfDB) and the Worldwide Fund for Nature (WWF) that took place between 2013 and 2015, is an example of how participatory scenario planning can benefit the continent. The process involved a series of analytical studies combined with workshops with policy makers and development partners with skill sets ranging from conservation and ecology to economics and international development. The outcomes of these workshops were four expert-developed and collectively-owned scenarios for the evolution of Africa's ecological resource base over the next 50 years (2015–2065).

In this paper, we synthesize the key learnings from the African Ecological Futures project, complemented by recent data and information on subsequent developments and trends since the release of this foundational piece of work. We (i) analyse the drivers, pressures and impacts of Africa's development trajectory on its environment; (ii) describe scenarios and response opportunities to define Africa's ecological future; and (iii) explore two case studies (see Boxes 1 and 2), which demonstrate important leverage points. Our hope is that our results provide information to support stakeholders committed to putting Africa on a sustainable development trajectory now, to guarantee an ecological society in the future. Recognising a continental set of scenarios may not be applicable to every national context, our ambition is that these scenarios be adapted to specific national and local contexts.

2. Approach and Methods

The approach taken for the African Ecological Futures process included three main steps. Firstly, the scientific basis for the African Ecological Futures process was provided by an in-depth analysis of six main sectors that are likely to shape Africa's development: energy, water, agriculture, extractives, trade and investment, and infrastructure. Analytical papers on each of these six areas, commissioned by WWF and AfDB through consultancies and internal experts, assessed the main challenges and potential development pathways that Africa could follow over the next 50 years. Each of these papers included an exploration of policy and development directions and their implications, as well as potential points of intervention for ensuring a sustainable, ecologically secure pathway of growth.

Secondly, these analyses were used to inform two scenario planning workshops in 2014 and 2015. The scenarios were constructed around two axes: (i) the level of centralisation versus decentralisation of decision making; and (ii) global orientation (i.e., global production and trade) or African orientation (i.e., intra-African production and trade) (Figure 2). These axes represent two broad-level dimensions of Africa's policy and development trajectory, underlying the biggest trends in Africa's development pathway. During the workshops, stakeholders undertook an exploratory analysis of possible scenarios, constituted along these two axes, to explore a wide range of potential futures across policy, management and planning domains. A deliberate attempt was made to avoid more desirable (best-case) or less desirable (worst-case) [19].

The first workshop in Cape Town, South Africa brought together 41 experts (both from within WWF and external) from across the continent. The workshop had two main functions: (i) to review and validate the six sector papers; and (ii) to apply the findings of these analyses to the participatory scenario planning exercise by reflecting on the implications of the different axes of development. The workshop resulted in a shared understanding of key drivers and pressures central to Africa's growth that have profound implications for the continent's ecological resources, and the parameters that could enable preferred outcomes for ecologically sustainable growth in Africa. A second workshop was held at the 15th African Ministerial Conference on the Environment (AMCEN) in Cairo, Egypt, and brought together partners at the forefront of development and environment challenges across Africa, including representatives from the African Development Bank, Albertine Rift Conservation Society (ARCOS), Birdlife International, the International Union for Conservation of Nature (IUCN), the New Partnership for Africa's Development (NEPAD), the United Nations Economic Commission for Africa (UNECA), and the United Nations Environment Programme (UNEP), as well as government representatives. This second workshop resulted in (i) a shared understanding among development partners of the dynamics and uncertainties that will determine Africa's ecological future; (ii) a series of scenario narratives that can assist decision makers in identifying areas of risk and opportunities for growth; and (iii) a general understanding of how decision makers and partners can influence development trajectories and manage emerging risks.

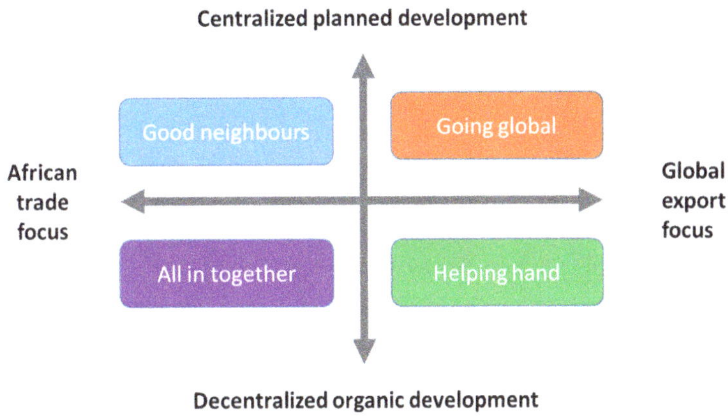

Figure 2. Scenarios developed during the African Ecological Futures process.

As a third step, the findings from the analysis and the participatory scenario planning workshops were used to construct an overall African Ecological Futures report, as a basis for continued dialogue with governments and development partners across Africa. For the purpose of this paper, the findings of this process are summarised in the form of an overview of the scenario narratives, an overall analysis of development trends and their consequences, based on the drivers, pressures, state, impacts, and responses (DPSIR) framework [33], as well as an overview of the key leverage points and response opportunities for securing Africa's ecological future. Furthermore, between 2015 and 2021, a number of more concrete case studies were undertaken as part of the follow-up process, to further analyse development trends and opportunities in a number of key sectors. The details of these case studies are not the subject of this paper, but summaries are presented for illustrative purposes (Boxes 1 and 2).

3. Results

3.1. Scenario Narratives

Four storylines or narratives—i.e., qualitative description of future developments—were developed to reflect distinct trajectories for Africa and African nations:

1. "Going global" where resource- rich regions take a planned, export-driven path to developing extractive and agricultural commodities, based on centralised decision making and connected economic infrastructure;
2. "Helping hands" where resource rich areas are the focus of extractive economic activities driven by local actors developing local resources for export through decentralised decision making and supported by local (off grid) infrastructure;
3. "All in together" where densely populated areas with renewable resources develop local agricultural industries through participatory decision making and local co-operative schemes driven by local actors; and
4. "Good neighbours" where the future is characterised by a strong drive for African-based development to increase intra-regional trade, where countries begin to take a coherent domestic view with regards to their production and consumption, and large regional infrastructure investments are needed.

The scenarios were primarily intended to instigate discussion on how different local and sectoral contexts influence the development pathways of countries and the continent, and the consequences thereof for the environment. In this regard, there is no 'ideal' scenario, nor is there the intention that these scenarios are fully reflective of reality. In fact, the four scenarios are not necessarily mutually exclusive for any country or region and different areas within any country may simultaneously exhibit different scenarios for different sectors.

3.2. *Analysis of Development Trends: Where Development Meets the Environment*

Here, we discuss the broad drivers of change on the continent, how these manifest on the ground as pressures or 'threats' to the environment, and what impact they are having now, and may have in the future, on Africa's natural resource base and people.

3.2.1. Drivers of Change

A broad set of complex, interconnected, and multi-scalar drivers have and will continue to drive these environmental changes (Figure 3). We discuss these drivers, the pressure they apply to the environment, and the impacts they are having.

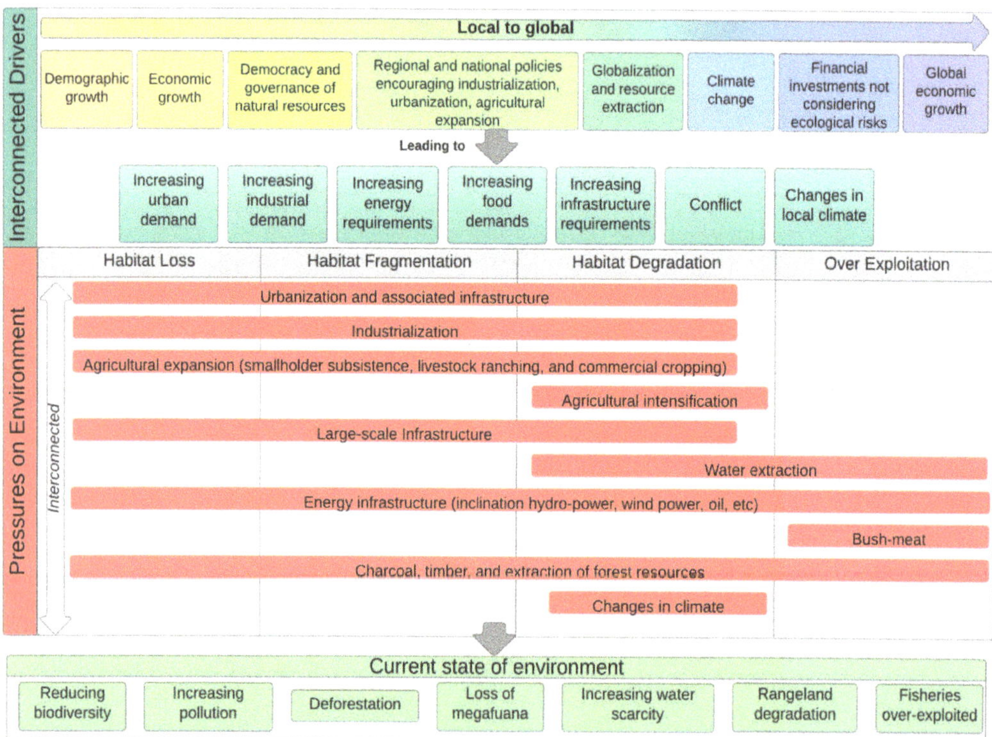

Figure 3. Conceptual model of drivers, pressures and impacts on the environment, as highlighted by stakeholders in the African Ecological Futures workshops. The drivers show a top tier of higher-level influences, with many of these interconnected and occurring at multiple scales from local to global. The second-tier highlights some of the common direct pressures that these create, leading to pressures on the environment, including over-exploitation, habitat loss, fragmentation, and degradation. These pressures lead to the current state of the environment.

Local and regional development factors: Africa's growth story is already one of the defining global narratives of the 20th and 21st century. Demographic change, changing lifestyles, and economic growth are key factors that drive growing pressures on the environment. By 2100, Africa is expected to have grown to between 3 and 4 billion people, while its middle class is projected to triple by 2060 to 1.1 billion [34]. GDP growth rates, which averaged 5% between 2000 and 2010 [35], are expected to continue. Despite a new set of challenges associated with growing inequality [36], the share of Africans living on less than US$1.90 a day has fallen substantially—from 54% in 1990 to 41% in 2015. That said, the actual number of poor people in Africa has increased from 278 to 413 million in the same period due to population growth [37]. Consequently, between 1961 and 2008, Africa's ecological footprint (i.e., the quantity of nature required to sustain a person) doubled—partly due to its rapidly growing population and middle class [1]. Although individuals' ecological footprint is still relatively small [38], this will likely increase due to greater demand for and consumption of goods and services, such as energy, housing, food, water, land use, and infrastructure.

National governance factors: Changes in national security and democratic civil society engagement will intensify environmental pressure. Countries without rigid democratic processes or with wide-spread conflict tend to have weaker natural resource governance [39,40]. Poor governance and an unstable political environment, in particular accompanied by a shrinking democratic space for civil society, can be major bottlenecks for sustainable development. Civil society organisations are key to strong governance institutions–using polycentric governance approaches that rely on participative approaches that build from the local level. Advocacy efforts can result in transformative change in the behaviour of business, governments, and consumers [41].

Global and external factors: Factors beyond the continent will also increase pressure on the African environment. Increasing food requirements globally will put pressure on Africa's arable lands, which represents roughly 60% of all globally uncultivated arable land. It is projected that global cropland will increased by 26%, or 3.35 million km^2, between 2010 and 2050—particularly throughout sub-Saharan Africa [42]. Furthermore, global shocks (e.g., oil commodity price downturns) will have downstream impacts on African nations, inducing changes to national priorities and policies, and altering local demand for natural resources. For example, the COVID-19 pandemic has increased pressure on the natural environment through reduced funding for conservation and restrictions on the operations of conservation agencies [4]. As the continent experiences rapid development, foreign direct investment inflows into Africa have significantly increased in the last decade, albeit contracting during the COVID-19 pandemic [18]. Investors' values and safeguards will influence whether potentially environmentally damaging development projects can access capital and licensing. Climate change induced impacts—such as droughts, flooding, storm surges, wildfires and receding glaciers—will put additional pressure on natural resources, and likely exacerbate the impact of other drivers [43].

3.2.2. Pressures on the Environment

Drivers of change manifest as interlinked pressures that ultimately impact the environment. These impacts occur in four broad but related categories: habitat conversion, habitat fragmentation, habitat degradation, and overexploitation of resources (Figure 3). Complete habitat conversion means an entire natural ecosystem with its biodiversity and ecosystem services is converted to a human land use. As habitat is converted, remaining intact patches become fragmented, which is particularly problematic for large migratory mammals, but also has profound implications for the survival of species in the context of climate change [44]. Degradation (e.g., through pollution, overgrazing, or deforestation) results in reduced habitat quality for biodiversity and impaired ecosystem services. Overexploitation, also called overharvesting, refers to harvesting a renewable resource to the point of diminishing returns (e.g., illegal and unregulated bushmeat trade, or overfishing) [45]. The pressures, that cause these four categories of impact include:

1. Urbanisation and industrialisation: Africa's urban areas have expanded at 5% per year for the last 20 years, and by 2050, 1.2 billion people will live in cities [46]. Urban expansion will result in the conversion of intact habitat that currently supports biodiversity and ecosystem services, either directly to urban areas, or indirectly, to provide the food, water, energy and other material that cities demand [47–49]. Some of this may be planned rezoning, while much will include illegal, unplanned encroachment into

urban green infrastructure and surrounding landscapes from unplanned peri-urban or informal settlements [50]. Urbanisation also requires supporting infrastructure such as roads and power lines, which will also have environmental impacts. Furthermore, unless sanitation and regulatory controls on industry are enforced, remaining habitats in or near urban areas will become polluted and degraded.

2. Agricultural expansion and/or intensification: Agricultural land conversion is among the largest global contributors to habitat loss, species extinction, and a major emitter of greenhouse gases [42,51]. Under current scenarios of agricultural production, vast areas of habitat across the continent will have to be cleared and farmed to meet the food requirements of a richer, larger population. Under a business-as-usual scenario it is predicted that vertebrate species in Africa will lose ~14.4% of their habitat on average by 2050 through land conversion [42]. In particular, increasing agricultural production threatens the most marginal land, causing the deterioration of soil and water resources. Pollution is also likely to increase due to intensified and untargeted use of fertilisers, herbicides, and pesticides, in part related to growing pesticide resistance, the use of hybrid seed and less fallowing and crop rotation [52].
3. Large-scale infrastructure: Thirty-three planned or existing transportation corridors exist across the continent that if completed, will total over 53,000 km in length [14]. The corridors involve large-scale expansion and construction of infrastructure such as roads, railroads, pipelines, and port facilities. These will open up extensive areas of land to new environmental pressures, and cause widespread fragmentation of ecosystems [53]. This is particularly concerning for projects that pass through important areas for biodiversity and wildlife, including the Southern Agricultural Growth Corridors (SAGCOT), the Lamu Southern Sudan Transport Corridor (LAPSSET), and many others [54].
4. Water extraction: Already, many freshwater basins and associated terrestrial ecosystems that rely on them are negatively impacted by industrial overexploitation, particularly irrigation. In most African countries, less than 50% of the population has access to improved sanitation facilities (30% in sub-Saharan Africa) and less than 75% have access to improved drinking water sources. Climate change will further exacerbate all aspects of water insecurity [55]. Increased volumes of water will be required, along with appropriate infrastructure, to feed a growing population [56–58].
5. Energy infrastructure: As many as 580 million Africans do not have access to electricity, and 900 million do not have clean cooking energy. Thus, substantial increases in electricity generation and transmission are required [59,60]. Even where renewable energy systems are created, there will likely be some level of environmental impact from energy development. For instance, hydropower dams are projected to increase by more than 23% by 2040 [59,61] and could lead to loss of habitats through damming. A growth in wind power turbines could causes collisions of birds and bats [62,63]. The impacts of global energy demands will exacerbate these issues, degrading habitats such as in Murchison Falls National Park, Uganda [64], and catalysing further investment in large scale infrastructure projects [65,66].
6. Bushmeat and overfishing: Large scale bushmeat overexploitation has led to defaunation of Africa's tropical forests [67], and much of its savannahs. Likewise, overfishing of both freshwater and marine species is driving some species to near extinction [68]. The ecological, nutritional, economic, and intrinsic values of wildlife are all at risk of being lost because present policies and practices cannot reconcile different values, nor manage resources sustainably. In some countries, changes in wealth or violent conflicts may increase the consumption of bushmeat [69–71].
7. Charcoal, timber, and extraction of forest and mineral resources: Sixty-five percent of the world's charcoal is produced in sub-Saharan Africa, mainly by smallholders [72], driven in particular by increasing demands from growing urban areas [11]. Furthermore, industrial logging has been the biggest driver of degradation within forested areas in Central Africa. While the growth of the African middle class is likely

to reduce domestic charcoal use over time, there is a risk that international export markets will remain [73]. Moreover, increased timber demand will likely continue to drive clearing of tropical, savannah, and coastal forests and woodlands. Mineral extraction, including materials such as cobalt used for solar photovoltaic batteries, sand for cement, aluminum, as well as diamonds, gold, natural gas, bauxite, iron ore, among other commodities will release toxic minerals with health impacts on local populations [74].

8. Changing climate: The short- and long-term impacts of climate change continue to unravel, with large scale negative impacts in urban and rural areas [43]. For instance, it is expected that 75–250 million people in the 2020s, and 350–600 million people by the 2050s, will be exposed to increased water stress as a result of changes in the frequency and intensity of extreme events [55,75]. Climate change will also impact livestock forage production, impacting the livelihoods of over 180 million people in rangelands [76]. Global predictions for biodiversity loss suggest that African biodiversity will lose considerable habitat with climate change [32,77].

3.3. Intervention Opportunities

The drivers of change identified during the African Ecological futures process revealed fundamental risk areas for the environment. The participatory scenarios planning workshop, subsequently, provided an opportunity to interrogate these challenges and risks under different conditions (scenarios) as well as to identify key leverage points and response opportunities (Figure 1). This process resulted in a number of recommended, cross-cutting approaches, as summarised below.

3.3.1. Effective Natural Resource Governance

Most response opportunities come under the overarching umbrella of 'effective natural resource governance'. This includes the legislation, regulations and informal rules. Over the last century, natural resource governance in Africa has been dominated by top-down models of state control [78,79], which are a largely ineffective but commonplace colonial artefact. However, over the past 50 years, concerted efforts have been made across Africa to devolve natural resource governance [80–82]. While coming with its own challenges, this process enables benefits to reach landowners and citizens, and if managed well, increases the legitimacy of natural resource governance efforts [83–85].

Streamlined governance at multiple scales, from global commitments such as SDGs 16 and 17, through to national policy and local resource management can leverage capacity and empower people to sustainably manage natural resources. Clear roles and responsibilities will ensure that various actors understand their rights over and access to natural resources. This also ensures there is more accountability, inclusion, and justice [86–88]. Devolving natural resource management increases sustainability because landowners and citizens are more likely to resist or mitigate pressures if they receive proportionally greater benefits from their natural resources and feel a greater sense of participation in governance [50].

There are persuasive examples of this in water use, fisheries, rangelands, forest products, and wildlife resources. For instance, communities living around the Maasai Mara National Reserve, Kenya, receive devolved financial benefits from ecotourism, which create an incentive strong enough for them to use the land for biodiversity conservation and livestock grazing, and to resist the potentially more profitable but less sustainable alternative of converting the area to croplands [89,90]. Another example is in Namibia, where community conservancies and community forests have facilitated the significant devolution of natural resource management authority from central government to local resource users since 1996. Communities now have the rights to manage common pool resources and realise the benefits of the management through income from employment in tourism and conservation hunting, the sale of indigenous plant products and crafts, local small and medium enterprises, and in-kind benefits such as the distribution of harvested game meat [91,92]. Associated to these efforts, elephant populations have been reported to have increased on communal land, from 7500 in 1995 to 22,800 in 2016 [93].

Voluntary standards and certification schemes on the sustainability of commodities, such as the Rainforest Alliance certification in Ghana, have shown to have beneficial outcomes in terms financial support, information and knowledge, technical assistance-conditioned by the presence of active farmer organisations and access to agricultural inputs and credit [94].

By strengthening institutions at various scales, including training and financial backing to ensure long term sustainability and reduce dependency on volunteerism, development agencies, among others, can play an important role in supporting African nations to shape their own social and ecological priorities. Likewise, social media and other new technologies (e.g., block chains, artificial intelligence) may play an important role in creating new opportunities to support participatory and effective governance [95,96].

3.3.2. Integrated Planning Capabilities

Strategic integrated planning facilitates judicious evidence-based decision making for coordinated land use management. Large scale social–ecological transformations, such as agricultural and infrastructure expansion, are inevitable in Africa's future. Currently, most development and land use planning does not sufficiently consider ecological impacts, is ad-hoc, and not coordinated across sectors [97]. There are immediate opportunities to address this through more integrated planning [98].

Understanding spatial and temporal ecological sensitivity, and the ways in which human activities can erode this, will allow decision makers to design and implement economic, industrial, agricultural, urban, and other forms of development in less environmentally damaging ways. Decision makers must have access to and use credible social and ecological data, account for long term ecological impacts of any actions, and deploy appropriate frameworks for assessing environmental impacts, rigorously following mitigation hierarchies to ensure "no net loss" of biodiversity and natural ecosystems [99]. Opportunities with developments that can be easily scaled up include (i) lengthening windows for public review, so costs and benefits of a development and any potential negative impacts are better discussed; and (ii) weighing up the social–ecological costs and benefits of several alternative development options rather than just one, which is the current norm. This is specifically relevant in the case of infrastructure development, which is often associated with undesirable social and environmental side effects (see Box 1). A good example of this is the "Aberdare road" project in Kenya, where an economic cost-benefit analysis identified several alternative options that benefited the economy and people more, while potentially impacting the environment less [100].

Integrated planning can be deployed across scales, including local, national and regional, as a tool to overcome ad-hoc and uncoordinated responses to ecological challenges, and pre-empt co-benefits and trade-offs, winners and losers in any scenario. For instance, considering that agriculture will undoubtedly expand across Africa (see Box 2) significant investment into sustainable intensification of both small and commercial producers to close yield gaps and conserve biodiversity will be crucial. Also, meeting this increased demand for food sustainably will require integrated spatial planning to maximise production and reduce land degradation. Cross-sectoral collaborations will need to plan for a land sparing-sharing continuum [101,102]. Approaches such as [103] principles for reconciling agriculture and conservation through a landscape approach can serve as useful guides [104]. An example of good practice is the Amboseli Ecosystem Management Plan in Kenya which integrates multiple land uses, including natural resource conservation and agriculture, for the greater good of all stakeholders. The plan has been gazetted by local stakeholders, county government, national government agencies and the Attorney General, following many of the guidelines outlined in the "effective natural resource governance" section above.

3.3.3. Investment Safeguards and Frameworks

Establishing appropriate investment safeguards can limit ecologically damaging projects. Firstly, this can be achieved through clear regulations which provide legislative recourse and create a disincentive for those transgressing codes of practice. Secondly, the adoption of voluntary codes and principles by major lending institutions can help establish clear and structured guidelines for assessing the social–ecological impacts of investments across their lifespan. For instance, the Equator Principles is a risk management framework established by the International Financial Corporation which provides a minimum standard for due diligence of risk and nature-related safeguards. Such national and international regulations can drive increased transparency, while reducing the capital available for those who do not meet performance criteria. In general, regulations have not been strategically or stringently applied in Africa to date, suggesting that simply following a code of practice is the first step (see also [28]).

Forming and applying regulations to maximise benefits and limit risks of Foreign Direct Investment and Overseas Development Assistance is also important given these investments represent such substantial proportions of the development expenditures of many African Nations. For example, in 2017 in the Republic of Congo, foreign investment accounted for 39.44% of GDP [105]. Regulatory frameworks should provide clear and enforceable standards for non-traditional and emerging market investors operating in Africa. Some lenders, for example, the European Union, have strict environmental regulations in their own jurisdictions, but these do not necessarily apply when they fund projects in Africa. For instance, the Export–Import Bank of China has provided more than USD149 billion to 1800 projects through the Belt and Road Initiative aiming to bolster a network of land and sea links with Southeast Asia, Central Asia, the Middle East, Europe, and Africa. China's lending has encountered criticism of being a "debt trap" which some believe will aggravate the financial vulnerability of developing countries with associated financial, geopolitical, and sovereignty risks [106]. Removing this double-standard would represent an immediate win for African sustainability, ensuring the achievement of Biodiversity Net Gain and Net Zero targets. The Forum on China-Africa Cooperation may provide an important opportunity for discussing solutions to these challenges.

In order to encourage the financial sector to evaluate and prioritise ecologically sound investments, enabling frameworks and tools which allow ecological concerns to be integrated in traditional financial risk assessments need to be established. By developing valuation methodologies that allow investors to respond to clear market signals, the value creation opportunities inherent in preserving and creating ecological and natural capital will emerge. In this regard, Natural Capital Accounting and Assessment is gaining rapid ground as an approach which brings ecological considerations to the forefront of policy, planning and decision making, particularly in the face of an uncertain future [107].

3.3.4. New Partnership Models

Many of the challenges of managing ecologically sensitive areas and assets are problems of collective action, while other examples show this can be overcome through effective institutional arrangements and partnerships [108]. To be more effective, there is a need for new partnership models that reconfigure the relationships between state, business, and civil society.

For instance, new institutional structures such as water user associations, which are increasingly being formalised across Africa, allow local communities to co-manage their own resources through polycentric governance with help from civil society, government and the private sector [109]. At a different scale, the "Nile Basin Initiative" is an example of a multi-country, multi-partner initiative that presents a platform facilitating discussions around the effective management of water and other resources in the Nile basin, promoting stability and sustainable resource governance [110].

Public–private partnerships are also likely to play an increasingly significant role in African sustainability. These can present important opportunities for domestic and international businesses to act as responsible stewards of natural resources to increase profit, enhance their reputation and create jobs. To achieve this, businesses will need to develop new skills and capabilities to engage as partners with communities, government, and other businesses, and more actively contribute to shaping public procurement processes and enabling policies (see the World Bank's Public Private Partnership Knowledge Lab https://pppknowledgelab.org/data, accessed on 4 August 2021).

A good example is the Akassa Brass Community Development Scheme run by Equinor (a Norwegian company), for over 20 years, which owns several strategic oil assets in Nigeria's Niger Delta region. Although it is part of the often-maligned extractives sector, by delivering social investments and infrastructure projects, and by placing an emphasis on shared values, this scheme involving community forestry on customary land appears to have fostered harmonious relations with host communities, private investors, and governments [111].

3.3.5. Clear Data, Management Information, and Decision Support Tools

Underpinning all of these intervention opportunities should be clear data, information management and decision support frameworks, each of which is vital to decision making which reflects the value of Africa's ecosystems and societies. Traditional decision support tools such as cost–benefit analysis are often inadequate in their considerations of ecological costs and benefits. This is in part because there are political judgments based on pre-existing assumptions, and in part because they focus on monetising benefits over considerations of the intrinsic values of ecological assets [107].

There are a number of alternative tools, such as strategic environmental assessments (SEAs). SEAs can facilitate strategic foresight and suggest monitoring mechanisms, determine effect size and severity of an investment, consider how it can withstand future climate impacts, and alternative routing, indirect redundancies, or severed wildlife migratory routes [28].

Likewise, conservation planning tools that allow for spatial analyses that allocate land parcels for multiple land uses such as agriculture and conservation in a manner that economically optimises both land uses and achieves pre-defined objectives for each are also available [112]. For example, there are tools which analyse mutual interdependencies, as well as trade-offs between sectors. For instance, in Burkina Faso and Ghana, the MAXUS tool is used to examine the spatial connections between energy development and food security [113]. In the Rufiji River Basin, Tanzania, spatial modelling has assessed water energy food trade-offs across infrastructural development scenarios [114].

African governments have also put natural capital accounting into practice. For instance, the Government of Botswana used natural capital accounting to construct water accounts from 1993–2012. Results from this were incorporated into the National Development Plan 11, and were used to guide: the raw water abstraction strategy project, catchment management committees, the National Water Master Plan review, and the Botswana National Water Conservation and Water Demand Management Strategy [115,116].

Additionally, the United Nation's System of Environmental-Economic Accounts is an international framework working to incorporate nature into the System of National Accounts. It includes the Central Framework which deals with, for example, water, energy, and mineral accounts, and offers guidelines for the next generation of standards under development known as Ecosystem Accounting [117]. There are other tools, however: the Green Growth Knowledge Partnership recently published a report which reviewed 28 data platforms and tools that have the potential to be used in integrating natural capital approaches in policy and planning processes [118].

Further work is needed to consolidate existing approaches and develop novel decision support tools that appropriately account for ecological considerations, without being technically complex and costly (e.g., [118]). Additionally, all decision support tools rely on access to appropriate credible data, such as the location of ecologically sensitive areas, data on the threats posed by economic activities, and data on possible solutions. Although there are efforts to enhance access to credible information, such as the conservation evidence project (www.conservationevidence.com, accessed on 4 August 2021), data sharing and availability is currently insufficient in many African countries [119–121]–emphasising the need for greater open access.

Box 1. Green Infrastructure, Green Cities.

Future estimates suggest that over 60% of the total population will reside in urban areas by 2060 [46]. The needs of these residents will need to be met. Given the magnitude and potential impacts of this task, investing in "green infrastructure" is an opportunity to improve sustainability. Green infrastructure is a network of natural or man-made environmental features that deliver ecosystem services within the built environment [122]. Green infrastructure is not simply open green spaces, but include practices such as infiltration, evapotranspiration, rainwater harvesting, bioretention, preserving, and restoring natural landscape features such as forests, floodplains, wetlands, waterways, and their banks, as well as planting site-specific features such as trees, green roads, road verges, permeable sidewalks, and cisterns. Green infrastructure can provide ecosystem services, including temperature regulation, augmenting water supply and improving water quality, while creating jobs, and mitigating and adapting to climate change, thereby avoiding significant costs. Indirectly, it also improves the aesthetic quality of cities, which can attract businesses, investment, and tourism and unlock financing mechanisms such as public–private partnerships, impact investment groups, and green funds. Green infrastructure can therefore contribute importantly towards achieving African Ecological Futures. However, currently, there is a limited understanding of the distinction between different types of infrastructure and in particular the potential benefits associated with green and blue infrastructure, as opposed to conventional, grey infrastructure. Consequently, planning and policy processes in urban areas often do not consider green alternatives for infrastructure; instead, often choosing less suitable designs that may not provide all intended benefits. Factoring in green infrastructures in strategic planning processes and having the right investment safeguards and frameworks in place will ensure a balance between developing and safeguarding sensitive ecosystems. There remain several barriers towards the large-scale application of green infrastructure, including: a limited understanding of green infrastructure and the economic, social and ecological benefits; and little technical guidance for how to implement green infrastructure in urban areas. Both strategic planning and investment are limited by lack of clear "data, management information and decision support frameworks". Quantifying the benefits of green infrastructure is not easy, and there is currently little Africa-specific data [123]. So, a robust evidence base for green infrastructure in Africa needs to be developed.

Box 2. Sustainable Agriculture.

Agriculture is the largest employer in Africa (including 175 million people in sub-Saharan Africa), and provides an important route to overcoming poverty [6]. Agriculture contributes 15% of Africa's GDP on average [124], 20% of merchandise exports [125], with crops constituting 85% of the total agricultural production value [124].

To meet growing demand for food (and other products), both extensification and intensification are required. These both can negatively affect ecosystems, especially considering that areas of high agricultural value are often also important for biodiversity conservation. Sustainable intensification, where agricultural yields are increased without adverse environmental impact and without the conversion of additional non-agricultural land, will be vital.

Sustainable intensification of agriculture in Africa is heavily dependent on 'effective natural resource governance' and 'integrated strategic planning'. It requires diverse and context-specific solutions which are tailored to both smallholders and large-scale commercial farming [126]. These may be in the form of advanced technologies (particularly for commercial farming), where there are also opportunities to learn from past failures [127]. There is also a need for solutions that include agro-ecological practices which address the immediate needs of smallholders [31], as well as globally agreed-on certification schemes which reward sustainable agricultural practices [128].

"Decision making frameworks" which support land policy development and implementation do exist, such as the AU/ECA/AfDB Land Policy Initiative [129,130]. However, there is scope for significant improvements, as well as developments of novel frameworks, that appropriately account for ecological considerations.

There is increasing investment into Africa's agricultural sector through Africa based farmer or private company investments, national public sector expenditure, foreign direct investment, and overseas development assistance. For instance, the number and scale of long-term leases or ownership of land in Africa by capital-rich private, government, or public–private sectors in particular, has increased dramatically over the past decade, motivated principally by the rise of commodity prices, food security, and biofuel production. Agricultural growth corridors, such as Tanzania's Southern Agricultural Growth corridor and Mozambique's Beira Agricultural Corridor, are aiming to dramatically expand land area under agriculture. However, significant challenges remain in ensuring these efforts work for local landowners and smallholders [131], as well as in reconciling the impacts of these agricultural corridors with areas that are deemed important for biodiversity conservation [132]. Therefore, 'safeguards on financial investment' in African agriculture are critical to limiting negative ecological impacts.

4. Conclusions

This work represents a synthesis of the findings of the African Ecological Futures project carried out between 2013 and 2015, while also integrating additional relevant scientific advances made since then. We reviewed the drivers, pressures, and impacts of Africa's development on its natural environment and described intervention opportunities that, if leveraged, could help define a continent-wide sustainable development trajectory. The aim of this work is to empower decision makers, investors, and implementers with information to support smart decisions now, to guarantee a sustainable, ecological future, ahead.

The scenarios process proved to be a valuable tool for analysing trade-offs and guiding discussions of what is an extremely complex system of interacting parameters that influence the potential development trajectories of Africa. By identifying a set of parameters that are more of a societal nature, as opposed to the more traditional "green" versus "brown" development scenarios, this analysis provides a framework for an open discussion about the core choices to be made by policy makers in Africa-including the dynamics and implications of external versus inward-looking development, and open versus closed societies.

This work further highlights the potential of participatory scenario planning and futures approaches for tackling the dynamic and evolving landscape of development challenges. While the process itself has proven effective, we recognize the limitations of an expert-based review and scenario development process. In particular, although efforts have been made to consider local stakeholders and the role of local governance in this work, we acknowledge that the recommendations presented are predominantly top-down and represent the views of 'experts' but not necessarily consensus from the broader 'community'. For instance, alternative views from landowners, citizens and local communities could help better understand the complexity of nature-people interactions at local level. In future, we acknowledge that ecological futures and scenario planning must include the knowledge and wisdom of these other voices, including but not limited to: rural and urban land owners, communities, and citizens [107,133].

Furthermore, in the approach taken in this process, it should be noted that scenarios derived from the African Ecological Futures process should not be seen as the end-result of this process. Rather, the process and deliberations were intended to instigate further discussion on how different local and sectoral contexts influence the development pathways of countries, and the continent, and the consequences thereof for the environment. In this regard, it is also clear that different scenarios may apply to different local and sectoral contexts, and therefore, that the planning process should be adapted to such specific contexts in an iterative manner.

It should also be noted that the 'global shock' presented by the COVID-19 pandemic has shown that extremely rare events can have widespread and severe impacts that jeopardise the success of implementing response opportunities. These extreme events can alter the relationships and impacts of global drivers of environmental change, highlighting vulnerabilities and disrupting progress towards achieving the sustainable development goals [4,134]. Such events are difficult to predict, but a resilience-based approach can allow social–ecological systems with the flexibility to respond and adapt within a changing environment [135].

The analysis and perspectives we share above are specific to Africa, but the entire planet faces an uncertain social–ecological future. We must develop and implement processes and approaches for innovating, analysing and communicating alternative development trajectories and sustainable futures. For instance, the recently completed review on The Economics of Biodiversity [107] identified three potential pathways to sustainability: (i) ensure that our demands on nature do not exceed supply, and that we increase nature's supply relative to its current level; (ii) change our measures of economic success to help guide us on a more sustainable path; and (iii) transform our institutions and systems to enable and sustain these changes for future generations. Exploring the implications of these pathways in the context of a scenarios process will be critical to identifying actionable and evidence-based development interventions to guide policy decisions.

The overall imperative of development across the continent, coupled with increased access to information, new decision support and planning tools, emerging new technologies, and an increasing interest in nature and sustainability provide a unique opportunity to redefine the continent's future. In particular, national and regional decision makers can learn from the short term and extractive development pathways followed by industrialised nations which dramatically depleted their own and others' ecosystems, wiping out biodiversity, undermining resilience and imperilling the planet in their unbridled pursuit of economic growth. The challenge is a development pathway that is just and equitable so that no one is left behind while simultaneously delivering benefits to people and nature.

African citizens and leaders have the ability to control their own ecological destiny-but it requires shared vision, robust evidence, and a committed and coordinated response. An inclusive African Ecological Futures process that recognises both endogenous and exogenous drivers, threats, and opportunities is a critical first step towards realising dynamic new development pathways for the continent. However, we cannot walk these pathways alone. Our future is tied to the future of the planet and we must work together as a continent and as a global community to embrace new trajectories, new approaches, new technologies, and new values—that recognise and strengthen nature's critical role in economic development and human wellbeing—as part of a sustainable global future that addresses human-induced change and ensures the resilience and stability of the earth system overall.

Author Contributions: Conceptualisation, P.S., J.P.R.T., T.C. and J.S.W.; Writing—original draft preparation, P.T., P.B. and J.R.A.; Writing—review and editing, P.S., P.T., P.B., J.R.A., J.P.R.T., Y.K., V.U. and J.S.W.; Visualisation, P.T. and P.B.; Supervision, P.S., T.C. and J.S.W.; Project administration, T.C.; Funding acquisition, P.S. All authors have read and agreed to the published version of the manuscript.

Funding: This work was funded by African Development Bank, the Worldwide Fund for Nature. Recent work was also funded in part by the UK Research and Innovation's Global Challenges Research Fund under the Development Corridors Partnership (grant no. ES/P011500/1).

Institutional Review Board Statement: Not applicable.

Informed Consent Statement: Not applicable.

Data Availability Statement: This article does not have any data to be availed.

Acknowledgments: We wish to thank the many experts from WWF, AfDB and partners that have contributed to the African Ecological Futures work over the past 8 years, a list that would be too long to mention here. In particular, we wish to acknowledge the leadership of Frederik Kwame Kumah, Deon Nell, and Laurent Some, who have spearheaded this initiative from the beginning, as well as Alice Ruhweza, who took over the role of champion on behalf of WWF since 2019. We also thank Rose Muthoni Muiyuro for her assistance.

Conflicts of Interest: The authors declare no conflict of interest.

References

1. AfDB and WWF. African Ecological Footprint Report: Green Infrastructure for Africa's Ecological Security; Gland, Switzerland. 2012. Available online: https://www.afdb.org/sites/default/files/documents/projects-and-operations/africa_ecological_footprint_report_-_green_infrastructure_for_africas_ecological_security.pdf (accessed on 24 May 2021).
2. Ripple, W.J.; Newsome, T.M.; Wolf, C.; Dirzo, R.; Everatt, K.T.; Galetti, M.; Hayward, M.W.; Kerley, G.I.H.; Levi, T.; Lindsey, P.A.; et al. Collapse of the world's largest herbivores. *Sci. Adv.* **2015**, *1*, e1400103. [CrossRef]
3. Wolf, C.; Ripple, W.J. Prey depletion as a threat to the world's large carnivores. *R. Soc. Open Sci.* **2016**, *3*, 160252. [CrossRef]
4. Lindsey, P.; Allan, J.; Brehony, P.; Dickman, A.; Robson, A.; Begg, C.; Bhammar, H.; Blanken, L.; Breuer, T.; Fitzgerald, K.; et al. Conserving Africa's wildlife and wildlands through the COVID-19 crisis and beyond. *Nat. Ecol. Evol.* **2020**, *4*, 1300–1310. [CrossRef]
5. Owen-Smith, N. Megafaunal Extinctions: The Conservation Message from 11,000 Years B.P. *Conserv. Biol.* **1989**, *3*, 405–412. [CrossRef] [PubMed]
6. AfDB Tracking Africa's Progress in Figures; Tunis, 2014. Available online: https://www.afdb.org/fileadmin/uploads/afdb/Documents/Publications/Tracking_Africa%E2%80%99s_Progress_in_Figures.pdf (accessed on 24 May 2021).
7. AfDB African Economic Outlook 2021; Côte d'Ivoire. Available online: https://www.afdb.org/en/knowledge/publications/african-economic-outlook (accessed on 4 June 2021).
8. ITU (International Telecommunications Union). *Measuring Digital Development: Facts and Figures 2019*; Geneva, Switzerland, 2019. Available online: https://news.itu.int/measuring-digital-development-facts-figures-2019/ (accessed on 5 November 2019).
9. Eberhard, R. *Access to Water and Sanitation in Sub-Saharan Africa*; Berlin, Germany, 2019. Available online: https://www.oecd.org/water/GIZ_2018_Access_Study_Part%20II_Narrative%20Report_Briefing_document.pdf (accessed on 24 May 2021).
10. Schoch, M.; Lakner, C. The Number of Poor People Continues to Rise in Sub-Saharan Africa, Despite a Slow Decline in the Poverty Rate. 2020. UNDP. Available online: https://blogs.worldbank.org/opendata/number-poor-people-continues-rise-sub-saharan-africa-despite-slow-decline-poverty-rate (accessed on 24 May 2021).
11. Rudel, T.K. The national determinants of deforestation in sub-Saharan Africa. *Philos. Trans. R. Soc. B Biol. Sci.* **2013**, *368*, 20120405. [CrossRef] [PubMed]
12. Belhabib, D.; Cheung, W.W.L.; Kroodsma, D.; Lam, V.W.Y.; Underwood, P.J.; Virdin, J. Catching industrial fishing incursions into inshore waters of Africa from space. *Fish Fish.* **2020**, *21*, 379–392. [CrossRef]
13. Asche, F.; Garlock, T.M.; Akpalu, W.; Amaechina, E.C.; Botta, R.; Chukwuone, N.A.; Eggert, H.; Hutchings, K.; Lokina, R.; Tibesigwa, B.; et al. Fisheries performance in Africa: An analysis based on data from 14 countries. *Mar. Policy* **2021**, *125*, 104263. [CrossRef]
14. Archer, E.; Dziba, L.; Mulongoy, K.J.; Maoela, M.A.; Walters, M. *Regional Assessment Report on Biodiversity and Ecosystem Services for Africa*; IPBES: Bonn, Germany, 2018. [CrossRef]
15. Laurance, W.F.; Sloan, S.; Weng, L.; Sayer, J.A. Estimating the Environmental Costs of Africa's Massive "development Corridors". *Curr. Biol.* **2015**, *25*, 3202–3208. [CrossRef]
16. Hallegatte, S.; Walsh, B. Available online: https://blogs.worldbank.org/climatechange/covid-climate-change-and-poverty-avoiding-worst-impacts (accessed on 24 May 2021).

17. Erdoğan, S.; Çakar, N.D.; Ulucak, R.; Danish; Kassouri, Y. The role of natural resources abundance and dependence in achieving environmental sustainability: Evidence from resource-based economies. *Sustain. Dev.* **2021**, *29*, 143–154. [CrossRef]
18. UNCTAD. Handbook of Statistics 2020—Population. Available online: https://unctad.org/webflyer/handbook-statistics-2020 (accessed on 24 May 2021).
19. UNICEF Home Page. Available online: https://www.unicef.org/press-releases/dividend-or-disaster-unicefs-new-report-population-growth-africa (accessed on 24 May 2021).
20. AfDB; WWF. African Ecological Futures 2015. Available online: https://www.afdb.org/fileadmin/uploads/afdb/Documents/Generic-Documents/xwwf_african_futures_report_english-lo-rez.pdf (accessed on 24 May 2021).
21. Moomaw, W.; Griffin, K.K.; Lomax, J. The Critical Role of Global Food Consumption Patterns in Achieving Sustainable Food Systems and Food for All. Available online: https://wedocs.unep.org/bitstream/handle/20.500.11822/25186/Food_Consumption_Patterns.pdf?sequence=1&isAllowed=y (accessed on 24 May 2021).
22. UN. Sustainable Development Goals. Available online: https://sustainabledevelopment.un.org/post2015/summit (accessed on 24 May 2021).
23. CBD. Strategic Plan for Biodiversity 2011–2020 and the Aichi Targets "Living in Harmony with Nature". Available online: https://www.cbd.int/doc/strategic-plan/2011-2020/Aichi-Targets-EN.pdf (accessed on 24 May 2021).
24. UNFCCC. Paris Agreement. Available online: https://unfccc.int/sites/default/files/english_paris_agreement.pdf. (accessed on 24 May 2021).
25. African Union Goals & Priority Areas of Agenda 2063. Available online: https://au.int/en/agenda2063/goals (accessed on 24 May 2021).
26. Sardar, Z. The Namesake: Futures; futures studies; futurology; futuristic; foresight—What's in a name? *Futures* **2010**, *42*, 177–184. [CrossRef]
27. van Vuuren, D.P.; Kok, M.T.J.; Girod, B.; Lucas, P.L.; de Vries, B. Scenarios in Global Environmental Assessments: Key characteristics and lessons for future use. *Glob. Environ. Chang.* **2012**, *22*, 884–895. [CrossRef]
28. Thorn, J.P.R.; Klein, J.A.; Steger, C.; Hopping, K.A.; Capitani, C.; Tucker, C.M.; Reid, R.S.; Marchant, R.A. Scenario archetypes reveal risks and opportunities for global mountain futures. *Glob. Environ. Chang.* **2021**, *69*, 102291. [CrossRef]
29. Vervoort, J.M.; Thornton, P.K.; Kristjanson, P.; Förch, W.; Ericksen, P.J.; Kok, K.; Ingram, J.S.I.; Herrero, M.; Palazzo, A.; Helfgott, A.E.S.; et al. Challenges to scenario-guided adaptive action on food security under climate change. *Glob. Environ. Chang.* **2014**, *28*, 383–394. [CrossRef]
30. Lavorel, S.; Colloff, M.J.; Locatelli, B.; Gorddard, R.; Prober, S.M.; Gabillet, M.; Devaux, C.; Laforgue, D.; Peyrache-Gadeau, V. Mustering the power of ecosystems for adaptation to climate change. *Environ. Sci. Policy* **2019**, *92*, 87–97. [CrossRef]
31. Thorn, J.P.R.; Klein, J.A.; Steger, C.; Hopping, K.A.; Capitani, C.; Tucker, C.M.; Nolin, A.W.; Reid, R.S.; Seidl, R.; Chitale, V.S.; et al. A systematic review of participatory scenario planning to envision mountain social–ecological systems futures. *Ecol. Soc.* **2020**, *25*, 1–55. [CrossRef]
32. IPCC; Rivera, A.; Bravo, C.; Buob, G. Summary for Policymakers. In *Climate Change and Land: An IPCC Special Report on Climate Change, Desertification, Land Degradation, Sustainable Land Management, Food Security, and Greenhouse Gas Fluxes in Terrestrial Ecosystems*; IPCC: Geneva, Switzerland, 2019; ISBN 9789291691548.
33. Rounsevell, M.D.A.; Dawson, T.P.; Harrison, P.A. A conceptual framework to assess the effects of environmental change on ecosystem services. *Biodivers. Conserv.* **2010**, *19*, 2823–2842. [CrossRef]
34. Ezeh, A.; Kissling, F.; Singer, P. Why sub-Saharan Africa might exceed its projected population size by 2100. *Lancet* **2020**, *396*, 1131–1133. [CrossRef]
35. AfDB. African Development Report. Available online: https://www.afdb.org/fileadmin/uploads/afdb/Documents/Publications/African_Development_Report_2012.pdf. (accessed on 24 May 2021).
36. Calvin, K.; Bond-Lamberty, B.; Clarke, L.; Edmonds, J.; Eom, J.; Hartin, C.; Kim, S.; Kyle, P.; Link, R.; Moss, R.; et al. The SSP4: A world of deepening inequality. *Glob. Environ. Chang.* **2017**, *42*, 284–296. [CrossRef]
37. Beegle, K.; Christiaensen, L. *Accelerating Poverty Reduction in Africa*; World Bank: Washington, DC, USA, 2019.
38. Zakari, R.; Zolfagharian, S.; Nourbakhsh, M.; Zin, R.M.; Gheisari, M. Ecological footprint of different nations. In Proceedings of the WCSE 2012—International Workshop on Computer Science and Engineering. *IJET* **2012**, *4*, 464–467.
39. USIP. Natural Resources, Conflict, and Conflict Resolution. Available online: https://www.usip.org/publications/2007/09/natural-resources-conflict-and-conflict-resolution (accessed on 24 May 2021).
40. Maphosa, S.B. Natural Resources and Conflict: Unlocking the Economic Dimension of Peace-Building in Africa. Africa Institute of South Africa (AISA). Available online: https://www.africaportal.org/publications/natural-resources-and-conflict-unlocking-the-economic-dimension-of-peace-building-in-africa/ (accessed on 24 May 2021).
41. The Economist Intelligence Unit. Democracy Index 2020: In Sickness and in Health? Available online: https://www.eiu.com/n/campaigns/democracy-index-2020/ (accessed on 24 May 2021).
42. Williams, D.R.; Clark, M.; Buchanan, G.M.; Ficetola, G.F.; Rondinini, C.; Tilman, D. Proactive conservation to prevent habitat losses to agricultural expansion. *Nat. Sustain.* **2020**, *4*, 314–322. [CrossRef]
43. IPCC. Global Warming of 1.5 °C An IPCC Special Report on the Impacts of Global Warming of 1.5 °C Above Pre-Industrial Levels and Related Global Greenhouse Gas Emission Pathways, in the Context of Strengthening the Global Response to the Threat of Climate Change. Available online: https://www.ipcc.ch/sr15/ (accessed on 24 May 2021).

44. Frishkoff, L.O.; Karp, D.S.; Flanders, J.R.; Zook, J.; Hadly, E.A.; Daily, G.C.; M'gonigle, L.K. Climate change and habitat conversion favour the same species. *Ecol. Lett.* **2016**, *19*, 1081–1090. [CrossRef]
45. Freedman, B. Chapter 12~Resources and Sustainable Development. Available online: https://ecampusontario.pressbooks.pub/environmentalscience/chapter/chapter-12-resources-and-sustainable-development/ (accessed on 24 May 2021).
46. UNECA. A Regional Approach to Leveraging Urbanization for Africa's Structural Transformation. Available online: https://www.uneca.org/regional-approach-leveraging-urbanization-africa%E2%80%99s-structural-transformation (accessed on 24 May 2021).
47. Effiong, E.L. Urbanization and Environmental Quality in Africa. Available online: https://mpra.ub.uni-muenchen.de/73224/1/MPRA_paper_73224.pdf (accessed on 24 May 2021).
48. Güneralp, B.; Lwasa, S.; Masundire, H.; Parnell, S.; Seto, K.C. Urbanization in Africa: Challenges and opportunities for conservation. *Environ. Res. Lett.* **2018**, *13*, 015002. [CrossRef]
49. Goyal, R.; Sy, A. Available online: https://www.brookings.edu/blog/africa-in-focus/2015/12/02/mobilizing-africas-rapid-urbanization-for-sustainable-climate-change/ (accessed on 24 May 2021).
50. Thorn, J.P.R.; Biancardi Aleu, R.; Wijesinghe, A.; Mdongwe, M.; Marchant, R.A.; Shackleton, S. Barriers and enablers to mainstreaming green infrastructure for climate adaptation in peri-urban settlements in Namibia and Tanzania. *Landsc. Urban Plan.* in press.
51. Ehrlich, P.R.; Ehrlich, A.H. Can a collapse of global civilization be avoided? *Proc. R. Soc. B Biol. Sci.* **2013**, *280*, 20122845. [CrossRef] [PubMed]
52. Thorn, J.P.R. Adaptation "from below" to changes in species distribution, habitat and climate in agro-ecosystems in the Terai Plains of Nepal. *Ambio* **2019**, *48*, 1482–1497. [CrossRef]
53. Kleinschroth, F.; Laporte, N.; Laurance, W.F.; Goetz, S.J.; Ghazoul, J. Road expansion and persistence in forests of the Congo Basin. *Nat. Sustain.* **2019**, *2*, 628–634. [CrossRef]
54. Juffe-Bignoli, D.; Bull, J.; Burgess, N.; Thorn, J.P.R.; Tam, C.; Hobbs, J. Development corridors, their impacts on biodiversity, and impact mitigation best practice: A global review. *Front. Ecol. Evolution.* **2021**, *9*, 683949. [CrossRef]
55. Diop, S.; Scheren, P.; Niang, A. *Climate Change and Water Resources in Africa*; Springer International Publishing: Berlin/Heidelberg, Germany, 2021.
56. Jacobsen, M.; Webster, M.; Vairavamoorthy, K. *The Future of Water in African Cities: Why Waste Water?* The World Bank: Washington, DC, USA, 2012.
57. Hassan Rashid, M.A.U.; Manzoor, M.M.; Mukhtar, S. Urbanization and its effects on water resources: An exploratory analysis. *Asian J. Water Environ. Pollut.* **2018**, *15*, 67–74. [CrossRef]
58. McGrane, S.J. Impacts of urbanisation on hydrological and water quality dynamics, and urban water management: A review. *Hydrol. Sci. J.* **2016**, *61*, 2295–2311. [CrossRef]
59. International Energy Agency. Africa Energy Outlook 2019: World Energy Outlook Special Report. Available online: https://www.iea.org/reports/africa-energy-outlook-2019 (accessed on 24 May 2021).
60. International Energy Agency Access to electricity—SDG7: Data and Projections—Analysis. Available online: https://www.iea.org/reports/sdg7-data-and-projections/access-to-electricity (accessed on 24 May 2021).
61. IRENA. Africa 2030: Roadmap for a Renewable Energy Future. Available online: https://irena.org/publications/2015/Oct/Africa-2030-Roadmap-for-a-Renewable-Energy-Future (accessed on 24 May 2021).
62. Katzner, T.E.; Nelson, D.M.; Diffendorfer, J.E.; Duerr, A.E.; Campbell, C.J.; Leslie, D.; Vander Zanden, H.B.; Yee, J.L.; Sur, M.; Huso, M.M.P.; et al. Wind energy: An ecological challenge. *Science* **2019**, *366*, 1206–1207.
63. Rehbein, J.A.; Watson, J.E.M.; Lane, J.L.; Sonter, L.J.; Venter, O.; Atkinson, S.C.; Allan, J.R. Renewable energy development threatens many globally important biodiversity areas. *Glob. Chang. Biol.* **2020**, *26*, 3040–3051. [CrossRef]
64. MacKenzie, C.A.; Fuda, R.K.; Ryan, S.J.; Hartter, J. Drilling through Conservation Policy: Oil Exploration in Murchison Falls Protected Area, Uganda. *Conserv. Soc.* **2017**, *15*, 322–333.
65. ESF Consultants Lokichar to Lamu Crude Oil Pipeline: Environmental and Social Impact Assessment. Available online: https://naturaljustice.org/wp-content/uploads/2021/01/Annex-I-A1-09Oct2019_Scoping-Report-and-TOR.pdf (accessed on 24 May 2021).
66. IEA World Energy Outlook 2015—Analysis. Available online: https://www.iea.org/reports/world-energy-outlook-2015 (accessed on 24 May 2021).
67. Fa, J.E.; Peres, C.A.; Meeuwig, J. Bushmeat Exploitation in Tropical Forests: An Intercontinental Comparison. *Conserv. Biol.* **2002**, *16*, 232–237. [CrossRef]
68. Yan, H.F.; Kyne, P.M.; Jabado, R.W.; Leeney, R.H.; Davidson, L.N.K.; Derrick, D.H.; Finucci, B.; Freckleton, R.P.; Fordham, S.V.; Dulvy, N.K. Overfishing and habitat loss drives range contraction of iconic marine fishes to near extinction. *Sci. Adv.* **2021**, *7*, 6026. [CrossRef] [PubMed]
69. De Merode, E.; Smith, K.H.; Homewood, K.; Pettifor, R.; Rowcliffe, M.; Cowlishaw, G. The impact of armed conflict on protected-area efficacy in Central Africa. *Biol. Lett.* **2007**, *3*, 299–301. [CrossRef]
70. Nackoney, J.; Molinario, G.; Potapov, P.; Turubanova, S.; Hansen, M.C.; Furuichi, T. Impacts of civil conflict on primary forest habitat in northern Democratic Republic of the Congo, 1990–2010. *Biol. Conserv.* **2014**, *170*, 321–328. [CrossRef]
71. Cawthorn, D.M.; Hoffman, L.C. The bushmeat and food security nexus: A global account of the contributions, conundrums and ethical collisions. *Food Res. Int.* **2015**, *76*, 906–925. [CrossRef]

72. Mensah, K.E.; Damnyag, L.; Kwabena, N.S. Analysis of charcoal production with recent developments in Sub-Sahara Africa: A review. *Afr. Geogr. Rev.* **2020**. [CrossRef]
73. Dos Santos, S.; Adams, E.A.; Neville, G.; Wada, Y.; de Sherbinin, A.; Mullin Bernhardt, E.; Adamo, S.B. Urban growth and water access in sub-Saharan Africa: Progress, challenges, and emerging research directions. *Sci. Total Environ.* **2017**, *607–608*, 497–508. [CrossRef]
74. Edwards, D.P.; Sloan, S.; Weng, L.; Dirks, P.; Sayer, J.; Laurance, W.F. Mining and the African environment. *Conserv. Lett.* **2014**, *7*, 302–311. [CrossRef]
75. Arnell, N.W. Climate change and global water resources: SRES emissions and socio-economic scenarios. *Glob. Environ. Chang.* **2004**, *14*, 31–52. [CrossRef]
76. Boone, R.B.; Conant, R.T.; Sircely, J.; Thornton, P.K.; Herrero, M. Climate change impacts on selected global rangeland ecosystem services. *Glob. Chang. Biol.* **2018**, *24*, 1382–1393. [CrossRef]
77. Newbold, T. Future effects of climate and land-use change on terrestrial vertebrate community diversity under different scenarios. *Proc. R. Soc. London B Biol. Sci.* **2018**, *285*. [CrossRef]
78. Brockington, D. *Fortress Conservation: The Preservation of the Mkomazi Game Reserve, Tanzania*; Indiana University Press: Bloomington, IN, USA, 2002.
79. Scott, J.C. *Seeing Like a State: How Certain Schemes to Improve the Human Condition Have Failed*; Yale University Press: New Haven, CT, USA, 1998.
80. IUCN Parks for Life: Report on the IVth World Congress on National Parks and Protected Areas. Available online: https://portals.iucn.org/library/sites/library/files/documents/1993-007.pdf (accessed on 24 May 2021).
81. Hulme, D.; Murphree, M. Communities, wildlife and the new conversation in Africa. *J. Int. Dev.* **1999**, *11*, 277–285. [CrossRef]
82. Adams, W.M.; Hulme, D. If community conservation is the answer in Africa, what is the Question? *Oryx* **2001**, *35*, 193–200. [CrossRef]
83. Folke, C.; Hahn, T.; Olsson, P.; Norberg, J. Adaptive Governance of Social–ecological Systems. *Annu. Rev. Environ. Resour.* **2005**, *30*, 441–473. [CrossRef]
84. Lemos, M.C.; Agrawal, A. Environmental governance. *Annu. Rev. Environ. Resour.* **2006**, *31*, 297–325. [CrossRef]
85. Persha, L.; Agrawal, A.; Chhatre, A. Social and ecological synergy: Local rulemaking, forest livelihoods, and biodiversity conservation. *Science* **2011**, *331*, 1606–1608. [CrossRef]
86. Lockwood, M.; Davidson, J.; Curtis, A.; Stratford, E.; Griffith, R. Governance principles for natural resource management. *Soc. Nat. Resour.* **2010**, *23*, 986–1001. [CrossRef]
87. Chene, M. Natural Resource Management Transparency and Governance: A Literature Review Focusing on Extractive Industries. Available online: https://www.u4.no/publications/natural-resource-management-transparency-and-governance (accessed on 24 May 2021).
88. CEESP; IUCN. IUCN Natural Resource Governance Framework (NRGF). Available online: https://www.iucn.org/commissions/commission-environmental-economic-and-social-policy/our-work/knowledge-baskets/natural-resource-governance (accessed on 24 May 2021).
89. Okech, R.; Bob, U. Sustainable Ecotourism Management in Kenya. *Ethiop. J. Environ. Stud. Manag.* **2009**, *2*. [CrossRef]
90. Juma, L.O.; Khademi-Vidra, A. Community-based tourism and sustainable development of rural regions in Kenya; Perceptions of the citizenry. *Sustainability* **2019**, *11*, 4733. [CrossRef]
91. White, P.A.; Belant, J.L. Provisioning of game meat to rural communities as a benefit of sport hunting in Zambia. *PLoS ONE* **2015**, *10*, e0117237. [CrossRef]
92. Community Conservation Namibia Sustainable Wildlife Utilisation. Available online: https://communityconservationnamibia.com/support-to-conservation/natural-resource-management/sustainable-wildlife-utilisation (accessed on 24 May 2021).
93. Thouless, C.; Dublin, H.T.; Blanc, J.; Skinner, D.; Daniel, T.; Taylor, R.; Bouche, P. African Elephant Status Report 2016. Available online: https://portals.iucn.org/library/sites/library/files/documents/SSC-OP-060_A.pdf (accessed on 24 May 2021).
94. Astrid Fenger, N.; Skovmand Bosselmann, A.; Asare, R.; de Neergaard, A. The impact of certification on the natural and financial capitals of Ghanaian cocoa farmers. *Agroecol. Sustain. Food Syst.* **2017**, *41*, 143–166. [CrossRef]
95. Browne, E. Social Media and Governance. Available online: https://www.comminit.com/content/social-media-and-governance (accessed on 24 May 2021).
96. Svidronova, M.M.; Kascakova, A.; Vrbicanova, V. Can Social Media be a Tool for Participatory Governance in Slovak Municipalities? *NISPAcee J. Public Adm. Policy* **2018**, *11*, 81–101. [CrossRef]
97. Laurance, W.F.; Peletier-Jellema, A.; Geenen, B.; Koster, H.; Verweij, P.; Van Dijck, P.; Lovejoy, T.E.; Schleicher, J.; Van Kuijk, M. Reducing the global environmental impacts of rapid infrastructure expansion. *Curr. Biol.* **2015**, *25*, R259–R262. [CrossRef]
98. Abuya, D.; Oyugi, M.O.; Oyaro, E. Management of the Effects of Land Use Changes on Urban Infrastructure Capacity: A Case Study of Ruaka Town, Kiambu County, Kenya. *J. Geogr. Inf. Syst.* **2019**, *8*, 158–190.
99. Arlidge, W.N.S.; Bull, J.W.; Addison, P.F.E.; Burgass, M.J.; Gianuca, D.; Gorham, T.M.; Jacob, C.D.S.; Shumway, N.; Sinclair, S.P.; Watson, J.E.M.; et al. A Global Mitigation Hierarchy for Nature Conservation. *Bioscience* **2018**, *68*, 336–347. [CrossRef]
100. Tyrell, P.; Allan, J. Evaluating the Socio-Economic Potential of Road Development Projects Around the Aberdare Range. A Report for Rhino Ark. 2020. Available online: https://pure.uva.nl/ws/files/51204676/TyrellAllan_2020_Evaluating_the_economic_potential_of_road_development_projects_around_the_Aberdare_range.pdf (accessed on 24 May 2021).

101. Ricciardi, V.; Mehrabi, Z.; Wittman, H.; James, D.; Ramankutty, N. Higher yields and more biodiversity on smaller farms. *Nat. Sustain.* **2021**, 1–7. [CrossRef]
102. Kremen, C.; Merenlender, A.M. Landscapes that work for biodiversity and people. *Science* **2018**, *362*, eaau6020. [CrossRef] [PubMed]
103. Sayer, J.; Sunderland, T.; Ghazoul, J.; Pfund, J.L.; Sheil, D.; Meijaard, E.; Venter, M.; Boedhihartono, A.K.; Day, M.; Garcia, C.; et al. Ten principles for a landscape approach to reconciling agriculture, conservation, and other competing land uses. *Proc. Natl. Acad. Sci. USA* **2013**, *110*, 8349–8356. [CrossRef]
104. Arts, B.; Buizer, M.; Horlings, L.; Ingram, V.; Van Oosten, C.; Opdam, P. Landscape Approaches: A State-of-the-Art Review. Available online: https://www.annualreviews.org/doi/abs/10.1146/annurev-environ-102016-060932 (accessed on 24 May 2021).
105. World Bank Foreign Direct Investment, Net Inflows (% of GDP)—Republic of Congo. Available online: https://data.worldbank.org/indicator/BX.KLT.DINV.WD.GD.ZS?locations=CG (accessed on 24 May 2021).
106. Reuters China's EximBank Provides More Than $149 Bln for Belt and Road Projects. Available online: https://www.reuters.com/article/china-bank-beltandroad/chinas-eximbank-provides-more-than-149-bln-for-belt-and-road-projects-idUSL3N2202TI (accessed on 24 May 2021).
107. Dasgupta, P. *The Economics of Biodiversity: The Dasgupta Review*; HM Treasury: London, UK, 2021; ISBN 9781911680307.
108. Ostrom, E. *Governing the Commons: The Evolution of Institutions for Collective Action*; Cambridge University Press: New York, NY, USA, 1990.
109. Bruns, B. Polycentric solutions for groundwater governance in Sub-Saharan Africa: Encouraging institutional artisanship in an extended ladder of participation. *Water* **2021**, *13*, 630. [CrossRef]
110. Nile Basin Initiative Nile Basin Initiative—About. Available online: https://www.rusumoproject.org/index.php/en/109-about-the-sub-basin-2/217-learn-about-the-nile-basin-initiative (accessed on 24 May 2021).
111. Joab-Peterside, S. Green Governance: The Case of Akassa Community Forests Management and Development Plan. Available online: http://geog.berkeley.edu/ProjectsResources/ND%20Website/NigerDelta/WP/19-Joab-Peterside.pdf (accessed on 24 May 2021).
112. Grantham, H.S.; Shapiro, A.; Bonfils, D.; Gond, V.; Goldman, E.; Maisels, F.; Plumptre, A.J.; Rayden, T.; Robinson, J.G.; Strindberg, S.; et al. Spatial priorities for conserving the most intact biodiverse forests within Central Africa. *Environ. Res. Lett.* **2020**, *15*, 940–945. [CrossRef]
113. Burger, R.E.A.; Abraham, E. Maximizing water–food–energy nexus synergies at basin scale. In *Advances in Science, Technology and Innovation*; Springer Nature: Berlin/Heidelberg, Germany, 2020; pp. 67–70.
114. Geressu, R.; Siderius, C.; Harou, J.J.; Kashaigili, J.; Pettinotti, L.; Conway, D. Assessing River Basin Development Given Water-Energy-Food-Environment Interdependencies. *Earth's Futur.* **2020**, *8*, e2019EF001464. [CrossRef]
115. IISD Botswana Natural Capital Accounting for Water Supports SDG 6 Monitoring. Available online: http://sdg.iisd.org/news/botswana-natural-capital-accounting-for-water-supports-sdg-6-monitoring/ (accessed on 24 May 2021).
116. Lange, G.-M.; Wodon, Q.; Carey, K. *Building a Sustainable Future: The Changing Wealth of Nations 2018*; World Bank Group: Washington, DC, USA, 2018.
117. SEEA Ecosystem Accounting | System of Environmental Economic Accounting. Available online: https://unstats.un.org/unsd/envaccounting/seearev/eea_final_en.pdf (accessed on 24 May 2021).
118. GGKP Measuring Nature' s Contribution to Economic Development towards a Framework of Indicators for National Natural Capital Reporting. Available online: https://www.greengrowthknowledge.org/research/measuring-nature%E2%80%99s-contribution-economic-development-towards-framework-indicators-national (accessed on 24 May 2021).
119. CBD. South Africa Declaration. Available online: https://www.cbd.int/idb/image/2015/celebrations/sa-declaration.pdf (accessed on 24 May 2021).
120. IISD Africa Rising Conference: Biodiversity Data for Sustainable Development. Available online: http://sdg.iisd.org/news/africa-rising-conference-focuses-on-biodiversity-data-for-sustainable-development/ (accessed on 24 May 2021).
121. Stephenson, P.J.; Bowles-Newark, N.; Regan, E.; Stanwell-Smith, D.; Diagana, M.; Höft, R.; Abarchi, H.; Abrahamse, T.; Akello, C.; Allison, H. Unblocking the flow of biodiversity data for decision-making in Africa. *Biol. Conserv.* **2017**, *213*, 335–340. [CrossRef]
122. Gulati, M.; Scholtz, L. The Case for Investment in Green Infrastructure in African Cities. Available online: https://africa.panda.org/?31761/The-case-for-investment-in-green-infrastructure-in-African-cities (accessed on 24 May 2021).
123. Thorn, J.P.R.; Biancardi Aleu, R.; Wijesinghe, A.; Mdongwe, M.; Marchant, R.A.; Shackleton, S. Mainstreaming Green Infrastructure for Climate Adaptation in Peri-urban Settlements in Namibia and Tanzania. *Landsc. Urban. Plan.*. In review.
124. OECD/FAO. OECD-FAO Agricultural Outlook 2016–2025. Available online: http://www.fao.org/3/I5778E/I5778E.pdf (accessed on 24 May 2021).
125. FAO. *Statistical Yearbook: Africa*; Statistics Division: Rome, Italy, 2014.
126. Adenle, A.A.; Wedig, K.; Azadi, H. Sustainable agriculture and food security in Africa: The role of innovative technologies and international organizations. *Technol. Soc.* **2019**, *58*, 101143. [CrossRef]
127. Dano, E.C. *Unmasking the New Green Revolution in Africa. Motives, Players and Dynamics*; Third World Network: Penang, Malaysia, 2007.
128. Pegasys. African Ecological Futures Phase 1: Agricultural Report. Available online: https://wedocs.unep.org/bitstream/handle/20.500.11822/20554/African_Ecological_Futures_Synthesis_Paper.pdf (accessed on 24 May 2021).

129. UNECA African Land Policy Centre—About. Available online: https://www.uneca.org/african-land-policy-centre (accessed on 24 May 2021).
130. AU-ECA-AfDB Consortium. Land Policy in Africa: A Framework to Strengthen Land Rights, Enhance Productivity and Secure Livelihoods. Available online: https://www.afdb.org/en/documents/document/framework-and-guidelines-on-land-policy-in-africa-27129 (accessed on 24 May 2021).
131. Watts, N. *Investing for Impact: Finance and Farming in the Southern Highlands of Tanzania*; University of Cambridge: Cambridge, UK, 2018.
132. Cisneros-Araujo, P.; Ramirez-Lopez, M.; Juffe-Bignoli, D.; Fensholt, R.; Muro, J.; Mateo-Sánchez, M.C.; Burgess, N.D. Remote sensing of wildlife connectivity networks and priority locations for conservation in the Southern Agricultural Growth Corridor (SAGCOT) in Tanzania. *Remote Sens. Ecol. Conserv.* **2021**. [CrossRef]
133. Ford, J.D.; Cameron, L.; Rubis, J.; Maillet, M.; Nakashima, D.; Willox, A.C.; Pearce, T. Including indigenous knowledge and experience in IPCC assessment reports. *Nat. Clim. Chang.* **2016**, *6*, 349–353. [CrossRef]
134. OECD. OECD Regional Outlook 2021: Addressing COVID-19 and Moving to Net Zero Greenhouse Gas Emissions. Available online: https://www.oecd.org/publications/oecd-regional-outlook-2021-17017efe-en.htm (accessed on 24 May 2021).
135. Folke, C. Resilience (Republished). *Ecol. Soc.* **2016**, *21*, 1–30. [CrossRef]

sustainability MDPI

Article

Mapping National Environmental Sustainability Distribution by Ecological Footprint: The Case of Italy

Silvio Franco, Barbara Pancino * and Angelo Martella

Department of Economics, Engineering, Society and Business Organizations, University of Tuscia, 01100 Viterbo, Italy; franco@unitus.it (S.F.); a.martella@unitus.it (A.M.)
* Correspondence: bpancino@unitus.it

Abstract: The paper proposes a possible way of spatially representing sustainability in Italy. For this purpose, the ecological footprint approach was used as a methodological framework to assess the level of sustainability of the 8092 Italian municipalities. For each municipality, the exploitation of ecosystem services, assessed by the ecological footprint indicator, and the corresponding availability of biological capacity, associated to an indicator, have been calculated and compared, thus generating a map representing the relative sustainability of Italian municipalities. The results show a very scattered distribution of ecological balance, wherein unsustainable conditions characterize more than 60% of the territory and almost 95% of the Italian population. Despite the limitations of the methodology and some assumptions regarding the ecological footprint assessment at the municipality level, the study represents an attempt to produce an innovating tool that, based on an operational definition of sustainability, can represent natural resource exploitation at the local level, and provide useful information to address coherent and targeted environmental policies of sustainability.

Keywords: ecological footprint; Italian municipalities; sustainability map

Citation: Franco, S.; Pancino, B.; Martella, A. Mapping National Environmental Sustainability Distribution by Ecological Footprint: The Case of Italy. *Sustainability* **2021**, *13*, 8671. https://doi.org/10.3390/su13158671

Academic Editors: Adam P. Hejnowicz and Jessica P. R. Thorn

Received: 6 June 2021
Accepted: 29 July 2021
Published: 3 August 2021

Publisher's Note: MDPI stays neutral with regard to jurisdictional claims in published maps and institutional affiliations.

Copyright: © 2021 by the authors. Licensee MDPI, Basel, Switzerland. This article is an open access article distributed under the terms and conditions of the Creative Commons Attribution (CC BY) license (https://creativecommons.org/licenses/by/4.0/).

1. Introduction

The concept of sustainability has increased in prevalence in the research sector as well as in public opinion. This situation has induced the widespread use of terms "sustainability" and "sustainable" in many different contexts, but, at the same time, it has generated vagueness about the real meaning of the concept [1,2]. However, from a scientific and economic perspective, the definition of sustainability should be clarified, and become an attribute that can characterize a specific object as sustainable.

To verify whether sustainability features are present in an object, certain conditions must exist [3]: a clear definition of sustainability; one (or more) indicator(s) to operationally apply the definition; the possibility of correctly assessing this (these) indicator(s) in a specific object (for instance, a product, a process, a firm or a region).

Regarding the first aspect, without entering the wide debate on the dimensions (environmental, economic, and social) of the concept of sustainability, their nature and their reciprocal interactions, in this work, we will focus on the relationship between the supply of natural resources and their demand from anthropogenic activities within a territory. In other words, a definition that looks at the environmental dimension of sustainability, intended as the preservation of natural capital, from an economic perspective will be adopted.

With reference to the assessment of sustainability, an indicator based on the ecological footprint approach will be used; a methodology that, as will be argued, is consistent with the definition of sustainability adopted in the study.

Our analysis focused on Italy, a country that, according to its current ecological footprint (4.41 gha per capita) [4], which is some five times higher than its resource-generating capacity (0.88 gha per capita), is strongly unsustainable. At the same time, the country is also characterized by very different local situations in terms of environmental pressures and natural resources availability. The "structural" unsustainability of the Italian economic system is evident from the trend in its ecological balance time series, which has been negative since the first evaluation in 1961 [4]. This is confirmed in a recent study [5], which reports that Italy "has the fourth-highest per capita EF in the EU 27 countries", and its footprint is mainly created by transportation and food consumption.

In this situation, it is interesting to assess how this unsustainability is spread throughout the national territory by adopting an innovative approach aiming at estimating the ecological balance at the municipality level in a synthetic way. Such information could be useful for constructing national and local policies capable of integrating economic development with the awareness of environmental issues, also through the use of Next Generation funds.

Moving from these considerations, the objective of this study is to build a map of the local sustainability in the Italian territory, assuming as the spatial reference a single municipality. The first section is devoted to a discussion of the assumed definition of sustainability and the theoretical approach, which can assess the existence of such conditions in a territory. In the second section, the methodology adopted for the analysis is presented. Then, the study's outcome, represented by the constructed sustainability map of Italian municipalities, is presented and discussed. The paper ends with some final remarks about the limitations of the study and its possible implications and future developments.

2. Background

For the development of our analysis, it was necessary to stick to a definition of environmental sustainability that is operational and can be assessed for a specific "object"; that is, in our case study, the Italian municipalities. The choice of this territorial scale enables us to provide an evaluation that considers as much detail as possible from an administrative point of view. Indeed, the administrative division of the Italian territory consists of 20 regions, 107 provinces and 8092 municipalities, the latter of which represents the most restricted level of administrative bodies.

2.1. Sustainability Definition

The concept of environmental sustainability considers natural capital—defined as the set of functions provided by the environment [6,7]—via two different approaches: possible substitution with man-made capital, and strict preservation. These two positions establish the difference between the paradigms of weak and strong sustainability [8].

For neoclassical economists, sustainability is a condition wherein the capital (in a broad sense) is maintained at a constant level [9]; to this end, natural capital can be substituted with man-made capital [10,11]. When the income of an economic activity is reinvested in manufactured or human capital, and its value is greater than the value of the natural capital lost in such an activity, a (weak) sustainability condition is established [9]. As Dietz and Neumayer [8] pointed out, the weak sustainability paradigm represents an extension of the neoclassical approach to economic growth, wherein natural resources were explicitly considered as a factor of production. Specifically, "the Hartwick–Solow models of the 1970s imputed non-renewable and renewable natural resources into a Cobb–Douglas production function, which is characterised by a constant and unitary elasticity of substitution between factors of production. This entailed the assumption that natural capital was similar to produced capital and could easily be substituted for it", [8] (p. 618). Consequently, from a weak sustainability perspective, there are no fundamental differences between the nature of the kinds of well-being that natural and man-made capital can generate [12].

On the other hand, the strong sustainability paradigm is based on the idea that natural capital accomplishes many functions, some of which are not replaceable [6,13]. The functions of natural capital associated with production and consumption processes, such as raw material provision and waste assimilation, can be partially substituted by man-made capital. The same happens for some amenity services, which represent another function of natural capital [8].

However, the basic life support function cannot be substituted [14]. This implies that "the global environmental and ecological system that provides us with the basic functions of food, water, breathable air and a stable climate should be subject to a strong sustainability rule", [8] (p. 619). In addition, some other reasons for natural capital non-substitutability must be considered—the consequences of its depletion are largely unknown and uncertain, and its loss is often irreversible [15,16].

In the strong sustainability approach, the possibility of replacing natural capital with man-made capital is not excluded. However, this option cannot be applied when the level of natural capital exploitation leads to the irreversible destruction of such capital [12]. This is true for those elements of natural capital that make an essential contribution to human well-being [6]. The need to preserve the consistency of these "critical" components of natural capital requires the adoption of a strong sustainability perspective in economics [17].

2.2. Sustainability Assessment Using Ecological Footprint

The definition of strong sustainability demands that natural capital be preserved in physical terms. It implies that every empirical analysis intending to evaluate a condition of strong sustainability must be based on a measurement of the physical dimension of the natural capital, in terms of its availability and possible exploitation by economic activities.

Regarding indicators able to perform such a measurement, the literature is extremely wide and diverse; for a discussion on this topic see, among others, [18–22]. In such reviews, many sustainability indicators are discussed, compared, and evaluated; one of the most important is ecological footprint [23]. Indeed, among scholars there is general agreement that the ecological footprint is an indicator that enables a strong sustainability measurement [7].

Consistent with Daly's two principles of strong sustainability [24], the ecological footprint methodology accounts for the demand and supply of the basic resources and ecosystem services that a community needs to support its lifestyle [25]. Monfreda et al. [26] state that the ecological footprint approach follows the core requirements of strong sustainability; Knaus et al. [27] claims that it reflects the principles of strong sustainability; Mori and Christodoulou [21] affirm that it is based on strong sustainability and Huang [28] asserts that ecological footprint is a strong sustainability indicator.

The ecological footprint approach accounts for the level of sustainability of a territory by first assessing the indicator ecological footprint (EF), which expresses the bioproductive area required by the local population to produce the renewable resources and ecological services it uses. This value is then compared with biocapacity (BC), which tracks the supply of renewable resources and ecological services provided by the local ecosystems [25,29].

Such comparison leads to the assessment of an indicator, ecological balance (EB), able to translate in quantitative terms the environmental surplus/deficit situation of a region, and hence to verify its strong sustainability condition. EB is calculated as the difference between the availability of resources available in a region, measured by BC, and the resources consumed by the activities of a local population, measured by EF. If EB is higher than zero, i.e., EF is lower than BC, the carrying capacity of the region is not exceeded, and the region is judged to be sustainable, under a strong sustainability approach [30].

Some authors raise extensive criticisms about both the ecological footprint approach in general, and its full suitability as an indicator of strong sustainability (see for example [30–33] for a discussion of critical and supporting points of view). However, this methodology has been used in different studies aiming at evaluating strong sustainability at the local level; for example, in Germany [27], Australia [34,35], Italy [36], Canada [37,38], the Mediterranean area [39], China [28] and Portugal [40]. All these studies, independently of their quantitative results, show how it is possible to assess the environmental impact of economic activities on a local scale using the ecological footprint.

It is worth considering that in studies evaluating sustainability at a local scale using the ecological footprint method, a different approach to the interpretation of BC can be adopted.

Indeed, the regional EF value can be compared with the average global biocapacity instead of the BC of the region itself [41]. This way of evaluating local sustainability refers to the idea that natural resource functions, in particular the absorption of CO_2 emissions, cannot be confined to local ecosystems.

In our case, a direct comparison between local BC and EF seems more coherent with the aim of the study, and consequently this approach will be applied to evaluate the strong sustainability of Italian municipalities.

3. Materials and Methods

The assessment, performed for all the (j = 8092) Italian municipalities, is based on the evaluation of the ecological balance (EB) per capita, obtained as the difference between biocapacity (BC) per capita and Ecological Footprint (EF) per capita:

$$EB_j = BC_j - EF_j \tag{1}$$

Although these concepts are now consolidated in the scientific literature, as regards the method of calculation of EF and BC and their meaning, please refer to [23] and to all the subsequent bibliography.

3.1. Calculation of EF_j

The estimation of the EF per capita at the municipal level (EF_j) was carried out considering the most updated value of Italian per capita ecological footprint (EF_N), which refers to the year 2017 [4].

Moving from this figure, the ecological footprint of the residents in each municipality was estimated by considering their relative level of consumption with respect to the national one. The assumption of a direct relationship between the level of consumption and the ecological footprint in a region, which is based on the idea that the quantity of purchased goods is strictly related to the exploitation of bioproductive resources demanded by their production, is supported by some studies [42,43].

The local level of consumption is affected by different drivers; in this study, two of them were explicitly considered: (i) the average income of municipality inhabitants and (ii) the regional consumer price index.

Likewise, a similar effect of the price index is quite evident, which has a significant impact on the possibilities of residents' purchases; this is particularly true in a country such as Italy, where economic differences between different areas (namely, the north and south) are quite noticeable. Other variables influencing the ecological footprint, such as the residents' purchasing power or the preferences in consumer expenditures for different products, were not considered in the calculation. Indeed, besides the difficulty of getting a reliable estimation of their value at the municipality level, the first one is strictly linked to the local price index and the second one has a limited influence in determining the aggregate level of consumption.

For each municipality, the index (I_Inc_j) defined as the ratio between the local and the national per capita income [44] was assessed.

Regarding the local price index, even if no data were available from official sources, it was possible to refer to a database created in a recent study [45]. As such data are calculated at the provincial level, the price index applied in our study (I_Pr_j) assumes the same value for all the municipalities within a province.

Then, the per capita ecological footprint in each municipality (EF_j) is estimated as follows:

$$EF_j = EF_N \times I\_Inc_j \times I\_Pr_j \tag{2}$$

3.2. Calculation of BC_j

Following the standard ecological footprint methodology [23], for each municipality (j), the land area S_{ij} that falls into each of the following i − 5 land-use classes was assessed:

1. Built-up land;
2. Cropland;
3. Grazing land;
4. Forest land;
5. Water.

By means of GIS software, the calculation was performed by overlapping the municipal borders network with the CORINE Land Cover (CLC) map.

The CORINE (Coordination of Information on the Environment) land cover database reports data on land use in European countries at a high spatial resolution. The CLC project started in 1985, and was coordinated by the European Environmental Agency to produce consistent and reproducible data concerning the state of the environment in the European Community [46]. The CLC databases for years 1990, 2000, 2006, 2012 and 2018 are available. The data are available on a spatial scale of 1:100,000 with a minimum mapping unit of 25 hectares for areal phenomena, and a minimum width of 100 m for linear phenomena [47].

Land cover is organized into 44 classes structured at three hierarchical levels [48]. To assess the area in each one of the five classes for the biocapacity evaluation, the second level of the CLC legend was considered. Table 1 shows how this reclassification was carried out.

Table 1. Reclassification of CLC level II classes into BC classes.

CLC—Level I	CLC—Level II	BC Classes
1. Artificial surfaces	1.1 Urban fabric	Built-up land
	1.2 Industrial, commercial and transport units	Built-up land
	1.3 Mine, dump, and construction sites	Built-up land
	1.4 Artificial, non-agricultural vegetated areas	Built-up land
2. Agricultural areas	2.1 Arable land	Crop land
	2.2 Permanent crops	Crop land
	2.3 Pastures	Grazing land
	2.4 Heterogeneous agricultural areas	Crop land
3. Forest and seminatural areas	3.1 Forest	Forest land
	3.2 Shrub and/or herbaceous vegetation	Grazing land
	3.3 Open spaces with little or no vegetation	Grazing land
4. Wetlands	4.1 Inland wetlands	Water (inland)
	4.2 Coastal wetlands	Not included
5. Water bodies	5.1 Inland waters	Water (inland)
	5.2 Marine waters	Not included

Source: Our elaboration on CLC (2018), Global Footprint Network (2021).

Each area was converted into a bioproductive surface, measured in global hectares, through the equivalence factor (EQF_i) and yield factor (Yw_i) coefficients for Italy in 2016 [4] (see Table 2). A global hectare (gha), which is the accounting unit for the EF and BC, is a hectare with the world average biological productivity for a given year [4]. Equivalent factors convert one of the five land types into a standard unit of biologically productive area, represented by one gha. A yield factor accounts for the level of productivity of a given land type in a specific country with respect to the average world productivity.

Table 2. Values of equivalence factor and yield factor.

Land-Use Type	Equivalence Factor (gha/ha)	Yield Factor
Built-up land	2.522	0.767
Crop land	2.522	0.767
Grazing land	0.457	1.908
Forest land	1.286	1.679
Water (inland)	0.368	0.897

Source: Global Footprint Network, 2020.

Then, the per capita bio-productive area BC_j of the municipality j is calculated by dividing the biocapacity of each municipality by its population, using the following equation:

$$BC_j = \frac{\sum_{i=1}^{5}(S_{ij} \times Yw_i \times EQF_j)}{Pop_j} \quad (3)$$

The resulting per capita values of EF_j and BC_j were compared to assess the EB_j per capita for each municipality. Municipalities with $EB_j < 0$ were considered unsustainable, while those with $EB_j > 0$ were marked as sustainable. The results of the sustainability assessment are graphically illustrated by means of a set of maps, wherein municipalities with different values of EF, BC and EB are marked with different colors to highlight the areas of the country where a strong sustainability condition is achieved or is lacking.

4. Results

Figures 1 and 2, respectively, show the spatial distribution of EF_j and BC_j among Italian municipalities.

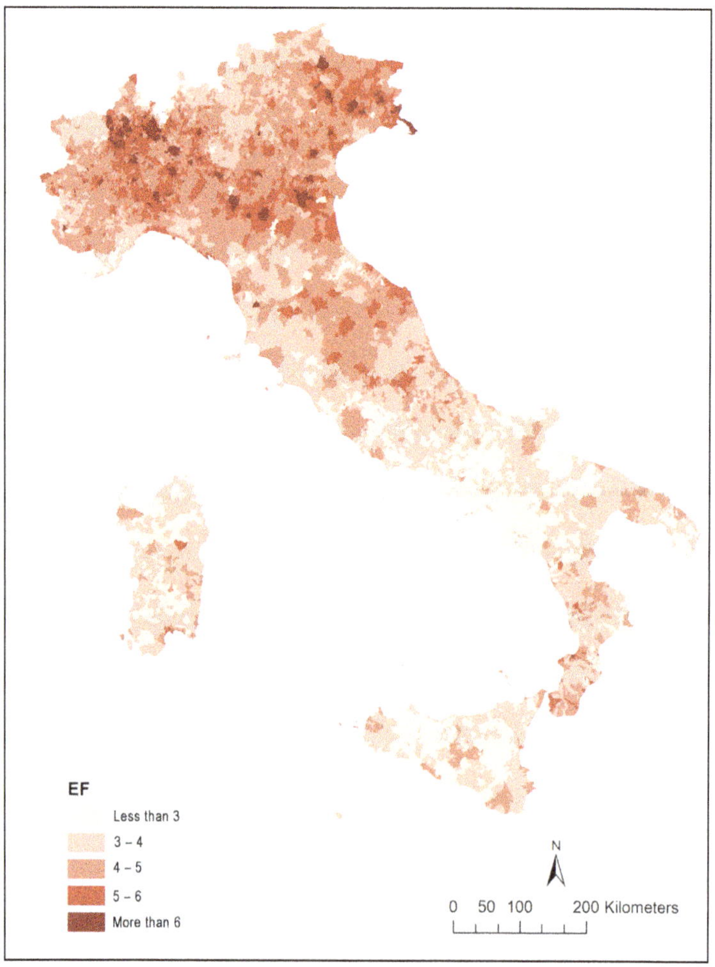

Figure 1. Distribution of per capita ecological footprint (EF_j).

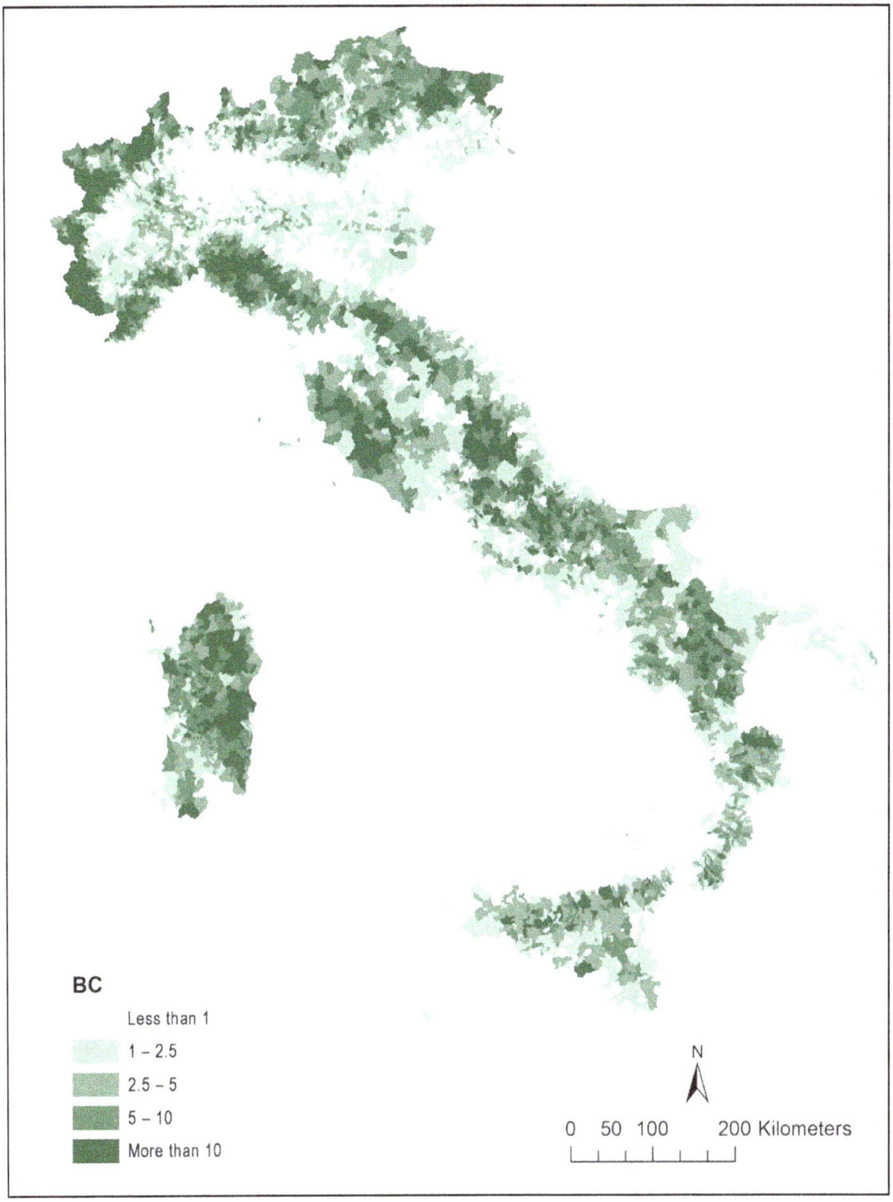

Figure 2. Distribution of per capita biocapacity (BC_j).

The EF_j distribution (Figure 1) reports in spatial terms the (potential) level of consumption of the local residents evaluated with respect to the national situation. In fact, following the posited assumptions, the combination of per capita income and price index is the weight factor used to estimate the average individual ecological footprints of municipalities' residents. To derive a better interpretation of this map, it should be considered that the Italian per capita EF in 2017 is 4.41 gha [4]. This suggests that the municipalities, which are in the range of 4–5 gha, more or less have an ecological footprint in line with the national one. Differently, in the first two classes, with an EFj less than 4 gha, are included those with a lower level of consumption, while the opposite situation characterizes those municipalities wherein the EFj is higher than 5 gha. The fact that, in general terms, the estimated levels of consumption are higher in northern Italy, despite the price levels being lower in the southern regions, confirms that the greatest effect on results is to be attributed to the average income of citizens.

The BC_j distribution (Figure 2) is directly related to the municipalities' population density weighted by the relative prevalence of the different land-use categories. Consequently, the higher values of biocapacity are located in correspondence with the main mountain ranges (Alps and Apennines) and in the large rural areas of central and southern Italy. In contrast, low levels of biocapacity characterize the urban areas, in particular the ones located around the main Italian cities (Rome, Naples and Milan), the northeast industrial district and the flat regions of Emilia–Romagna, Tuscany and Puglia. In addition, it should be considered that the average Italian biocapacity in 2017 was 0.88 gha per capita [4], and, consequently, only the municipalities falling in the first class have a lower value of BC. This implies that, even if less than one-third of Italian municipalities have a below-average level of biocapacity, in some areas the availability of natural resources is so scarce as to influence the national figure.

Figure 3, which represents the main outcome of the study, shows the distribution of the ecological balance indicator, and can be interpreted as the map of Italian sustainability. The municipalities colored in shades of orange-red are unsustainable ($EF_j > BC_j$), and the ones in shades of yellow-green are sustainable ($EF_j < BC_j$). Looking at the map, it can be observed that the distribution of sustainable/unsustainable municipalities broadly follows the BC_j distribution; the high level of association between BC and EB is confirmed by a correlation coefficient of 0.891. This is a consequence of the different scales of variation in EF (which ranges approximately from 2 to 8 gha) and BC (which ranges from 0 to more than 100); this gives BC greater influence in determining the final EB value.

Figure 4 shows the frequency distribution of Italian municipalities with respect to their EB value. The distribution is very asymmetric, and a long tail on the right is observed, related to the municipalities with very high values of BC. Consequently, the mean (−0.18 gha) and the median (−2.04 gha) of the EB_j distribution provide two quite different indications of the general tendency of Italian municipalities' sustainability.

Further inferences of the outcomes of this study can be derived from the synthetic figures reported in Table 3.

Table 3. Data on sustainable/unsustainable municipalities, area, and population in Italy.

Condition	Municipalities		Area		Population	
	Number	%	km^2	%	Million	%
Sustainable	2314	28.6%	118,654	39.3%	3.159	5.2%
Unsustainable	5778	71.4%	183,419	60.7%	57.638	94.8%
Total	8092	100.0%	302,073	100.0%	60.797	100.0%

The negative value of the median implies that more than half the Italian municipalities are not sustainable; nevertheless, the fact that the share of municipalities with $EB_j < 0$ is 71.4% is quite impressive. As regards the spatial dimension, more than 39% of the Italian territories are sustainable municipalities, showing a positive value of EB. As far as the Italian population is concerned, the vast majority (almost 95%) live in municipalities characterized by unsustainable conditions.

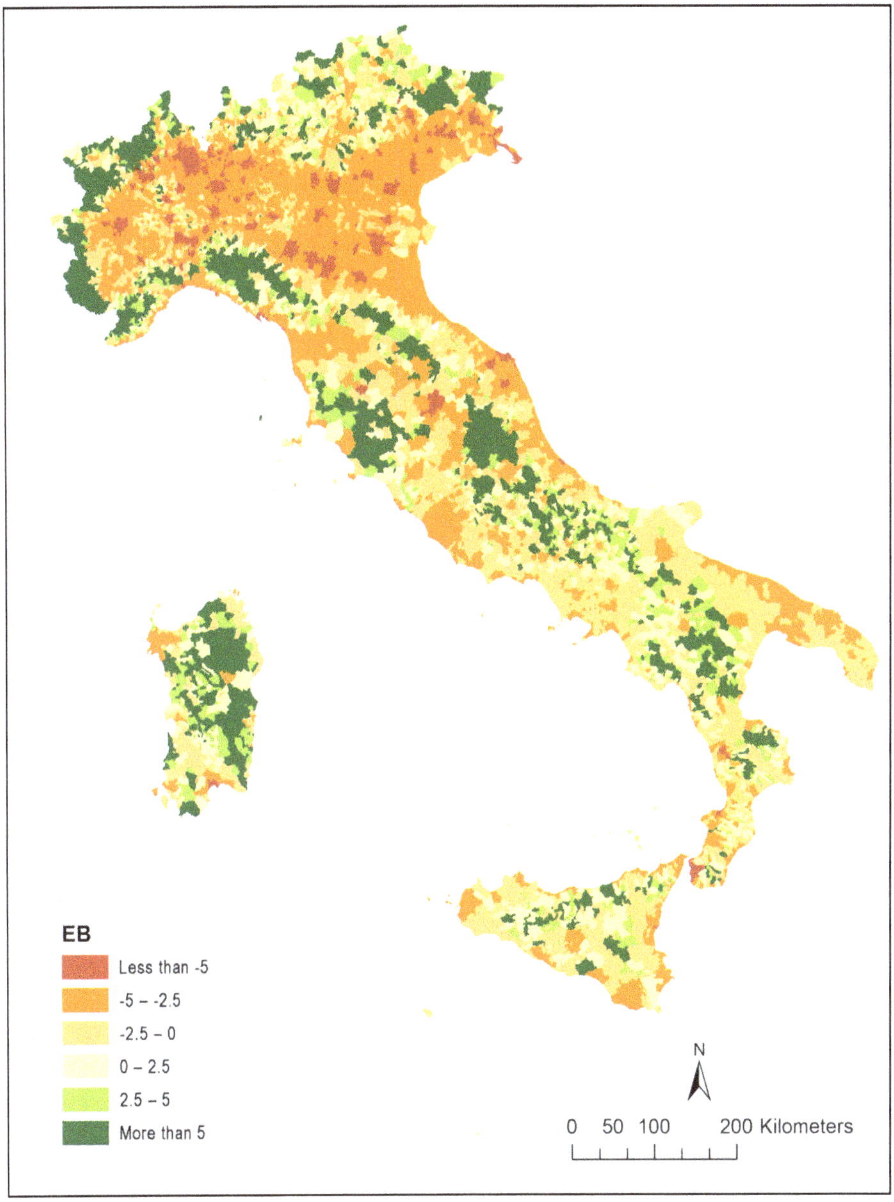

Figure 3. Distribution of ecological balance—EB_j (gha).

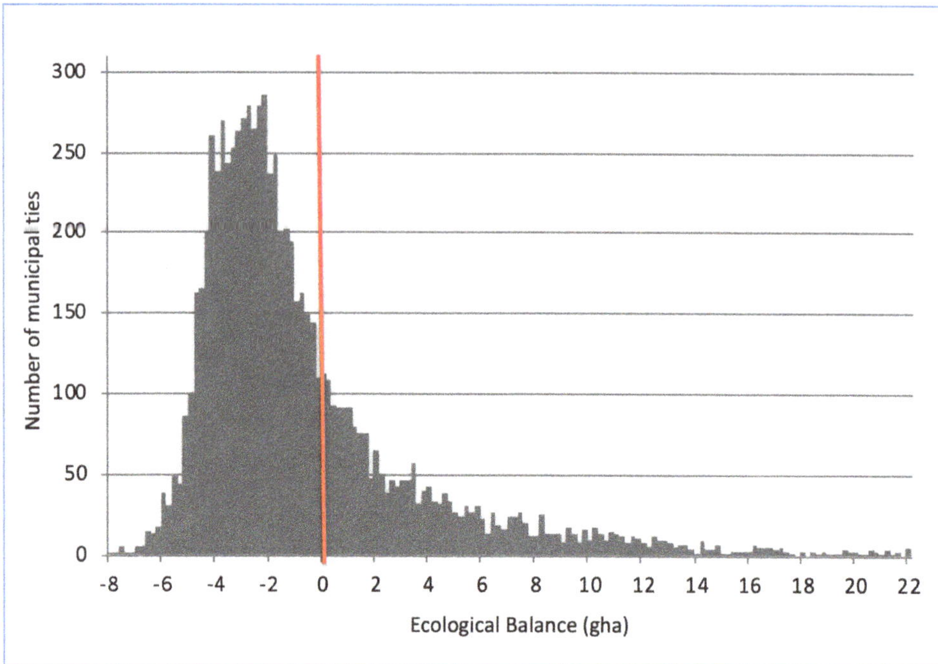

Figure 4. Frequency distribution of the municipalities by per capita EB (gha).

5. Discussion

These results are closely linked to the ecological footprint approach and to the methodology applied in this study to evaluate the EB. In this analysis, the role of the spatial scale is crucial, as it greatly influences the correct interpretation of the sustainability condition. Focusing on a small scale, such as the municipality, offers more detailed information about the territory.

The ecological balance of Italy, as a whole country, has a value of −3.53 gha per capita [4]. This study shows that this synthetic figure hides a wide set of different situations, with possible implications for better addressing environmental policies. To this end, two aspects should be considered. The first point is represented by the fact that almost 95% of the Italian population live in unsustainable municipalities, a figure that evidences the widespread local anthropogenic pressure on natural resources. The second point is linked to the possibility of identifying areas that produce positive externalities, thus partially compensating for the unsustainability of other territories.

The distribution of sustainable municipalities is more scattered, especially in northern Italy, where the role of the Alps and the less industrialized areas is evident in increasing the sustainability of the municipalities nearby. The sustainable areas grow in size as one travels south along the Apennines. A "green" spot emerges in the heart of Tuscany, in a region called Maremma, characterized by a historic vocation of extensive agriculture. The situation in Calabria, Sicily, and Sardinia is quite peculiar, as the inner part of each region is mostly sustainable, while the municipalities along the coast are not. This is linked to the fact that the inland areas of these regions are characterized by the presence of mountainous reliefs and vast agricultural areas. This determines, on the one hand, the concentration of the population along the coasts (flatter and better served by communication routes) and, on the other hand, the greater biocapacity of inland areas linked to land uses (forest, crop, grazing land).

From this point of view, ecological footprint accountability provides useful information in planning environmental policies in different stages of the decision-making process [49]. This was reported also by [50], who adapted the policy cycle of [51], highlighting the usefulness of the ecological footprint in each phase of the policy-making process. Indeed, EF can be very useful in the early warning phase, allowing the identification of ecological hot spots that need to be addressed. In our study, the identification of the most unstainable areas could drive national environmental policies to more targeted interventions.

Similarly, in the monitoring phase, where there is a need to assess the evolution of the environmental problems and the eventual effect of the adopted policies, EF can make a positive contribution, highlighting the possible effects of the implemented policies.

Furthermore, EF—given the immediacy with which it gives results—can also be useful in the headline and issue-framing phase, where it might be necessary to make a comparison among regions and to raise stakeholder awareness (in our case mainly at the local level).

A lower significance can be assigned to the policy development phase, linking the specific environmental policies with the general strategic policy framework, while for the implementation phase, EF appears to not be useful at all.

Figure 5 summarizes the relation between the steps of the policy cycle and the usefulness of information provided by the EF accountability approach.

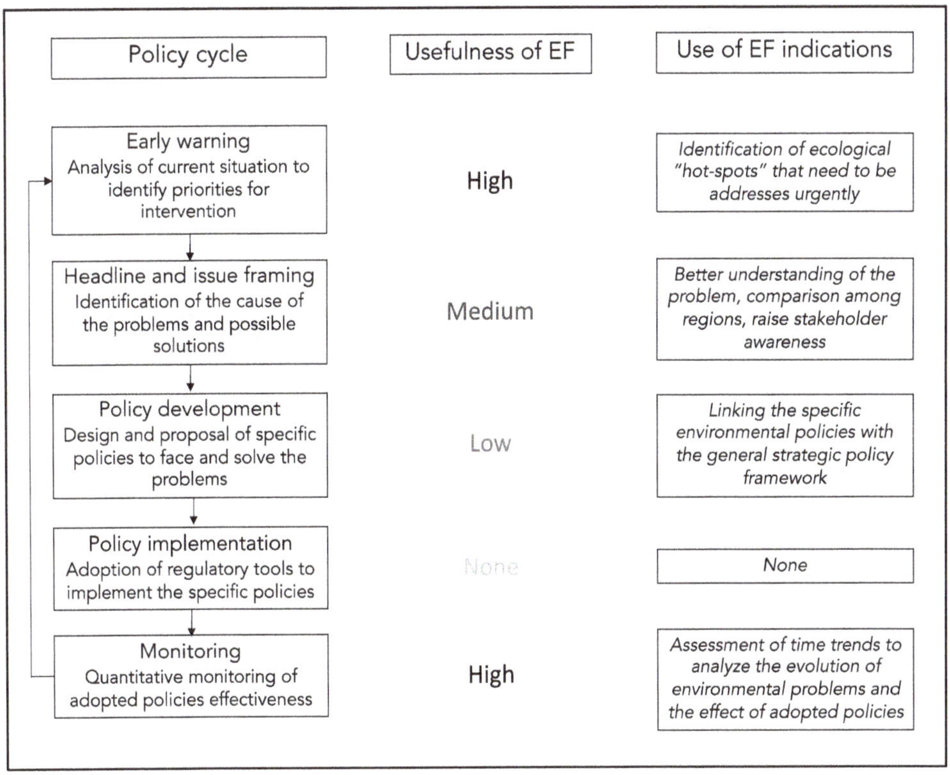

Figure 5. Policy usefulness of the ecological footprint based on the scheme proposed by [50].

6. Limitations

The outcomes of the study may be criticized for a few reasons.

The first point concerns the methodological approach; in fact, as pointed out in the paper, the ecological footprint, even if it is recognized by many scholars as a reliable indicator for sustainability evaluation, has some conceptual limitations and, especially when applied on a local scale, requires quantitative simplification; for instance, the interconnectivity between municipalities, and how footprints within a municipality might be (more) influenced by adjacent populations and municipalities, particularly if resources are more heavily consumed elsewhere.

The second aspect is linked to the hypothesis underlying the local EF calculation, which is based on the assumption that the average purchase power of a municipality population directly relates to its level of consumption, which, in turn, determines the EF value. As pointed out in the methodology, this assumption, even if it is quite reasonable and supported by some studies, introduces a simplification into the local ecological balance calculation, and then into the Italian map of sustainability. It is evident, indeed, that EF not only depends on the quantity of consumption, but also on the typologies of purchased goods, and the income and price levels do not exactly reflect the amount of consumption.

Another limitation of the study that should be highlighted is the way in which BC is calculated. Indeed, the assessment of BC per capita in each municipality was based on the land types' bioproductivity (yield factors), assessed at the national level. In other words, the assessment of municipal ecological balance does not consider the variability in the local bioproductivity of cropland, forest land and grazing land. Such a limitation could be overcome in future studies by calculating the average productivity of the different land types on a regional scale.

7. Conclusions

This study aimed at mapping sustainability, using the ecological footprint approach as a tool to design a framework for environmental policy planning. Its outcomes are to be considered from a technical perspective. Despite the presence of numerous international policies that pay attention to environmental aspects, including the New Green Deal and the Sustainable Development Goals, there is a dearth of operational instruments that can help policy-makers in carrying out their function.

Despite the intrinsic limits of both the descriptive capacity and the calculation method of the ecological footprint, as raised by many authors, we consider that this indicator of sustainability has the advantage of being applicable at any scale, and it can show in quantitative terms the ecological balance of a territory. Consequently, it can provide a useful framework to identify specific areas of intervention, as it clearly shows where the highest anthropogenic pressure on an ecosystem occurs.

Going beyond considerations of the generalized unsustainability of the Italian territory and, consequently, the urgent need to promote specific environmental policies at the national level, the study highlighted substantial differences in the spatial distribution of the demand and supply of natural resources. In the applied methodology, the demand for natural resources is linked to national consumption styles, weighted by the local situation in terms of residents' real purchase power, while the supply is defined by municipalities' bio-productivity, divided by their population.

Environmental sustainability has potentially great implications for human welfare, and hence it represents a key goal of local policies. In the definition of these policies, indicators are becoming an increasingly essential tool. Their use is no longer limited to monitoring the progress of policies' implementation; rather, they assume crucial importance in the policy planning and decision phases.

This study, while providing preliminary results and presenting important limitations, emphasizes that the environmental sustainability of anthropogenic activities at the local level is affected by three main drivers: population density, residents' lifestyles and the bioproductivity of different land uses. This suggests that, if policy-makers actually intend to pursue the goal of environmental sustainability, their interventions should cover different areas; among these, as suggested by this study, priority should be assigned to policies aimed at redefining residential models, consumption behaviors and, last but not least, land-use patterns.

Author Contributions: Conceptualization, S.F. and B.P.; Data curation, A.M.; Methodology, S.F. and A.M.; Writing—original draft, S.F.; Writing—review & editing, B.P. All authors have read and agreed to the published version of the manuscript.

Funding: This research received no external funding.

Institutional Review Board Statement: The research is compliant with the Ethical Code of the University of Viterbo "La Tuscia".

Informed Consent Statement: Not applicable.

Data Availability Statement: Not applicable.

Conflicts of Interest: The authors declare no conflict of interest.

References

1. Mebratu, D. Sustainability and sustainable development: Historical and conceptual review. *Environ. Impact Assess. Rev.* **1998**, *18*, 493–520. [CrossRef]
2. Bolis, I.; Morioka, S.; Sznelwar, L.I. When sustainable development risks losing its meaning. Delimiting the concept with a comprehensive literature review and a conceptual model. *J. Clean. Prod.* **2014**, *83*, 7–20. [CrossRef]
3. Rennings, K.; Wiggering, H. Steps towards indicators of sustainable development: Linking economic and ecological concepts. *Ecol. Econ.* **1997**, *20*, 25–36. [CrossRef]
4. Global Footprint Network. 2021. Available online: https://www.footprintnetwork.org (accessed on 12 May 2021).
5. Ozcan, B.; Khan, D.; Bozoklu, S. Dynamics of ecological balance in OECD countries: Sustainable or unsustainable? *Sustain. Prod. Consum.* **2021**, *26*, 638–647. [CrossRef]
6. Ekins, P.; Simon, S.; Deutsch, L.; Folke, C.; De Groot, R. A framework for the practical application of the concepts of critical natural capital and strong sustainability. *Ecol. Econ.* **2003**, *44*, 165–185. [CrossRef]
7. Barbier, E.B. The concept of natural capital. *Oxf. Rev. Econ. Policy* **2019**, *35*, 14–36. [CrossRef]
8. Dietz, S.; Neumayer, E. Weak and strong sustainability in the SEEA: Concepts and measurement. *Ecol. Econ.* **2007**, *61*, 617–626. [CrossRef]
9. Ayres, R.U.; Gowdy, J.M. Strong versus weak sustainability: Economics, natural sciences and "consilience". *Environ. Ethics* **2001**, *23*, 155–168. [CrossRef]
10. Costanza, R.; Daly, H.E. Natural Capital and Sustainable Development. *Conserv. Biol.* **1992**, *6*, 37–46. [CrossRef]
11. Pearce, D.; Hamilton, K.; Atkinson, G. Measuring sustainable development: Progress on indicators. *Environ. Dev. Econ.* **1996**, *1*, 85–101. [CrossRef]
12. Pelenc, J.; Ballet, J. Strong sustainability, critical natural capital and the capability approach. *Ecol. Econ.* **2015**, *112*, 36–44. [CrossRef]
13. Segerson, K.; Pearce, D.W.; Turner, R.K. Economics of Natural Resources and the Environment. *Land Econ.* **1991**, *67*, 272. [CrossRef]
14. Barbier, E.B.; Burgess, J.C.; Folke, C. *Paradise Lost? The Ecological Economics of Biodiversity*; Earthscan: London, UK, 1994.
15. Turner, R.K.; Pearce, D.W. *Sustainable Development: Ethics and Economics*; Centre for Social and Economic Research on the Global Environment: Norwich/London, UK, 1992; Working Paper NO. PA 92-09.
16. Ang, F.; Van Passel, S. Beyond the Environmentalist's Paradox and the Debate on Weak versus Strong Sustainability. *BioScience* **2012**, *62*, 251–259.
17. Barbier, E.B.; Burgess, J.C. Natural Resource Economics, Planetary Boundaries and Strong Sustainability. *Sustainability* **2017**, *9*, 1858. [CrossRef]
18. McCool, S.F.; Stankey, G.H. Indicators of Sustainability: Challenges and Opportunities at the Interface of Science and Policy. *Environ. Manag.* **2004**, *33*, 294–305. [CrossRef]
19. Hezri, A.A.; Dovers, S.R. Sustainability indicators, policy and governance: Issues for ecological economics. *Ecol. Econ.* **2006**, *60*, 86–99. [CrossRef]
20. Moldan, B.; Janoušková, S.; Hak, T. How to understand and measure environmental sustainability: Indicators and targets. *Ecol. Indic.* **2012**, *17*, 4–13. [CrossRef]
21. Mori, K.; Christodoulou, A. Review of sustainability indices and indicators: Towards a new City Sustainability Index (CSI). *Environ. Impact Assess. Rev.* **2012**, *32*, 94–106. [CrossRef]
22. Singh, R.K.; Murty, H.; Gupta, S.; Dikshit, A.K. An overview of sustainability assessment methodologies. *Ecol. Indic.* **2012**, *15*, 281–299. [CrossRef]
23. Wackernagel, M.; Rees, W.E. *Our Ecological Footprint: Reducing Human Impact on the Earth*; New Society Publishers: Gabriola Island, CO, Canada, 1996.
24. Daly, H.E. *Steady-State Economics—Second Edition with New Essays*; Earthscan: London, UK, 1992.
25. Mancini, M.S.; Galli, A.; Niccolucci, V.; Lin, D.; Hanscom, L.; Wackernagel, M.; Bastianoni, S.; Marchettini, N. Stocks and flows of natural capital: Implications for Ecological Footprint. *Ecol. Indic.* **2017**, *77*, 123–128. [CrossRef]

26. Monfreda, C.; Wackernagel, M.; Deumling, D. Establishing national natural capital accounts based on detailed Ecological Footprint and biological capacity assessments. *Land Use Policy* **2004**, *21*, 231–246. [CrossRef]
27. Knaus, M.; Löhr, D.; O'Regan, B. Valuation of ecological impacts—A regional approach using the ecological footprint concept. *Environ. Impact Assess. Rev.* **2006**, *26*, 156–169. [CrossRef]
28. Huang, L. Exploring the Strengths and Limits of Strong and Weak Sustainability Indicators: A Case Study of the Assessment of China's Megacities with EF and GPI. *Sustainability* **2018**, *10*, 349. [CrossRef]
29. Galli, A.; Wackernagel, M.; Iha, K.; Lazarus, E. Ecological Footprint: Implications for biodiversity. *Biol. Conserv.* **2014**, *173*, 121–132. [CrossRef]
30. Neumayer, E. *Weak Versus Strong Sustainability: Exploring the Limits of Two Opposing Paradigms*, 4th ed.; Edward Elgar: Cheltenham, UK, 2013.
31. van den Bergh, J.C.J.M.; Verbruggen, H. Spatial sustainability, trade and indicators: An evaluation of the 'ecological footprint'. *Ecol. Econ.* **1999**, *29*, 61–72. [CrossRef]
32. Fiala, N. Measuring sustainability: Why the ecological footprint is bad economics and bad environmental science. *Ecol. Econ.* **2008**, *67*, 519–525. [CrossRef]
33. Galli, A.; Giampietro, M.; Goldfingerd, S.; Lazarus, E.; Lind, D.; Saltelli, A.; Wackernagel, M.; Müller, F. Questioning the Ecological Footprint. *Ecol. Indic.* **2016**, *69*, 224–232. [CrossRef]
34. Graymore, M.L.M.; Sipe, N.G.; Rickson, R.E. Regional sustainability: How useful are current tools of sustainability assessment at the regional scale? *Ecol. Econ.* **2008**, *67*, 362–372. [CrossRef]
35. Graymore, M.; Sipe, N.; Rickson, R.E. Sustaining Human Carrying Capacity: A tool for regional sustainability assessment. *Ecol. Econ.* **2010**, *69*, 459–468. [CrossRef]
36. Scotti, M.; Bondavalli, C.; Bodini, A. Ecological Footprint as a tool for local sustainability: The municipality of Piacenza (Italy) as a case study. *Environ. Impact Assess. Rev.* **2009**, *29*, 39–50. [CrossRef]
37. Wilson, J.; Grant, J.L. Calculating ecological footprints at the municipal level: What is a reasonable approach for Canada? *Local Environ.* **2009**, *14*, 963–979. [CrossRef]
38. Isman, M.; Archambault, M.; Racette, P.; Konga, C.N.; Llaque, R.M.; Lin, D.; Iha, K.; Ouellet-Plamondon, C.M. Ecological Footprint assessment for targeting climate change mitigation in cities: A case study of 15 Canadian cities according to census metropolitan areas. *J. Clean. Prod.* **2018**, *174*, 1032–1043. [CrossRef]
39. Baabou, W.; Grunewald, N.; Ouellet-Plamondon, C.; Gressot, M.; Galli, A. The Ecological Footprint of Mediterranean cities: Awareness creation and policy implications. *Environ. Sci. Policy* **2017**, *69*, 94–104. [CrossRef]
40. Galli, A.; Iha, K.; Pires, S.M.; Mancini, M.S.; Alves, A.A.; Zokai, G.; Lin, D.; Murthy, A.; Wackernagel, M. Assessing the Ecological Footprint and biocapacity of Portuguese cities: Critical results for environmental awareness and local management. *Cities* **2020**, *96*, 102442. [CrossRef]
41. Syrovátka, M. On sustainability interpretations of the Ecological Footprint. *Ecol. Econ.* **2020**, *169*, 106543. [CrossRef]
42. Niccolucci, V.; Tiezzi, E.; Pulselli, F.; Capineri, C. Biocapacity vs Ecological Footprint of world regions: A geopolitical interpretation. *Ecol. Indic.* **2012**, *16*, 23–30. [CrossRef]
43. Uddin, G.A.; Salahuddin, M.; Alam, K.; Gow, J. Ecological footprint and real income: Panel data evidence from the 27 highest emitting countries. *Ecol. Indic.* **2017**, *77*, 166–175. [CrossRef]
44. Italian Ministry of Economy and Finance. 2019. Available online: http://www1.finanze.gov.it, (accessed on 12 December 2019).
45. Ichino, A.; Boeri, T.; Moretti, E.; Posch, J. *Wage Equalization and Regional Misallocation: Evidence from Italian and German Provinces*; Centre for Economic Policy Research: London, UK, 2020; Discussion Paper DP13545.
46. EEA. European Environmental Agency, Corine Land Cover Documentation. 1995. Available online: http://www.eea.europa.eu/publications/COR0-landcover (accessed on 27 November 2020).
47. CLMS. Copernicus Land Monitoring Service Datasets. 2018. Available online: http://land.copernicus.eu/pan-european/corine-land-cover. (accessed on 27 November 2020).
48. Bossard, M.; Feranec, J.; Otahel, J. *CORINE Land Cover Technical Guide–Addendum 2000*; European Environment Agency: Copenhagen, Denmark, 2000.
49. Stoeglehner, G.; Narodoslawsky, M. Implementing ecological footprinting in decision-making processes. *Land Use Policy* **2008**, *25*, 421–431. [CrossRef]
50. Galli, A. On the rationale and policy usefulness of Ecological Footprint Accounting: The case of Morocco. *Environ. Sci. Policy* **2015**, *48*, 210–224. [CrossRef]
51. Knill, C.; Tosun, J. Policy making. In *Comparative Politics*; Caramani, D., Ed.; Oxford University Press: Oxford, UK, 2008; pp. 495–519.

sustainability

MDPI

Article

Linking Environmental Regulation and Financial Performance: The Mediating Role of Green Dynamic Capability and Sustainable Innovation

Xinpeng Xing [1], Tiansen Liu [2,3], Lin Shen [1] and Jianhua Wang [1,*]

1 School of Business, Jiangnan University, Wuxi 214122, China; xinpeng@jiangnan.edu.cn (X.X.); 8201609118@jiangnan.edu.cn (L.S.)
2 School of Economics and Management, Harbin Engineering University, Harbin 150001, China; tiansen0328@hrbeu.edu.cn
3 School of Social and Behavioral Sciences, Nanjing University, Nanjing 210023, China
* Correspondence: jianhua.w@jiangnan.edu.cn; Tel.: +86-182-5154-9736

Received: 16 December 2019; Accepted: 21 January 2020; Published: 31 January 2020

Abstract: This study develops a multiple mediating model for exploring the link between environmental regulation and financial performance through green dynamic capability, sustainability exploration/exploitation innovation, based on the data from 355 Chinese manufacturing firms. Empirical results support a mediating role of green dynamic capability and sustainability exploration/exploitation in the link between environmental regulation and financial performance, respectively. What's more, our findings indicate that environmental regulation can help improve financial performance via two multiple mediating paths, i.e., green dynamic capability and sustainability exploration innovation, as well as green dynamic capability and sustainability exploitation innovation. These key findings will help to understand how important green dynamic capability and sustainable innovation is when Chinese manufacturing firms establish a business-politics tie.

Keywords: environmental regulation; green dynamic capability; sustainability exploration innovation; sustainability exploitation innovation; financial performance

---

1. Introduction

Climate change and environmental pollution have become inevitable problems in the pursuit of economic growth [1,2]. In this context, sustainable development is viewed as a key way to mitigate the pressure caused by economic development and environmental protection, and requires the joint efforts from multi-stakeholders, including government, business organization, and public [3,4]. As a feasible approach for government to normalize environmental behaviors of economic participants, environmental regulation plays an increasingly important role in decreasing the negative effect of firms' operations process on the natural environment [5,6]. At the same time, technology innovation that can be viewed as a key driver of economic growth is also responsible for balancing economic development and environmental protection [7–9]. Therefore, it is a key to identify the link among environmental regulation, technology innovation, and economic output, thus designing more paths for firms' sustainable development.

To reveal this link, quite a lot of literature has emerged with a typical research finding as the Porter Hypothesis [10]. The early view in terms of economic growth and environmental protection argued that strict environmental regulations require firms to increase investment in environmental protection and green technology innovation [11]. Nevertheless, the rise in the compliance cost effect caused by environmental regulation is bound to decrease financial performance [12]. The Porter

Hypothesis posits that reasonably designing environmental policies can motivate innovation and productivity gains, partially/fully offset compliance cost, and even generate net benefits [13,14]. This may improve financial performance through creating absolute advantage, which mainly comes from "innovation advantage" and "first-mover advantage" [15]. However, there still exist some conflicts on how governments develop policy regulation that can improve the comprehensive performance of firms [2,16,17]. The empirical results of [16,18] illustrated that environmental regulation is usually negatively linked with financial performance, while by contrast, some studies suggested designing the environmental regulation that will not only lead the cost-saving innovation that can compensate for the compliance cost, but also help firms to gain low-cost and differential competitive edges, thereby improving financial performance [13,19,20]. However, the finding of [21,22] argued that environmental regulation positively affects firms' financial performance, thus verifying their fluctuating links.

The inconsistency between environmental regulation and financial performance has attracted a focus from subsequent research, and their transmission mechanism has been also characterized from multi-perspectives [23,24]. From the perspective of subdivision of environmental regulations, some research focused on such regulation from the aspects of administrative orders and market incentives. For instance, results of [20] presented that market incentives can help improve financial performance, but [25] indicated that this incentive has a negative effect. While on the whole, prior research held the same view that there is no significant relationship between command-controlled environmental regulation and financial performance.

From the perspective of segmented technology innovation, the effect of environmental regulation on economic output via technology innovation was explored [26,27]. For instance, [22] found that environmental commitment and sustainability exploitation innovation significantly mediate the link of environmental regulation and financial performance, while sustainability exploration innovation cannot motivate the effect of this regulation on financial performance. Additionally, this regulation can also positively affect financial performance through voluntary efficiency-oriented innovations, while it is not applicable to the regulation-oriented innovation. What's more, some literature suggested that internal organizational factors, e.g., CSR-contingent executive compensation, dynamic capability, slack resources, and strategy design, need to be considered in the "strong" Porter Hypothesis [28–31]. However, only a few literature incorporated internal factors of organizations into the research framework in terms of linking these three variables.

Overall, there are still two gaps in existing literature. First, the effect of existing environmental regulation on financial performance through technology innovation is controversial, and it needs to subdivide the specific type of environmental regulation and technology innovation [23,32]. Second, it needs to identify what are key internal factors that can lead the transmission mechanism of environmental regulation, sustainable dual-innovation, and financial performance [24,33]. In order to fill in these two gaps, this study will incorporate both green dynamic capability and sustainable innovation into the Porter Hypothesis framework, and then explore the potential path from environmental regulation to financial performance through green dynamic capability and sustainable innovation.

To fill the research gaps in the existing literature, our study aims to provide insights into the implementation path of the strong Porter Hypothesis by exploring the meditating effect of green dynamic capability and sustainability exploration/exploitation innovation in linking environmental regulation and financial performance. Specifically, this study extends the understanding of the relationship between environmental regulation, sustainable innovation, and firm performance by dividing sustainable innovation into sustainability exploration innovation and sustainability exploitation innovation. Furthermore, this study strives to new insights into the link between environmental regulation and financial performance by inserting green dynamic capability into the conceptual framework of Porter Hypothesis. Meanwhile, this study tries to add empirical evidence for the theoretical framework, proposed by prior studies [33], which emphasizes the importance of internal capability and sustainable innovation for the Porter Hypothesis. Overall, we expect to provide a better understanding of strong Porter Hypothesis by answering how firms establish green capability, and then develop sustainable innovation mode to improve economic growth in the specific environmental regulation.

The rest of this study is structured as follows. Section 2 organizes a theoretical analysis and then develops our research hypotheses. Section 3 introduces data collection and research methodology. Section 4 presents the key empirical results. Section 5 further discusses our key findings and thus proposes research implications. Section 6 concludes this study and presents our research limitations and future research.

2. Theoretical Underpinnings and Hypotheses Development

2.1. Environmental Regulation

Enlightened by the Resource-based View, maximizing the value of resources is the key to create competitive edges of firms or industries [25]. Accordingly, there are two sides about environmental regulation that can be discussed as follows. On the one hand, environmental policy may restrict the circulation of resources in the markets or inter-firms, thus resulting in the lack of resources. On the other hand, limiting the use of materials in production process will make firms explore innovative production modes, and thus enhance the competitive edges in terms of innovation aspects. This is similar to the Innovation Compensation Effect proposed by Porter. Based on the Porter Hypothesis, environmental regulation is not only a policy tool for protecting the natural environment, but also an arguable facet to promote firms' innovation practices. However, there still exists a considerable debate in terms of the efficiency and internal transmission mechanism of environmental regulation [16,18]. In view of the existence of environmental total factor productivity that was proposed in recent years, different types of environmental regulations have different influence on such productivity in different industrial sectors [23].

Therefore, researchers are committed to reveal the effect of different regulations on sustainable operations. Enlightened by the "narrow" Porter Hypothesis, some research divided environmental regulation into mandatory regulation and voluntary regulation, and drew the effect of these two on firms' performance levels [20,34]. Then [35] divided it into three aspects, i.e., command-controlled regulation, voluntary regulation, and market-oriented regulation. These regulations require firms to increase environmental management investment that aims to consolidate the technology innovation that will help environmental protection.

The interaction among environmental regulation, internal factors, sustainable innovation, and financial performance needs to be systematically explored. Among them, environmental regulation, as the starting point of Porter Hypothesis, is considered to be the key part of theoretical framework. Although a large number of studies analyzed the theoretical path from environmental regulation to firms' performance via technology innovation, few of them involved how to select an efficient path for both government and firms. Further, prior studies less focused on the inter-firms' factors in the research framework, e.g., operational capability, management attitudes, and strategic positions, which are affected by external policy tools [16]. Therefore, in order to find an efficient way to address these problems, we aim to analyze the internal and external effects that environmental regulation put on financial performance.

2.2. Green Dynamic Capability

Dynamic capability was defined as a kind of process in terms of firms' resources use, especially focusing on the process of rearranging or reconfiguring resources, and this process emphasizes how firms use the limited resources to adapt to market environments [36]. Subsequent studies proposed various views about it in terms of organization internality and effectiveness of resource use. [37] emphasized the integration of knowledge and technology resources of dynamic capability, and argued that this capability can be portrayed as a series of combine capabilities that promote firms to create knowledge.

From the economic perspective, some prior studies [38,39] viewed dynamic capability as a kind of organizational guideline that is represented by managers' capability to integrate opportunities to achieve high financial performance and establish firms' competitive edge. With the further research of dynamic capability in firms' sustainable innovation and high-quality development, [40] proposed the view of "green development" into the theory system of dynamic capability, and put forward the concept of green dynamic capability. It was interpreted as the capability of firms to develop their green organizational capability to respond to market changes by using existing resources and a range of knowledge renewal activities. This study takes the influence path of green dynamic capability as the internal factors in the organizations.

However, the literature about green dynamic capability tends to focus on a certain aspect about firm sustainability. For instance, [41] explored the positive internal mediating role of dynamic capability (sensing, seizing, and reconfiguring) and innovation towards firms' sustainability. [42] developed a framework of how dynamic capability works in the business network. What's more, [43] investigated how green dynamic capability improves the competitive edges from the green development perspective. Furthermore, few researches regarded green dynamic capability as the mediating role in the model of strong view of Porter Hypothesis. Therefore, to fulfill this gap, we explore the functions of green dynamic capability in the multiple paths involving environmental and economic factors to reinforce the researches related to the Porter Hypothesis.

2.3. Sustainability Exploitation/Exploration Innovation

Sustainable innovation was defined by [44,45] as innovation which has enabled firms to improve their financial performance while reducing negative environmental and social impacts. In the view of [46], sustainable practices are essential to firms' further development. And the reasons why firms perform innovation activities involve two aspects: For one thing, firms with high innovative capability are likely to tackle with dynamic environment problems; for another, individual or organizational innovation to obtain competitive edges.

To explore the effect of different types of sustainable innovation, the definition of sustainability exploitation innovation and sustainability exploration innovation conceptually was proposed by [47]. Sustainability exploitation innovation illustrates that organizational on-going incremental improvements are essential to tackle with the scarcity of resources, e.g., appropriate mechanism to resources reduction and renting rules to other firms [47,48]. In contrast, sustainability exploration innovation focuses on reducing environmental costs by enhancing exploration innovation activities, and it is designed to achieve business sustainability, future prosperity, and competitive edges [49]. At the same time, by analyzing the impact of green innovation in Taiwan's manufacturing industry in China, [50] concluded that green innovation will create competitive edges to manufacturing firms, of which green product innovation played the most important effect.

In the Instrumental Stakeholder Theory, organizations need to give the priority of shareholders' influence on sustainable development. It emphasizes that managing the link with shareholders smoothly when operating innovative practices contributes to significant improvement of financial performance. What's more, [51] found that organizations strive to gain competitive edges through sustainability exploitation innovation by successfully addressing stakeholder expectations. But the Sustainability-oriented Theory presents the initiatives of organizations when they confront with changing markets. It means that their purposes to make innovation are not only the pressure from government and shareholders, but also from their commitment to sustainable economic success [49]. Accordingly, establishing a conceptual model on exploitation/exploration innovation towards economic performance is necessary for practical improvement.

However, only a few studies involving sustainability exploitation innovation and sustainability exploration innovation function in a conceptual model, which contains both internal and external factors, e.g., organizational leadership and green dynamic capability. Hence, this study concentrates on firms' preferences between these two sustainable innovations when firms are influenced by the external environmental regulations and internal green dynamic capability, and moreover, whether their choices will contribute to shareholders' profits or not.

2.4. The Effect of Environmental Regulation on Sustainable Innovation

The main goal of government environmental regulation is to achieve harmonious development of environment and industry. Thus, the government needs to formulate relevant environmental policies according to the development situation, organizational structure, and existing resources. Based on Porter Hypothesis, appropriate environmental regulation can promote firms' innovation.

Considering the link of environmental regulation and innovation performance, the results are categorized into two aspects. Firstly, from the perspective of Cost Compensation Theory, a group of scholars explain that the implementation of environmental regulation increases the management cost of firms, thereby reducing the investment in innovation activities, thus inhibiting the innovative production of firms [52]. Secondly, in the view of innovation compensation, scholars argued that in order to gain competitive edges, these firms will be autonomous in their innovation activities to achieve efficient production, while the government and society will also take measures to stimulate and induce firms to carry out green innovation [53]. Furthermore, governments' environmental standards punish firms that violate the rules, thus encouraging other firms to comply with regulations that help firms to develop innovation activities. The sustainability exploitation innovation requires firms to emphasize the use of existing innovation resources and innovation practices [47]. Thus, sustainability exploration innovation and sustainability exploitation innovation can achieve the balance between economic development and environmental protection via enhancing sustainable innovation [54].

In view of innovation, exploration, and exploitation, innovations portray two different methods to attain innovative results for business organizations. Thus, environmental regulation promotes technology innovation. For instance, improving the efficiency of sustainability exploitation innovation tools and methods. From cost compensation perspective, environmental regulation force firms to implement innovative practices, thus, organizations may orchestrate their production resources and improve technology to meet the governments' needs. This helps develop following hypotheses:

Hypothesis 1a. *Environmental regulation positively affects sustainability exploitation innovation.*

Hypothesis 1b. *Environmental regulation positively affects sustainability exploration innovation.*

2.5. The Effect on Environmental Regulation on Green Dynamaic Capability

Environmental regulation requires firms to implement environmental management, and increase investment in innovation related to environmental protection and pollution reduction [25]. Environmental regulation requires firms, which aim to avoid green risks, to adjust green producing strategy in a certain period [35]. Furthermore, environmental regulation highlights the importance of reconfiguring internal resources, enhancing green produce line, and orchestrating human structure [55]. Environmental regulation has led to environmental changes, enabling firms to face a new market environment, in which they pay more attention to green innovation, environmental protection, pollutant emission reduction, energy, and material consumption [56].

According to the Dynamic Capability Theory, when facing the challenges from the dynamic environment, firms will integrate, establish, and reconfigure competences to gain competitiveness [57,58]. Relative to dynamic capability, green dynamic capability are regarded as the capability of a firm to integrate and reconfigure its existing resources and knowledge to renew and develop its green organizational capability [40,43]. Green dynamic capability is critical to firm success, in response to environmental protection, and shortages of resources and energy. To obtain green competitive edge, firms are motivated to take environmentally friendly productions by improving the capability to sense, seize, and orchestrate opportunities in the markets.

Thus, we proposed the following hypothesis.

Hypothesis 1c. *Environmental regulation positively affects green dynamic capability.*

2.6. The Effect of Green Dynamci Capability

The core point of sustainable development is integrating financial performance, environment, society, and other government control factors appropriately [59]. The improvement of firms' green dynamic capability allows for firms to make efficient adjustment to adapt to the changes of the green environment. On the one hand, it reorganizes the existing resources to achieve the optimal allocation. On the other hand, it also recognizes and seizes the opportunities of environmental chances to create the competitive edge of competitors. The establishment of competitive edges is positively associated with the sustainable production and innovation of products.

Sustainable practices will help firms achieve a high level of financial performance [60]. At the sensing level of dynamic capability, firms will undertake a series of innovation activities to capture customer needs, competitor information, and explore innovation opportunities [61,62] found that the lack of clean production technology and measures to address cleaner production will be the main obstacle to high-quality development for firms. Therefore, the capability to obtain information in the external environment quickly is inaccessible for firms to move to high-quality development. At the seizing level of dynamic capability, organizations will mobilize existing resources to capture opportunities and challenges. Since the high-quality development of the firms is affected by the internal organizational structure and external stakeholders, it is beneficial to the sustainable development activities if they can coordinate internal and external resources and fully mobilize it [63]. By constantly adjusting the organizational resources, the firms can enhance sustainable competitive edges. Further, in order to achieve high financial performance, organizations should learn how to update, use, and allocate resources to a rapidly changing external environment [44,64]. This leads us to suggest the following hypothesis:

Hypothesis 2. *Green dynamic capability positively affects financial performance.*

Green dynamic capability that emphasizes the use of existing green resources and capability, helps firms to meet the changing needs of the external environment by constructing and developing new organizational capability [65]. On the one hand, green dynamic capability helps firms to achieve a timely understanding of environmental protection laws, regulations, and policies, and quickly perceive and capture the subtle dynamic changes in the market by conducting an extensive information search [43]. Firms with strong green dynamic capability are more likely to fully grasp consumers' green demands and the future development trend of the industry. On the other hand, firms with strong green dynamic capability can adjust the production process effectively according to changes in actual demand [66]. Therefore, we believe that green dynamic capability can help firms to promote sustainable innovation, and improve the probability of success of green innovation by overcoming technical research and development difficulties.

Therefore, we propose following hypotheses:

Hypothesis 3a. *Green dynamic capability positively affects financial performance via sustainability exploitation innovation.*

Hypothesis 3b. *Green dynamic capability positively affects financial performance via sustainability exploration innovation.*

2.7. The Effect of Sustainable Innovation on Financial Performance

The impact of innovation activities on firms' economic output is also considered within two aspects. Some scholars insisted that sustainability exploratory innovation and sustainability exploitation innovation promoted the financial performance of firms to varying degrees [67]. In contrast, other scholars believed that the relationship between the two kinds of sustainable innovation and corporate financial performance was not directly related. For instance, [68] concluded that exploratory practice and financial performance in sustainable innovation were inverted due to the intensity of research and development and the influence of the firms' industry. [47] recommended that the main goal of sustainability exploration innovation is to new climate-friendly products and technology by conducting the exploration innovation.

Sustainability exploration innovation covers both sustainable development and exploration innovation, which aims to reduce the environmental impact of the product cycle via conducting exploration innovation [24,69]. New green products, technologies, and knowledge created by sustainability exploration innovation are beneficial to reduce material consumption and increase recycling rates [70]. Therefore, we believe that sustainability exploration innovation is able to explore new markets and differentiate the business from its competitors via providing new green products and green processes [54,71,72]. Additionally, sustainability exploitation innovation aims to continually reduce the use of materials and energy to enhance existing competitive edge eventually through exploitation innovation that pays attention to improving the existing products and designs [49]. It implies that sustainability exploitation innovation ensures the organization's viability against competitors by providing lower prices and higher quality products. Therefore, the sustainability exploitation innovation will improve business performance by developing the nature of innovation. Thus, we proposed following hypotheses:

Hypothesis 4a. *Sustainability exploitation innovation positively affects financial performance.*

Hypothesis 4b. *Sustainability exploratory innovation positively affects financial performance.*

2.8. The Mediating Effect of Sustainable Innovation

We proposed in Hypotheses H1a and H1b that environmental regulation positively influences two types of sustainable innovation: Sustainability exploitation innovation and sustainability exploration innovation, respectively. After that, we furthermore proposed in Hypotheses H3a and H3b that sustainability exploitation innovation and sustainability exploration innovation can improve the financial performance of manufacturing firms. Therefore, according to Hypotheses H1a and H3a, we argue that environmental regulation forces firms to enhance sustainability exploitation innovation to improve economic development and enhance the environmental protection which can improve their financial performance. Further, following the views proposed in Hypotheses H1b and H3b, we also believe that environmental regulation forces firms to enhance sustainability exploration innovation to achieve the balance between economic development and environmental protection, which can improve their financial performance. Thus, the following hypotheses are proposed:

Hypothesis 5a. *The link between environmental regulation and financial performance is mediated by sustainability exploitation innovation.*

Hypothesis 5b. *The link between environmental regulation and financial performance is mediated by sustainability exploration innovation.*

2.9. The Mediating Effect of Green Dynamic Capability

Besides exploring the internal mechanism of sustainable innovation, scholars begin to take the influence of the organizational factors of firms into account [16]. According to the Porter Hypothesis, environmental regulation promotes innovation activities, and how to implement and identify innovation activities depends on the dynamic control of internal organizational environment.

Firms tend to enhance the capability to identify innovation opportunities, control risks, and integrate resources when considering the high-level cost caused by a series of factors, e.g., resource shortage of human capital and technology backwardness. At the same time, taking the pressure of stakeholders into account, environmental regulation is likely to affect internal organization factors, e.g., leaders' judgment, skills, processes, production techniques, which are key elements of green dynamic capability. Thus, the improvement of dynamic capability depends on flexible and suitable regulations [8], which contributes to the higher economic output.

Meanwhile, considering the influence of environmental regulation, firms improve the effectiveness of innovation activities by changing the uncertain external factors, and reconfigure the factors of organizational rules within the organization. Thus, the effective use of organizational resources and the improvement of organizational structure flexibility are likely to affect the sustainable application of dynamic capability. By then, it makes firms quickly respond to the green markets' changes. Thus, we propose the following hypothesis:

Hypothesis 6. *The effect of environmental regulation on financial performance is mediated by green dynamic capability.*

2.10. The Multiple Mediating Role of Green Dynamic Capability and Sustainable Innovation

Green dynamic capability presents the capability of firms to adapt to external changes by using internal and external green resources, and firms with higher class of green dynamic capability can obtain great merits in realizing green innovation and development. According to Hypotheses H1c, H3a, and H4a, firms need to enhance their green dynamic capability to cope with changes caused by environmental regulation, and the strengthening of green dynamic capability may promote the implement of sustainability exploitation innovation, which is critical driver to improve firms' financial performance.

As proposed in hypothesis H1c, environmental regulation is positively associated with green dynamic capability. According to H3b and H4b, green dynamic capability have a positive impact on sustainability exploration innovation via improving reconfiguration and opportunity recognition capability of a firm, and sustainability exploration innovation can improve firms' financial performance. Therefore, based on the H1c, H3b, and H4b, this study believes that environmental regulation can enhance firms' financial performance via green dynamic capability and sustainability exploration innovation.

Based on the above discussion, we therefore hypothesize:

Hypothesis 7a. *The influence of environmental regulation on financial performance is subsequently mediated by green dynamic capability and sustainability exploitation innovation.*

Hypothesis 1b. *The influence of environmental regulation on financial performance is subsequently mediated by green dynamic capability and sustainability exploration innovation.*

According to hypotheses above, the conceptual framework proposed in this study is shown in Figure 1.

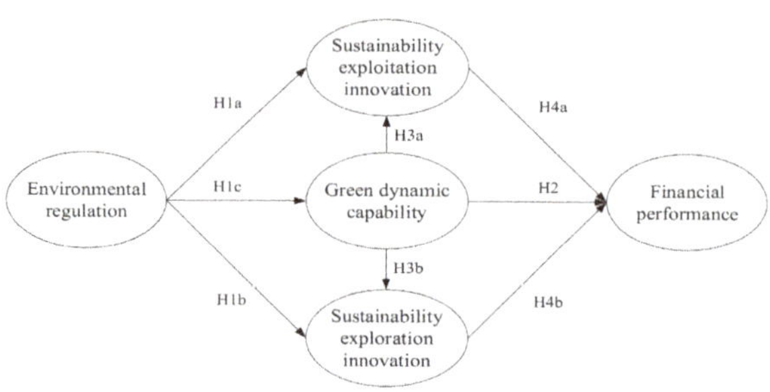

Figure 1. The conceptual framework

3. Research Methodology

Following previous studies [24,41,73], the research methodology is conducted in the following three stages. First, these survey questionnaires were firstly designed based on previous studies and confirmed by three professors and senior managers. Second, the questionnaires were randomly sent to a sample of senior managers in manufacturing firms. Third, the validity and reliability of the measurement in this study were assessed to ensure their adequacy for PLS-SEM analysis.

3.1. Data Collection

Our research subjects are manufacturing firms in China, and the questionnaires are filled in anonymously by the senior executives of the firm under investigation. The survey, which lasted from March to May in 2018, included the eastern, central, western, and north-eastern regions of China. The questionnaire in this study consists of two parts. The first part includes basic information of firms, and the second part mainly investigates the financial status of firms, the state of sustainable innovation of firms, and the firms' green dynamic capability. To ensure the validity of the questionnaire, three experts from universities and five top managers from manufacturing firms in Wuxi were invited to conduct a preliminary survey, and according to the professional and business executive feedback, the questionnaire was revised and improved.

This study issues 877 questionnaires in the study, e-mail, and online forms, a total of 585 questionnaires were recovered, the recovery rate of 66.70%. Among all received questionnaires, 230 invalid questionnaires that have obvious problems, e.g., incomplete filling, selecting one result for all questions, and obvious regularity were eliminated. Finally, 355 valid questionnaires were obtained, with an efficiency of 60.68%.

According to the preliminary analysis, firm scales cover micro-firms (less than 20 employees; 2.82%), small firms (from 20 to 299 employees; 36.06%), medium-sized firms (from 300 to 999 employees; 27.89%), and large firms (more than 1000 employees; 33.24%). In terms of the establishment of firms, the establishment of 1 to 5 years of the firm accounted for 7.04%, the establishment of 6 to 10 years of firms accounted for 22.25%, the establishment of 11 years to 15 years of firm exhibition 14.93%, the establishment of 16 years to 20 years of firms accounted for 22.25%, the establishment of more than 20 years of firms accounted for 33.52%.

In order to reduce the influence of common method variance (CMV), this study distributes the composition measurements of the questionnaire among different subject modules by reference to the method proposed by [74]. In order to verify whether there is a common method deviation, this study uses Harman single factor test to carry out exploratory factor analysis, extracting factors with characteristic values greater than 1 without rotation. The results present that the first factor extracted can only explain the variation of 31.41% (<40%), which indicates that there is no homologous deviation in the obtained data.

3.2. Variables

3.2.1. Dependent Variables

Financial Performance. Financial performance is defined as the indicator staking the business situation of a firm, which can directly reflect the profitability and business risk. Referring to the prior study [75], this study tested financial performance with 3 items (sales growth, profitability, and market share in the main product market) in which, managers were asked to rate financial performance relative to that of their major competitors. Respondents self-reported all the items through 7-point Likert scales ranging from 1 (strongly disagreement) to 7 (strongly agreement).

3.2.2. Independent Variables

Environmental Regulation. Environmental regulation represents a series of policies for governments to achieve environmental innovation in order to achieve environmental protection and economic development [76]. Reference to multi-item construct used by prior studies [77], this study measured environmental regulation through four items through a 7-point Likert scale ranging from 1 (strongly disagreement) to 7 (strongly agreement). Managers were asked to explain the extent to which environmental regulations affected his or her firms, including (1) strict criteria, (2) suitability for China's situation, (3) being aware of their impact, and (4) whether environmental issues can be effectively addressed.

3.2.3. Mediating Variables

Green Dynamic Capability. This capability refers to the capability to integrate and reconfigure internal and external competences to meet the requirements of changing environments and green product design [40,43]. Following a 7-item construct proposed by [43], we used measured green dynamic capability using the following five items: (1) The firm's capability to quickly monitor the environment, (2) the capability to learn and create, (3) the capability to integrate resources and expertise, (4) the capability to coordinate employees' capability to develop green technologies, and (5) successfully allocate resources to develop the capability to innovate in green.

Sustainability Exploitation Innovation and Sustainability Exploration Innovation. Sustainable innovation is a series of innovative activities to achieve sustainable development, in which sustainability exploitation innovation emphasizes the capability of a firm to participate in and improve the production process [71]. Sustainability exploration innovation refers to the process by which firms establish divergent and innovative adaptive mechanisms through their own practices [47]. Based on prior studies [49], sustainability exploitation innovation was measured using the following six items: (1) Our firm promptly responds to issues proposed by existing stakeholders; (2) our firm often focuses on evaluating external factors related to the interests of key stakeholders; (3) our firm conducts the proactive and flexible business procedures, giving support to meet the various demands of key stakeholders; (4) our firm's product/service design and development often reflect the requirement of key customers and suppliers; (5) our firm often takes advantage of various management measures to avoid risks of variability in key processes; (6) our firm often encourages members to meet sustainability standards via setting a series of key performance indicators. Each of these items was measured using a 7-point Likert scale ranging from 1 (totally disagree) to 7 (totally agree).

According to prior studies [24,51], sustainable product and process development (SER-SPPD) and sustainable-oriented learning (SER-SOL) constitute sustainability exploration innovation. SER-SPPD usually refers to green process engineering and product innovation. SOL usually refers to the capability to develop sustainability- related innovations. Reference to prior studies [24,71], SER-SPPD was measured using the following four items: (1) Our firm is likely to develop products or services in a radical way to face environmental changes; (2) our firm strives to reduce the negativity of external environment via the improvement of the products or services; (3) our firm reconfigures the business procedures periodically to meet green needs; (4) our firm often obtains environmentally friendly innovation to build green environment. SER-SOL was measured using the following four items: (1) Our firm often attaches great importance to train the knowledge and skills of workers to make sustainability practices more effectively; (2) our firm strives to form a sustainable organizational culture by enhancing innovative learning; (3) our firm is good at giving sustainable examples to enhance the capability of workers involving knowledge and practical skills; (4) our firm identifies external opportunities from partners to obtain ideas toward sustainability. Each of these items was measured using a 7-point Likert scale ranging from 1 (strongly disagreement) to 7 (strongly agreement).

3.2.4. Control Variables

Prior literature indicated that same unobservable determinants may jointly influence financial performance [78,79]. Following them, we control observable variables to mitigate bias caused by unobservable determinants. The existing literature presents that the basic features of firm, e.g., firm size and age, can influence the green innovation behavior of firm, and this study designs firm size and firm age as control variables [80]. Referring to [54,81], we use the natural arithmetic of business years to portray firm age, and use the natural arithmetic of the average number of staff in the past three years to portray firm size.

3.3. Reliability Test

To calculate the validity and reliability of the measurement, PLS-SEM model and SmartPLS 3.0 software were employed in this study. Following to prior studies [82,83], we measure the convergent validity of the scales following three principles (all indicator loadings > 0.70; Composite Reliabilities > 0.60; the average variance extracted > 0.50). All outer factor loadings for constructed items range from 0.742 to 0.861 (> 0.70) as shown in Table 1, the Composite Reliabilities values range from 0.879 to 0.913 (> 0.60), and the average variance extracted ranges from 0.603 to 0.708 (> 0.50). All three conditions for convergent validity thus hold. Additionally, Table 1 presents that Cronbach's α values range from 0.794 to 0.880, which exceeded the suggested threshold value of 0.70. The results of the reliability test show that a set of indicators in this study can explain a single latent construct [84].

Table 1. Reliability and convergent validity (N = 355).

Variable	Measurement Items	Factor Loadings	Cronbach's α	CR	AVE
Environmental regulation (ER)	ER1	0.825	0.847	0.897	0.685
	ER2	0.839			
	ER3	0.804			
	ER4	0.842			
Sustainability exploitation innovation (SEI)	SEI1	0.782	0.868	0.901	0.603
	SEI2	0.781			
	SEI3	0.771			
	SEI4	0.781			
	SEI5	0.801			
	SEI6	0.742			
Sustainability-oriented learning (SER-SOL)	SER-SOL1	0.831	0.838	0.892	0.673
	SER-SOL2	0.833			
	SER-SOL3	0.798			
	SER-SOL4	0.821			
Sustainable product and process development (SER-SPPD)	SER-SPPD1	0.857	0.857	0.903	0.701
	SER-SPPD2	0.833			
	SER-SPPD3	0.824			
	SER-SPPD4	0.834			
Green dynamic capability (GDC)	GDC1	0.859	0.880	0.913	0.677
	GDC2	0.800			
	GDC3	0.846			
	GDC4	0.794			
	GDC5	0.811			
Financial performance (FP)	FP1	0.831	0.794	0.879	0.708
	FP2	0.861			
	FP3	0.833			

Note: AVE: average variance extracted; CR: composite reliability.

3.4. Descriptive Statistics and Correlation Analysis

To examine the discriminant validity, we compare the square root of the AVE with the coefficients of correlation between the variable and other variables. The results of correlation statistics and discriminant validity are presented in Table 2 that presents the mean, standard deviation, the square root of AVE, and the correlation between variables. The square root of AVE for each variable is higher than the correlation coefficient between the variable and the other potential variables, which indicate that the discriminant validity of all variables is acceptable.

Table 2. Descriptive statistics and discriminant validity.

Variable	Mean	Standard Deviation	1	2	3	4	5	6
1. ER	5.612	0.912	**0.828**					
2. SEI	5.325	0.851	0.428 **	**0.777**				
3. SER-SOL	5.492	0.884	0.330 **	0.622 **	**0.820**			
4. SER-SPPD	5.364	1.021	0.347 **	0.558 **	0.438 **	**0.837**		
5. GDC	5.086	1.134	0.537 **	0.572 **	0.472 **	0.484 **	**0.823**	
6. FP	5.438	1.026	0.332 **	0.571 **	0.557 **	0.464 **	0.601 **	**0.841**

Note: * $p < 0.05$; ** $p < 0.01$. The square roots of AVE are shown in diagonal line. ER: Environmental regulation; SEI: Sustainability exploitation innovation; SER-SOL: Sustainability-oriented learning; SER-SPPD: Sustainable product and process development; GDC: Green dynamic capability; FP: Financial performance.

4. Results

To examine the research hypotheses, the partial least squares approach to structural equation modeling (PLS-SEM) was applied. Following the PLS specific evaluation process [41], we evaluate the PLS-SEM model by using the SmartPLS 3.0 software, and then interpreted the path coefficients and evaluated the significances by bootstrapping (1000 subsamples and individual-level changes reprocessing).

4.1. Direct Effects

To examine H1a and H1b, the results in Table 3 and Figure 2 indicate that the direct effects of environmental regulation on sustainability exploitation innovation and sustainability exploration innovation are different. Environmental regulation has a positive effect on sustainability exploitation innovation (β = 0.053, p < 0.001). However, the effect of environmental regulation on SER-SOL (β = 0.115, p > 0.100) and SER-SPPD (β = 0.136, p > 0.050) are not significant. Therefore, H1a is supported, but H1b is not supported. What's more, our results present that the effect of environmental regulation on green dynamic capability is significant (β = 0.538, p < 0.100), and the 95% bias-corrected confidence interval ranges from 0.446 to 0.612, which does include 0. Therefore, H1c is supported.

Table 3. Results for the direct effects (N = 355).

Corresponding Hypothesis	Direct Effect	Boot Effect	Bootstrapping Percentile 95% CI		Result
			LLCI	ULCI	
H1a	SEI ER	0.174 **	0.055	0.300	Supported
H1b	SER-SOL ER	0.115	−0.029	0.246	Not supported
	SER-SPPD ER	0.136	−0.017	0.256	Not supported
H1c	GDC ER	0.538 **	0.446	0.612	Supported
H2	FP GDC	0.366 **	0.245	0.481	Supported
H3a	SEI GDC	0.481 **	0.376	0.583	Supported
H3b	SER-SPPD GDC	0.409 **	0.293	0.510	Supported
	SER-SOL GDC	0.408 **	0.288	0.519	Supported
H4a	FP SEI	0.181 **	0.047	0.320	Supported
H4b	FP SER-SOL	0.250 **	0.124	0.387	Supported
	FP SER-SPPD	0.103	−0.003	0.199	Not supported

Note: * $p < 0.05$; ** $p < 0.01$. ER: Environmental regulation; SEI: Sustainability exploitation innovation; SER-SOL: Sustainability-oriented learning; SER-SPPD: Sustainable product and process development; GDC: Green dynamic capability; FP: Financial performance.

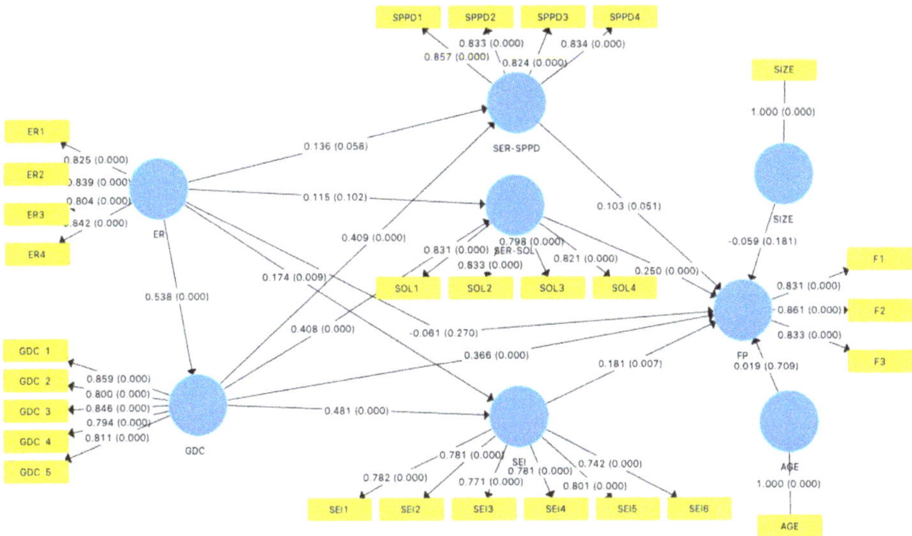

Figure 2. Results of multiple mediation model. Note: ER: Environmental regulation; SEI: Sustainability exploitation innovation; SER-SOL: Sustainability-oriented learning; SER-SPPD: Sustainable product and process development; GDC: Green dynamic capability; FP: Financial performance.

Accordingly, it is found that green dynamic capability positively affects the financial performance ($\beta = 0.366$, $p < 0.001$), the 95% bias-corrected confidence interval does not include 0 (LLCI = 0.245, ULCI = 0.481), thus H2 is empirically supported. To assess the effect of green dynamic capability on two kinds of sustainable innovation, the results present that there is a significant relationship between green dynamic capability and sustainability exploitation innovation ($\beta = 0.481$, $p < 0.001$). Additionally, the results also indicate that green dynamic capability can significantly affect sustainability exploration innovation from two dimensions: SER-SOL ($\beta = 0.408$, $p < 0.001$) and SER-SSPD ($\beta = 0.409$, $p < 0.001$). Therefore, H3a and H3b are supported.

Since the effects of sustainability exploitation innovation ($\beta = 0.181$, $p < 0.001$) and SER-SOL ($\beta = 0.250$, $p < 0.001$) on financial performance are significant, the effect of SER-SPPD ($\beta = 0.103$, $p > 0.005$) presents an opposite result, and the 95% bias-corrected confidence interval includes 0 (LLCI = −0.003, ULCI = 0.199). Therefore, H4a is empirically supported, while H4b is partially supported.

4.2. Indirect Effects

To test the mediating effects of sustainability exploitation/exploration innovation and green dynamic capability in line with H5 and H7, we use the bootstrapping method in SmartPLS 3.0. It is clear from Table 4 that sustainability exploitation innovation plays a meditate effect between environmental regulation and financial performance ($\beta = 0.032$, $p < 0.010$), the effect is significant with the 95% bias-corrected confidence interval is between 0.007 and 0.075. Thus, H5a can be proved. However, the meditating effects of SOL ($\beta = 0.029$, $p > 0.100$) and SPPD ($\beta = 0.014$, $p > 0.100$) on the relationship between environmental regulation and financial are insignificant, which means the mediating role of sustainability exploration innovation in the link of environmental regulation and financial performance is not significant. Thus, the results do not support H5b.

Table 4. Results for the indirect effects (N = 355).

Corresponding Hypothesis	Indirect Effect	Boot Effect	Bootstrapping Percentile 95% CI		Result
			LLCI	ULCI	
H5a	FP SEI ER	0.032 *	0.007	0.075	Supported
H5b	FP SER-SOL ER	0.029	−0.002	0.073	Not supported
	FP SER-SPPD ER	0.014	−0.001	0.042	Not supported
H6	FP GDC ER	0.197 **	0.129	0.271	Supported
H7a	FP SEI GDC ER	0.047 *	0.011	0.088	Supported
H7b	FP SER-SOL GDC ER	0.055 *	0.023	0.097	Supported
	FP SER-SPPD GDC ER	0.023	0.000	0.049	Not supported

Note: * $p < 0.05$; ** $p < 0.01$. ER: Environmental regulation; SEI: Sustainability exploitation innovation; SER-SOL: Sustainability-oriented learning; SER-SPPD: Sustainable product and process development; GDC: Green dynamic capability; FP: Financial performance.

Regarding H6, the result presents that green dynamic capability positively mediates the relationship between environmental regulation and financial performance, which has a significant value of 0.197 and the 95% bias-corrected confidence interval is between 0.129 and 0.271, which does not include 0. Thus, we conclude that environmental regulation can promote the improvement of firms' green dynamic capability, and thus improve the financial performance.

As can be seen in Table 4, different types of innovation have different influence on the relationship between environmental regulation and financial performance via green dynamic capability. According to H7a, sustainability exploitation innovation plays a significant multi-mediation role in the path of environmental regulation to financial performance ($\beta = 0.047$, $p < 0.050$). The results show that the value is proven to be significant with the 95% bias-corrected confidence interval between 0.011 and 0.088, which does not include 0. Thus, H7a is supported. At the same time, the 95% bias-corrected confidence interval of the multi-mediation effect composed of green dynamic capability and SER-SOL does not include 0 (LLCI = 0.023, ULCI = 0.097), which indicates that the coefficient of mediating effect is significant ($\beta = 0.055$, $p < 0.050$). By contrast, green dynamic capability and SER-SPPD cannot mediate the link between environmental regulation and financial performance ($\beta = 0.023$, $p > 0.005$). Therefore, H7b is partially supported.

5. Discussions

Prior studies that noted the importance of environmental regulation in promoting sustainable development have focused on the link between environmental regulation, technology innovation, and firm performance based on the research framework of Porter Hypothesis [2,85]. However, the effect of environmental regulation on innovation and firm performance remains still inconclusive [86]. Suggested by [20,24,33], some key factors within business organizations, e.g., internal capability, organizational slack, and stakeholder pressures etc., be addressed into framework of Porter Hypothesis to promote firms to integrate economic development and environmental protection. Few studies has addressed the green dynamic capability into the original Porter Hypothesis and provided empirical evidence that environmental regulation can affect financial performance via green dynamic capability and sustainable innovation. To extend previous studies on Porter Hypothesis, and provide new insights into the framework of Porter Hypothesis consisting of factors within the organization, the existing research was designed to determine the mediating role of green dynamic capability and sustainable innovation in the link between environmental regulation and financial performance.

(1) The meditating role of sustainability exploration/exploitation innovation in the link between environmental regulation and financial performance. Different from the existing literature [87], the most interesting result is that sustainability exploitation plays a mediating role in the relationship between environmental regulation and financial performance, while sustainability exploration innovation does not. These results are in accord with recent studies indicating that environmental regulation affects financial performance via sustainable innovation [88]. Our findings further supported prior research, which perceived that the effect of environmental regulation on financial performance not only depends on different kinds of environment regulations, but also depends on different types of innovation [24,85].

(2) The meditating role of green dynamic capability in the relationship between environmental regulation and financial performance. Our empirical results indicate that green dynamic capability significantly mediates the link between environmental regulation and financial performance. As for the mediating effect of green dynamic capability in the link between environmental regulation and financial performance, our key findings confirm that green significantly mediates the relationship between environmental regulation and financial performance. This provides empirical evidence for the conceptual framework proposed by prior studies [33], that the mediating effect of some internal factors should be considered in testing Porter Hypothesis, e.g., capability, slack resource, and environmental strategy. Additionally, this study expands the dynamic capability literature by introducing green dynamic capability into the framework of Porter Hypothesis.

(3) The multiple moderating effects of green dynamic capability and sustainable innovation. The results present that environmental regulation can not only promote financial performance through green dynamic capability and sustainability exploitation innovation, but also through green dynamic capability and sustainability oriented learning. These results can support prior literature concerning the complex interconnections among environmental regulation, technology innovation, and firm performance [16,20,24]. Our findings confirm the conceptual framework proposed by [16,22] and provide empirical evidence for strong vision of Porter Hypothesis. Different from the findings of [53,87], that technology innovation induced by environmental regulation can lead to a higher firm performance, our findings extend the Porter Hypothesis literature by revealing two possible paths to achieve the strong Porter Hypothesis. Additionally, this study also extends prior research [24,55] by the mediating effect of green dynamic capability in the relationship among environmental regulation, innovation, and firm performance.

6. Conclusions

Drawing on the Dynamic Capability Theory and the Porter Hypothesis framework, this study introduces a model that tends our knowledge regarding the relationship of environmental regulation and financial performance with multiple mediations of green dynamic capability, and sustainable innovation in manufacturing firms. The main results are as following:

(1) The indirect effect of environmental regulation on financial performance. The result of the analysis of survey data from 355 respondents demonstrate that there is not a significant link between environmental regulation and financial performance, but environmental regulation indirectly affect financial performance via green dynamic capability and sustainability exploitation innovation, respectively.

(2) The multiple mediating effect of green dynamic capability and sustainability exploitation innovation. Our findings indicate that green dynamic capability and sustainability

exploitation innovation subsequently mediated the link between environmental regulation and financial performance.

(3) The multiple mediating effect of green dynamic capability and sustainability exploration innovating. Environmental regulation also significantly affects financial performance partly through green capability and sustainability exploration innovation. Our findings provide some intriguing insights, especially when compared to the results of the existing literature.

6.1. Theoretical Implications

Our study makes contributions to the Porter Hypothesis literature in three ways. First, we established a multiple mediating model involving environmental regulation, green dynamic capability sustainability exploitation/exploration innovation, and financial performance together. We provide a holistic view in the multiple indirect effects of environmental regulation on financial performance through two types of sustainable innovation. Our findings indicated that strong Porter Hypothesis not only depends different types of environmental regulations and industry heterogeneity [89], but also depends on different types of sustainable innovation. These findings provide an insight into the role of sustainability exploitation innovating in the implement of strong Porter Hypothesis.

Second, our findings suggest that the strong vision of Porter Hypothesis not only depends on different kind of environmental regulations and technology innovation, but is also related to internal organization factors. The results of this study further enrich the research on green dynamic capability by uncovering the mediating effect of green dynamic capability in the link between environmental regulation and financial performance.

Third, prior studies related to the Porter Hypothesis have focused on the link between environmental regulation and technology innovation, and neglected the effect of factors internal to firms. This study enriches the literature by arguing the multiple mediating effects of green dynamic capability and sustainability exploration/exploitation innovations on the link between environmental regulation and financial performance. Our findings indicate that green dynamic capability respond the internal and external model when organizations complying with the environmental regulations.

6.2. Practical Implications

The empirical evidences offer following implications for practitioners. We find the optimize way innovation for firms to gain economic output. Environmental regulation will bring firms the actual development risks, and at the same time, low dynamic capability is likely to result in lack of competitive edge for firms [90]. With the rapid development of China's economy, environmental regulation not only directly affects sustainable innovation, but also sets up an invisible "access standards" for firms' green dynamic capability. It can be seen that the green dynamic capability play an important effect in the mechanism of environmental regulation. As small and medium-sized firms, organizing sustainability exploitation innovation activities may lead to more efficient innovation benefits, e.g., leasing innovative equipment and outsourcing innovative production. By contrast, for innovative firms, it is clear that perfecting their organizational structure will gain first priority. They can implement exploration innovation practices on the basis of maintaining a certain level of sustainability exploitation innovation, e.g., the production innovation, human resource cultivation, and management model innovation.

When it comes to government measures, we provide theoretical suggestions to high-quality development in line with Chinese manufacturing firms. Properly transmitting information to firms benefit firms to identify opportunities, control risks, and orchestrate relevant resources. At the same time, governments need to coordinate regulatory systems and enhance sustainability exploitation innovation models, including leasing systems and outsourcing systems. In that case, it will be easier to maintain the fairness and competitiveness of activities related to sustainability exploitation innovation. Most importantly, the government had better encourage firms towards sustainability exploration innovation appropriately, since it may be helpful to organizations' long-term goals [51].

As an inevitable element when analyzing the internal influence mechanism of environmental regulation, green dynamic capability portray firms' original capability react to the environmental policies. The results of this study point out green dynamic capability take an important part in the relationship between environmental regulation and firm performance. First, the high level of green dynamic capability of firms directly facilitates the financial performance, for the reason that firms are able to rapidly make risk identification, capital control, and market monitoring. Second, environmental regulation improves the efficiency of the organization's management pattern by changing the internal structure of the organization. Finally, the environmental regulation affects the internal organizational structure through changes in external market environments and innovative activities in the internal organizations, which contributes to efficient production lines and new products. The process enhances the reputation of firms to a certain extent, and contributes to high financial performance of firms [91,92].

With respect to sustainable innovation, it is found that sustainability exploitation innovation is more in line with the current trend of Chinese manufacturing firms than sustainability exploration innovation. It means that firms prefer to improve their performance by purchasing patents, and outsourcing production. This is because that the success rate of exploratory product innovation is relatively low along with high financial risks, so most firms that have difficulty transforming exploration innovation avoid radical innovation to reduce potential economic lose. At the same time, sustainability exploitation innovation and green dynamic capability have a multiple mediating effect on the link between environmental regulation and financial performance, since the government's requirements for high-quality development affect the capability of firms to perceive opportunities, control resource structure, and reconfigure resources, thus changing the innovation model. In terms of sustainability exploration innovation, sustainability-oriented learning plays a more obvious effect than sustainable product and process development. We also find the intermediary effect of sustainable product and process innovation is insignificant, thus, firms prefer to adopt sustainability exploitation innovation and sustainable orientation innovation to create innovative value. Accordingly, firms need to carry out more sustainable learning activities when resources are limited.

6.3. Limitation and Future Research

There are some limitations in this study which need to be further explored. First, this study only gives insight into the performance of firms at a certain point in time, since it is much more complex to collect long-term and stable firm data. In order to explore the long-term rewards of sustainability exploitation innovation, future research should be analyzed in long-term scenarios. Second, there may be different types of environmental regulation, e.g., mandatory and voluntary regulations, and it is preferable to investigate the different functions of them on green dynamic capability and sustainable innovation activities. Additionally, this study explored the effect of environmental regulation on financial performance via green dynamic capability, and sustainable innovation using the data sets were obtained through questionnaires. This leads to the lack of consideration of objective indicators. Therefore, future studies should be replicated by using the objective measurement methods of financial performance (e.g., ROA, profits, Tobin's Q) and innovation performance (e.g., patents, R&D input, and R&D personnel) to explore the effect different types of environmental regulations on financial performance. Third, this study takes 355 manufacturing firms as a sample and tests the inherent mechanism of environmental regulation, green dynamic capability, sustainability exploration/exploitation innovation, and financial performance. It is worthwhile to identify the universality in other industrials. Finally, the existing literature has confirmed the important role of some internal dynamics (e.g., social responsibility, stakeholder pressure) in promoting the implement of sustainable innovation and improving financial performance. Therefore, future research can properly explore how to introduce these internal dynamics into the framework of Porter Hypothesis to explore potential channels to achieve sustainable development.

Author Contributions: Conceptualization, J.W.; methodology, X.X. and L.S.; software, X.X.; investigation, X.X. and T.L.; data curation, X.X. and L.S.; writing—original draft preparation, X.X. and T.L.; writing—review and editing, J.W. and T.L.; supervision, J.W. All authors have read and agreed to the published version of the manuscript.

Funding: This study is funded by the National Natural Science Foundation of China (Grant No. 71673115), the Jiangsu Social Science Fund (Grant No. 19GLC004), and the Philosophy and Social Science Research Planning Project of Heilongjiang Province (Grant No. 19JLC117).

Conflicts of Interest: The authors declare no conflict of interest.

References

1. Wang, X.; Sun, C.; Wang, S.; Zhang, Z.; Zou, W. Going green or going away? A spatial empirical examination of the relationship between environmental regulations, biased technological progress, and green total factor productivity. *Int. J. Environ. Res. Public Health* **2018**, *15*, 1917. [CrossRef] [PubMed]
2. Hashmi, R.; Alam, K. Dynamic relationship among environmental regulation, innovation, CO_2 emissions, population, and economic growth in OECD countries: A panel investigation. *J. Clean. Prod.* **2019**, *231*, 1100–1109. [CrossRef]
3. Searchinger, T.D.; Wirsenius, S.; Beringer, T.; Dumas, P. Assessing the efficiency of changes in land use for mitigating climate change. *Nature* **2018**, *564*, 249–253. [CrossRef] [PubMed]
4. Ashford, N.A.; Hall, R.P. Achieving global climate and environmental goals by governmental regulatory targeting. *Ecol. Econ.* **2018**, *152*, 246–259. [CrossRef]
5. Costa-Campi, M.T.; García-Quevedo, J.; Martínez-Ros, E. What are the determinants of investment in environmental R&D? *Energ Policy* **2017**, *104*, 455–465.
6. Khan, Z.; Sisi, Z.; Siqun, Y. Environmental regulations an option: Asymmetry effect of environmental regulations on carbon emissions using non-linear ARDL. *Energy Sources Part A Recover. Util. Environ. Eff.* **2019**, *41*, 137–155. [CrossRef]
7. Inigo, E.A.; Albareda, L. Sustainability oriented innovation dynamics: Levels of dynamic capabilities and their path-dependent and self-reinforcing logics. *Technol. Forecast. Soc. Change* **2019**, *139*, 334–351. [CrossRef]
8. Bitencourt, C.C.; de Oliveira Santini, F.; Ladeira, W.J.; Santos, A.C.; Teixeira, E.K. The extended dynamic capabilities model: A meta-analysis. *Eur. Manag. J.* **2019**. [CrossRef]
9. Cristina De Stefano, M.; Montes-Sancho, M.J.; Busch, T. A natural resource-based view of climate change: Innovation challenges in the automobile industry. *J. Clean. Prod.* **2016**, *139*, 1436–1448. [CrossRef]
10. Liu, X.; Wang, E.; Cai, D. Environmental regulation and corporate financing-quasi-natural experiment evidence from China. *Sustainability* **2018**, *10*, 4028. [CrossRef]
11. Ford, J.A.; Steen, J.; Verreynne, M.L. How environmental regulations affect innovation in the Australian oil and gas industry: Going beyond the Porter Hypothesis. *J. Clean. Prod.* **2014**, *84*, 204–213. [CrossRef]
12. Zhao, X.; Liu, C.; Yang, M. The effects of environmental regulation on China's total factor productivity: An empirical study of carbon-intensive industries. *J. Clean. Prod.* **2018**, *179*, 325–334. [CrossRef]
13. Porter, M.E.; Van Der Linde, C. Toward a new conception of the environment-competitiveness relationship. *J. Econ. Perspect.* **1995**, *9*, 97–118. [CrossRef]
14. Lanoie, P.; Laurent-Lucchetti, J.; Johnstone, N.; Ambec, S. Environmental policy, innovation and performance: New insights on the porter hypothesis. *J. Econ. Manag. Strateg.* **2011**, *20*, 803–842. [CrossRef]
15. Li, M.; Dong, L.; Luan, J.; Wang, P. Do environmental regulations affect investors? Evidence from China's action plan for air pollution prevention. *J. Clean. Prod.* **2020**, *244*. [CrossRef]
16. Nor, N.M.; Bahari, N.A.S.; Adnan, N.A.; Kamal, S.M.Q.A.S.; Ali, I.M. The Effects of Environmental Disclosure on Financial Performance in Malaysia. *Procedia Econ. Financ.* **2016**, *35*, 117–126. [CrossRef]
17. Dou, J.; Han, X. How does the industry mobility affect pollution industry transfer in China: Empirical test on Pollution Haven Hypothesis and Porter Hypothesis. *J. Clean. Prod.* **2019**, *217*, 105–115. [CrossRef]
18. Kalt, J.P. The impact of domestic environmental regulatory policies on US international competitiveness. *Recur. Nat.* **1985**, *2*, 312–330.
19. Ambec, S.; Cohen, M.A.; Elgie, S.; Lanoie, P. The Porter hypothesis at 20: Can environmental regulation enhance innovation and competitiveness? *Rev. Environ. Econ. Policy* **2013**, *7*, 2–22. [CrossRef]

20. López-Gamero, M.D.; Molina-Azorín, J.F.; Claver-Cortés, E. The potential of environmental regulation to change managerial perception, environmental management, competitiveness and financial performance. *J. Clean. Prod.* **2010**, *18*, 963–974. [CrossRef]
21. Yuan, B.; Ren, S.; Chen, X. Can environmental regulation promote the coordinated development of economy and environment in China's manufacturing industry?–A panel data analysis of 28 sub-sectors. *J. Clean. Prod.* **2017**, *149*, 11–24. [CrossRef]
22. Mi, Z.; Zeng, G.; Xin, X.; Shang, Y.; Hai, J. The extension of the Porter hypothesis: Can the role of environmental regulation on economic development be affected by other dimensional regulations? *J. Clean. Prod.* **2018**, *203*, 933–942.
23. Shen, N.; Liao, H.; Deng, R.; Wang, Q. Different types of environmental regulations and the heterogeneous influence on the environmental total factor productivity: Empirical analysis of China's industry. *J. Clean. Prod.* **2019**, *211*, 171–184. [CrossRef]
24. Xing, X.; Liu, T.; Wang, J.; Shen, L.; Zhu, Y. Environmental regulation, environmental commitment, sustainability exploration/exploitation innovation, and firm sustainable development. *Sustainability* **2019**, *11*, 6001. [CrossRef]
25. Testa, F.; Iraldo, F.; Frey, M. The effect of environmental regulation on firms' competitive performance: The case of the building & construction sector in some EU regions. *J. Environ. Manage.* **2011**, *92*, 2136–2144.
26. Castellacci, F.; Lie, C.M. A taxonomy of green innovators: Empirical evidence from South Korea. *J. Clean. Prod.* **2017**, *143*, 1036–1047. [CrossRef]
27. Zhou, X.; Feng, C. The impact of environmental regulation on fossil energy consumption in China: Direct and indirect effects. *J. Clean. Prod.* **2017**, *142*, 3174–3183. [CrossRef]
28. Liu, T.; Ren, Z.; Zhang, Y.; Feng, B.; Lin, H.; Xiao, J.; Zeng, W.; Li, X.; Li, Z.; Rutherford, S. Modification effects of population expansion, ageing, and adaptation on heat-related mortality risks under different climate change scenarios in Guangzhou, China. *Int. J. Environ. Res. Public Health* **2019**, *16*, 376. [CrossRef]
29. Del Río González, P. The empirical analysis of the determinants for environmental technological change: A research agenda. *Ecol. Econ.* **2009**, *68*, 861–878. [CrossRef]
30. Li, Z.F.; Thibodeau, C. CSR-Contingent executive compensation incentive and earnings management. *Sustainability* **2019**, *11*, 3421. [CrossRef]
31. Ikram, A.; Li, Z.F.; Minor, D. CSR-contingent executive compensation contracts. *J. Bank. Financ.* **2019**. [CrossRef]
32. Song, H.; Zhao, C.; Zeng, J. Can environmental management improve financial performance: An empirical study of A-shares listed companies in China. *J. Clean. Prod.* **2017**, *141*, 1051–1056. [CrossRef]
33. Ramanathan, R.; He, Q.; Black, A.; Ghobadian, A.; Gallear, D. Environmental regulations, innovation and firm performance: A revisit of the Porter hypothesis. *J. Clean. Prod.* **2017**, *155*, 79–92. [CrossRef]
34. Sharma, S. Different strokes: Regulatory styles and environmental strategy in the North-American. *Bus. Strateg. Environ.* **2001**, *10*, 344–364. [CrossRef]
35. Ren, S.; Li, X.; Yuan, B.; Li, D.; Chen, X. The effects of three types of environmental regulation on eco-efficiency: A cross-region analysis in China. *J. Clean. Prod.* **2018**, *173*, 245–255. [CrossRef]
36. Teece, D.J.; Pisano, G.; Shuen, A. Dynamic capabilities and strategic management. *Strateg. Manag. J.* **1997**, *18*, 509–533. [CrossRef]
37. Kogut, B.; Zander, U. Knowledge of the firm, combinative capabilities, and the replication of technology. *Organ. Sci.* **1992**, *3*, 383–397. [CrossRef]
38. Dougherty, D. Interpretive barriers to successful product innovation in large firms. *Organ. Sci.* **1992**, *3*, 179–202. [CrossRef]
39. Helfat, C.E.; Raubitschek, R.S. Product sequencing: co-evolution of knowledge, capabilities and products. *Strateg. Manag. J.* **2000**, *21*, 961–979. [CrossRef]
40. Chen, Y.S.; Chang, C.H. The determinants of green product development performance: Green dynamic capabilities, green transformational leadership, and green creativity. *J. Bus. Ethics* **2013**, *116*, 107–119. [CrossRef]
41. Mousavi, S.; Bossink, B.; van Vliet, M. Dynamic capabilities and organizational routines for managing innovation towards sustainability. *J. Clean. Prod.* **2018**, *203*, 224–239. [CrossRef]
42. Brink, T. Orchestration of dynamic capabilities for competitive advantage. *Int. J. Energy Sect. Manag.* **2019**, *13*, 960–976. [CrossRef]

43. Lin, Y.H.; Chen, Y.S. Determinants of green competitive advantage: the roles of green knowledge sharing, green dynamic capabilities, and green service innovation. *Qual. Quant.* **2017**, *51*, 1663–1685. [CrossRef]
44. Boons, F.; Montalvo, C.; Quist, J.; Wagner, M. Sustainable innovation, business models and economic performance: An overview. *J. Clean. Prod.* **2013**, *45*, 1–8. [CrossRef]
45. Rashid, N.; Jabar, J.; Yahya, S.; Shami, S. Dynamic eco innovation practices: A systematic review of state of the art and future direction for eco innovation study. *Asian Soc. Sci.* **2015**, *11*, 8–21. [CrossRef]
46. Delai, I.; Takahashi, S. Corporate sustainability in emerging markets: Insights from the practices reported by the Brazilian retailers. *J. Clean. Prod.* **2013**, *47*, 211–221. [CrossRef]
47. Maletič, M.; Maletič, D.; Dahlgaard, J.J.; Dahlgaard-Park, S.M.; Gomišček, B. Sustainability exploration and sustainability exploitation: From a literature review towards a conceptual framework. *J. Clean. Prod.* **2014**, *79*, 182–194. [CrossRef]
48. March, J.G. Exploration and exploitation in organizational learning. *Organ. Sci.* **1991**, *2*, 71–87. [CrossRef]
49. Maletič, M.; Maletič, D.; Dahlgaard, J.J.; Dahlgaard-Park, S.M.; Gomišček, B. Effect of sustainability-oriented innovation practices on the overall organisational performance: An empirical examination. *Total Qual. Manag. Bus. Excell.* **2016**, *27*, 1171–1190. [CrossRef]
50. Chang, C.H. The influence of corporate environmental ethics on competitive advantage: The mediation role of green innovation. *J. Bus. Ethics* **2011**, *104*, 361–370. [CrossRef]
51. Maletič, M.; Maletič, D.; Gomišček, B. The impact of sustainability exploration and sustainability exploitation practices on the organisational performance: a cross-country comparison. *J. Clean. Prod.* **2016**, *138*, 158–169. [CrossRef]
52. You, D.; Zhang, Y.; Yuan, B. Environmental regulation and firm eco-innovation: Evidence of moderating effects of fiscal decentralization and political competition from listed Chinese industrial companies. *J. Clean. Prod.* **2019**, *207*, 1072–1083. [CrossRef]
53. Zhao, X.; Sun, B. The influence of Chinese environmental regulation on corporation innovation and competitiveness. *J. Clean. Prod.* **2016**, *112*, 1528–1536. [CrossRef]
54. Xing, X.; Wang, J.; Tou, L. The relationship between green organization identity and corporate environmental performance: The mediating role of sustainability exploration and exploitation innovation. *Int. J. Environ. Res. Public Health* **2019**, *16*, 921. [CrossRef]
55. Xie, R.H.; Yuan, Y.J.; Huang, J.J. Different types of environmental regulations and heterogeneous influence on "green" productivity: Evidence from China. *Ecol. Econ.* **2017**, *132*, 104–112. [CrossRef]
56. Zhang, Y.; Wang, J.; Xue, Y.; Yang, J. Impact of environmental regulations on green technological innovative behavior: An empirical study in China. *J. Clean. Prod.* **2018**, *188*, 763–773. [CrossRef]
57. Zollo, M.; Winter, S.G. Deliberate learning and the evolution of dynamic capabilities. *Organ. Sci.* **2002**, *13*, 339–351. [CrossRef]
58. Eisenhardt, K.M.; Martin, J.A. Dynamic capabilities: What are they? *Strateg. Manag. J.* **2000**, *21*, 1105–1121. [CrossRef]
59. Artiach, T.; Lee, D.; Nelson, D.; Walker, J. The determinants of corporate sustainability performance. *Account. Financ.* **2010**, *50*, 31–51. [CrossRef]
60. Ameer, R.; Othman, R. Sustainability practices and corporate financial performance: A study based on the top global corporations. *J. Bus. Ethics* **2012**, *108*, 61–79. [CrossRef]
61. Wales, W.J.; Parida, V.; Patel, P.C. Owners on both sides of the deal: Mergers and acquisitions and overlapping institutional ownership. *Strateg. Manag. J.* **2013**, *28*, 1319–1350.
62. Oliveira Neto, G.C.; Leite, R.R.; Shibao, F.Y.; Lucato, W.C. Framework to overcome barriers in the implementation of cleaner production in small and medium-sized enterprises: Multiple case studies in Brazil. *J. Clean. Prod.* **2017**, *142*, 50–62. [CrossRef]
63. De Marchi, V. Environmental innovation and R&D cooperation: Empirical evidence from Spanish manufacturing firms. *Res. Policy* **2012**, *41*, 614–623.
64. Adner, R.; Helfat, C.E. Corporate effects and dynamic managerial capabilities. *Strateg. Manag. J.* **2003**, *24*, 1011–1025. [CrossRef]
65. Chen, Y.S.; Lin, Y.H.; Lin, C.Y.; Chang, C.W. Enhancing green absorptive capacity, green dynamic capacities and green service innovation to improve firm performance: An analysis of Structural Equation Modeling (SEM). *Sustainability* **2015**, *7*, 15674–15692. [CrossRef]

66. Zhou, Y.; Hong, J.; Zhu, K.; Yang, Y.; Zhao, D. Dynamic capability matters: Uncovering its fundamental role in decision making of environmental innovation. *J. Clean. Prod.* **2018**, *177*, 516–526. [CrossRef]
67. Jansen, J.J.P.; Van Den Bosch, F.A.J.; Volberda, H.W. Exploratory innovation, exploitative innovation, and performance: Effects of organizational antecedents and environmental moderators. *Manage. Sci.* **2006**, *52*, 1661–1674. [CrossRef]
68. Uotila, J.; Maula, M.; Keil, T.; Zahra, S.A. Exploration, exploitation, and financial performance: analysis of S&P 500 corporations. *Strateg. Manag. J.* **2009**, *30*, 221–231.
69. Maletič, M.; Maletič, D.; Dahlgaard, J.J.; Dahlgaard-Park, S.M.; Gomišček, B. The Relationship between sustainability– oriented innovationpractices and organizational performance: empiricalevidence from slovenian organizations. *Organizacija* **2014**, *47*, 3–13. [CrossRef]
70. Tomšič, N.; Bojnec, Š.; Simčič, B. Corporate sustainability and economic performance in small and medium sized enterprises. *J. Clean. Prod.* **2015**, *108*, 603–612. [CrossRef]
71. Maletič, M.; Maletič, D.; Gomišček, B. The role of contingency factors on the relationship between sustainability practices and organizational performance. *J. Clean. Prod.* **2018**, *171*, 423–433. [CrossRef]
72. Jansen, J.J.P.; Van den Bosch, F.A.J.; Volberda, H.W. Exploratory innovation, exploitative innovation, and ambidexterity: The impact of environmental and organizational antecedents. *Schmalenbach Bus. Rev.* **2005**, *57*, 351–363. [CrossRef]
73. Aboelmaged, M. The drivers of sustainable manufacturing practices in Egyptian SMEs and their impact on competitive capabilities: A PLS-SEM model. *J. Clean. Prod.* **2018**, *175*, 207–221. [CrossRef]
74. Podsakoff, P.M.; MacKenzie, S.B.; Lee, J.Y.; Podsakoff, N.P. Common method biases in behavioral research: A critical review of the literature and recommended remedies. *J. Appl. Psychol.* **2003**, *88*, 879–903. [CrossRef]
75. Weng, H.H.R.; Chen, J.S.; Chen, P.C. Effects of green innovation on environmental and corporate performance: A stakeholder perspective. *Sustainability* **2015**, *7*, 4997–5026. [CrossRef]
76. Cohen, M.A.; Tubb, A. The impact of environmental regulation on firm and country competitiveness: A meta-analysis of the porter hypothesis. *J. Assoc. Environ. Resour. Econ.* **2018**, *5*, 371–399. [CrossRef]
77. Eiadat, Y.; Kelly, A.; Roche, F.; Eyadat, H. Green and competitive? An empirical test of the mediating role of environmental innovation strategy. *J. World Bus.* **2008**, *43*, 131–145. [CrossRef]
78. Li, F. Endogeneity in CEO power: A survey and experiment. *Invest. Anal. J.* **2016**, *45*, 149–162. [CrossRef]
79. Wang, H.; Lu, W.; Ye, M.; Chau, K.W.; Zhang, X. The curvilinear relationship between corporate social performance and corporate financial performance: Evidence from the international construction industry. *J. Clean. Prod.* **2016**, *137*, 1313–1322. [CrossRef]
80. Voss, G.B.; Voss, Z.G. Strategic ambidexterity in small and medium-sized enterprises: Implementing exploration and exploitation in product and market domains. *Organ. Sci.* **2013**, *24*, 1459–1477. [CrossRef]
81. Vanpoucke, E.; Vereecke, A.; Wetzels, M. Developing supplier integration capabilities for sustainable competitive advantage: A dynamic capabilities approach. *J. Oper. Manag.* **2014**, *32*, 446–461. [CrossRef]
82. Martinez-Conesa, I.; Soto-Acosta, P.; Palacios-Manzano, M. Corporate social responsibility and its effect on innovation and firm performance: An empirical research in SMEs. *J. Clean. Prod.* **2017**, *142*, 2374–2383. [CrossRef]
83. Fornell, C.; Larcker, D.F. Evaluating structural equation models with unobservable variables and measurement error. *J. Mark. Res.* **1981**, *18*, 39–50. [CrossRef]
84. Ağan, Y.; Kuzey, C.; Acar, M.F.; Açikgöz, A. The relationships between corporate social responsibility, environmental supplier development, and firm performance. *J. Clean. Prod.* **2016**, *112*, 1872–1881. [CrossRef]
85. Hu, D.; Wang, Y.; Huang, J.; Huang, H. How do different innovation forms mediate the relationship between environmental regulation and performance? *J. Clean. Prod.* **2017**, *161*, 466–476. [CrossRef]
86. Sheng, J.; Zhou, W.; Zhang, S. The role of the intensity of environmental regulation and corruption in the employment of manufacturing enterprises: Evidence from China. *J. Clean. Prod.* **2019**, *219*, 244–257. [CrossRef]
87. Rubashkina, Y.; Galeotti, M.; Verdolini, E. Environmental regulation and competitiveness: Empirical evidence on the Porter Hypothesis from European manufacturing sectors. *Energ Policy* **2015**, *83*, 288–300. [CrossRef]
88. Yu, W.; Ramanathan, R.; Nath, P. Environmental pressures and performance: An analysis of the roles of environmental innovation strategy and marketing capability. *Technol. Forecast. Soc. Change* **2017**, *117*, 160–169. [CrossRef]

89. Yuan, B.; Ren, S.; Chen, X. Can environmental regulation promote the coordinated development of economy and environment in China's manufacturing industry?–A panel data analysis of 28 sub-sectors. *J. Clean. Prod.* **2017**, *149*, 11–24. [CrossRef]
90. Zhao, X.; Zhao, Y.; Zeng, S.; Zhang, S. Corporate behavior and competitiveness: Impact of environmental regulation on Chinese firms. *J. Clean. Prod.* **2015**, *86*, 311–322. [CrossRef]
91. Collins, E.; Lawrence, S.; Pavlovich, K.; Ryan, C. Business networks and the uptake of sustainability practices: the case of New Zealand. *J. Clean. Prod.* **2007**, *15*, 729–740. [CrossRef]
92. Wolf, J. The relationship between sustainable supply chain management, stakeholder pressure and corporate sustainability performance. *J. Bus. Ethics* **2014**, *119*, 317–328. [CrossRef]

© 2020 by the authors. Licensee MDPI, Basel, Switzerland. This article is an open access article distributed under the terms and conditions of the Creative Commons Attribution (CC BY) license (http://creativecommons.org/licenses/by/4.0/).

MDPI
St. Alban-Anlage 66
4052 Basel
Switzerland
Tel. +41 61 683 77 34
Fax +41 61 302 89 18
www.mdpi.com

Sustainability Editorial Office
E-mail: sustainability@mdpi.com
www.mdpi.com/journal/sustainability

www.ingramcontent.com/pod-product-compliance
Lightning Source LLC
Chambersburg PA
CBHW041947260926
51850CB00018B/1734
9783036551852